The Siamang in Malaya

Contributions to Primatology

Vol. 4

S. Karger · Basel · München · Paris · London · New York · Sydney

The Siamang in Malaya

A Field Study of a Primate in Tropical Rain Forest

DAVID JOHN CHIVERS

103 figures, 21 plates and 57 tables, 1974

S. Karger · Basel · München · Paris · London · New York · Sydney

Contributions to Primatology

(Sucessor to 'Bibliotheca Primatologica')

Vol. 1: McGuire, M.T. (Los Angeles, Calif.): The St. Kitts Vervet. XII + 202 p., 8 fig., 30 tab., 14 maps, 1974. ISBN 3-8055-1692-4.

Vol. 2: Chevalier-Skolinkoff, Suzanne (San Francisco, Calif.): Ontogeny of Communication in the Stumptail Macaque *(Macaca arctoides)*. X + 174 p., 31 fig,. 24 tab., 1974. ISBN 3-8055-1647-9.

Vol. 3: Luckett, P. (New York, N.Y.), ed.: Reproductive Biology of the Primates. XII + 288 p., 193 fig., 17 tab., 1974. ISBN 3-8055-1671-1.

The previously published titles within the series 'Bibliotheca Primatologica' are listed on the third cover page.

S. Karger · Basel · München · Paris · London · New York · Sydney
Arnold-Böcklin-Strasse 25, CH-4011 Basel (Switzerland)

Printed in Switzerland by Buchdruckerei Maihof, Luzern
ISBN 3-8055-1668-1

Contents

Preface XI
Acknowledgements XII

Chapter 1: Introduction

Background 1
Aims and Scope 3
Malay Peninsula 4
Study Areas 7
1. Ulu Sempam 8
2. Kuala Lompat 14
3. Ulu Gombak 15
Study Methods 16
1. Habituation 20
2. Data Collection 22
A Siamang Day 23
Weather and Seasonality in Central Malaysia (1969 – 1970) 24

Chapter 2: Ranging Behaviour and Territory

Introduction 27
Measures of Movement 28
Methods 28
Results 29
1. Temporal Variation in Movement 31
2. Effect of Weather on Movement 34
Ranging Patterns 36

Sectional Analysis 40
Quadrat Analysis 41
Extent and Rate of Ranging 45
Night Sleeping Trees 49
Maintenance of Territory 52
Territory Size 56
Discussion 57
Summary 61

Chapter 3: Diet and Feeding Behaviour

Introduction 63
Vegetation Analysis 64
Diet 69
Food Categories 70
1. Food Trees 70
2. Feeding Visits 76
3. Feeding Time 78
4. Feeding Visit Duration 83
Food Tree Identity 86
1. Kuala Lompat 86
2. Ulu Sempam 91
Principal Food Trees 92
1. Kuala Lompat 99
2. Ulu Sempam 105
Feeding Behaviour 107
Dispersion of Food Trees 107
1. Kuala Lompat 107
2. Ulu Sempam 110
Drinking Behaviour 113
Feeding Habits and Posture 114
Feeding Rates 116
Movements within Food Tree 118
Discussion 119
Summary 126

Chapter 4: Daily Activity Patterns

Introduction 128
Daily Activity Time and Day Length 129
Differential Use of Forest Levels 133
Activity Bouts 137
Behavioural Synchrony in Group Activities 137

Daily Pattern of Activity Bouts ... 140
Periodicity of Maintenance Activities ... 140
Drinking Behaviour ... 146
Activity Time ... 148
Kuala Lompat (TS1) ... 148
1. Group Behaviour ... 148
2. Individual Behaviour ... 151
Ulu Sempam (RS2) ... 155
Weather and Daily Activity Patterns ... 157
Behaviour and Weather ... 158
Discussion ... 163
Summary ... 169

Chapter 5: Social Organisation

Introduction ... 172
Spatial Relations ... 172
Group Cohesion ... 173
Group Spread ... 178
Inter-Individual Distance ... 181
Spacing of RS2 in Ulu Sempam ... 183
Spacing and Weather ... 184
Interim Discussion ... 186
Group Progressions ... 188
Start and Finish of Daily Activities ... 188
Travel Order ... 189
Travel Sequence ... 192
Infant Behaviour ... 193
Grooming Behaviour ... 197
Group Behaviour ... 199
Individual Behaviour ... 205
1. Initiation of Pair Association and Grooming ... 205
2. Duration and Frequency of Grooming Bouts and Sequences ... 207
3. Daily Patterns for Each Pairing ... 209
4. Grooming Relations between Each Individual ... 211
Play and Maturation ... 214
Sexual Behaviour ... 219
Aggression and Group Formation ... 222
Discussion ... 227
Summary ... 232

Chapter 6: Calling Behaviour

Introduction 235
Pattern of Group Calls 237
Siamang 237
1. Male Calls 238
2. Female Calls 238
3. Inter-Group Conflicts 241
4. Alarm 241
White-Handed Gibbon 242
Group Calling Behaviour 245
Ulu Gombak (G) 247
Kuala Lompat (T) 249
1. Siamang 249
2. Gibbon 250
Ulu Sempam (R) 252
1. Siamang 252
2. Gibbon 255
Calling in Local Populations 256
Time of First Calls 256
Daily Number of Call Hours 258
Daily Number of Group Calls 259
Duration of Call Bouts 262
Calling and Weather 263
Discussion 264
Summary 267

Chapter 7: Distribution and Density

Introduction 269
Distribution 269
Siamang 269
Gibbons 271
1. White-Handed *(lar)* 271
2. Black-Handed *(agilis)* 272
Altitudinal Zonation 272
Group Size 274
Population Density 274
Methods 275
Results 275
1. Siamang 282
2. Gibbons 283
Discussion 284
Summary 287

Chapter 8: Concluding Discussion

Comparative Socio-Ecology ... 289
Malayan Forest Primates ... 291
The Siamang in Tropical Rain Forest ... 293
Siamang and the Great Apes ... 294
Summary ... 296
Postscript ... 299
References ... 301

Appendices

A. Patterns of Fieldwork and Primate Observation ... 306
B. Tree Species in Study Areas ... 309
C. Patterns of Territory Use by TS1 ... 313
D. Daily Patterns of Weather and Behaviour ... 318
E. Distribution and Density of Hylobatid Groups in Survey Localities in West Malaysia ... 325
F. Notes on Colour Phases and Female 'Great Calls' in Gibbons ... 330

Author Index ... 332
Subject Index ... 334

Preface

The aims of this work are to present a comprehensive quantitative description of the behaviour of a higher primate in the tropical rain forest environment, and to initiate detailed synecological comparisons between closely related species, following a two-year field study in West Malaysia. A fuller interpretation must await the collation of long-term observations, which will be collected along lines that can now be more clearly defined.

The study of forest primates has been beset by problems of observation. With the need to obtain such information, as the study of free-ranging primates broadens in scope and complexity, techniques have been developed to help resolve these problems. Selected social groups were observed continuously from dawn to dusk for long periods. Data were collected on ranging (and calling), diet and feeding, and social organisation by different – in many cases complementary – methods, and analysed from varied viewpoints. Thus it was possible to extract the maximum of information from the data, and thereby elucidate the full characteristics, and more subtle components, of each behaviour.

There are problems in communicating results in a multi-disciplinary field such as primatology, and it is hoped that the differing approaches of readers have been catered for herein. While it is becoming increasingly important to standardise procedures for comparative purposes, remaining problems concern the value of contrasting approaches, which are difficult (and perhaps undesirable) to incorporate into a unified scheme.

Acknowledgments

This field study of the siamang was financed by a Malaysian Commonwealth Scholarship (which included a substantial allowance for my wife), a Science Research Council Overseas Studentship, and a Goldsmiths' Company Travelling Studentship (which generously included a loan for the purchase of a Landrover). Grants for equipment and field assistance were received from the Boise Fund (Oxford), the Emslie Horniman Scholarship Fund (Royal Anthropological Institute, London), the New York Zoological Society, and the Merchant Thaylors' Company, London.

Departmental facilities were kindly provided initially by the Duckworth Laboratory of Physical Anthropology, University of Cambridge, subsequently by the School of Biological Sciences, University of Malaya, and finally by the Sub-department of Veterinary Anatomy, University of Cambridge. Special thanks are due to Mr. BERNARD DENSTON of the Duckworth Laboratory, to the Visual Aids Unit in the Department of Anatomy for photographic assistance, and to the Sub-department of Animal Behaviour for the use of their PDP-12 computer.

Thanks are due to M.Prager Ltd. (Kuala Lumpur and Singapore) for keeping my electronic equipment in running order while in the field, to Humphrey-Saunders Ltd. (Pinner, Middx.) for handling exposed film and advising me on the results, and to Mrs. C. BAYFIELD for the efficient way she handled the complexities of typing this manuscript.

In Malaya valuable help was received from the Chief Game Warden, Mr. BERNARD THONG, and all the state Game Wardens and their staff, especially those in Pahang where both the main study areas were situated. It was through ENCHE ABDUL JALIL BIN AHMAD that prolonged observation of siamang was possible in the Krau Game Reserve, from the rest house at

Kuala Lompat Ranger Post. Dr. E. Soepadmo of the University of Malaya, and Dr. Tim Whitmore of the Forest Research Institute, Kepong, kindly identified botanical specimens during visits to the study areas. Prof. E.J.H. Corner of the Botany School, Cambridge, identified species of fig on my return to England; assistance was also received from the Herbarium, Kew.

I am grateful to Dr. Lee Peng Choo of the Economic Planning Unit for important information on the forests of West Malaysia, and to the Government Departments who kindly provided me with maps, aerial photographs, and meteorological equipment. We were very grateful to the University of Malaya for opportunities to use their Field Studies Centre in Ulu Gombak, and to Lord and Lady Medway for their hospitality on our regular visits there. Valuable assistance was given by Mr. Yong Ghong Chong in setting up the two main study areas.

This study was initiated under the guidance of Dr. David Pilbeam and Dr. Alison Jolly, and then supervised by Lord Medway for the last four years – both in Malaya and England. I am especially grateful to Lord Medway for advice and encouragement whilst in Malaya, and for the time and trouble he took in making critical comments during the analysis and writing up of the results. Dr. Michael Simpson, Dr. Joe Herbert, and Mr. John Fleagle have read parts of the manuscript and contributed useful comments. I am grateful to Dr. Simpson for his advice on data analysis and non-parametric statistical tests, and to all those from whom I have gained benefit during discussions of primates and behaviour. The extensive comments of Prof. R. A. Hinde and Dr. R. D. Martin were particularly helpful for the revision of the original manuscript.

We are indebted to all those people and animals who made our two years in Malaysia so memorable; and to our cats Moppitt and the late Mr. Butterworth for their companion-ship during our isolation, and for the assistance they gave in spotting primates and in sampling the forest fauna. I shall always be indebted to my wife, Sarah, for the important and unselfish contribution she has made to *every* aspect of this long study.

Chapter 1: Introduction

Background

Gibbons *(Hylobates)* figure prominently in the art and literature of China from 1500 BC until 1644 AD [VAN GULIK, 1967]. While the macaque was the popular image of human restless curiosity, inconstancy, and trickery (as in Indian monkey lore), the gibbon became the purely Chinese symbol of the worldly ideals of poet and philosopher, and of the mysterious link between man and nature. Because of the unique place that the gibbon occupied in far-eastern culture, VAN GULIK found it possible to trace the extent of its habitat, its appearance and mannerisms over more than 2,000 years.

The gibbon was first brought to the notice of the West by the Venetian traveller Marco Polo in the 13th century. Linnaeus described the gibbon as *Homo lar* in 1771, and as *Simia lar* in 1788, following the description by BUFFON [1766] of the 'grand' and 'petit' gibbon. It was at this time that the siamang was first described – from Bencoolen (Sumatra) as *Simia gibbon* by MILLER [1778] – but it did not receive wide recognition until the time of RAFFLES [1821]. The gibbons were reviewed as four species of *Simia* by HARLAN [1834], and as eight species of *Hylobates* by MARTIN [1841]. It was in this year that GLOGER [1841] gave generic distinction to the siamang as *Symphalangus syndactylus*. The debate about whether the siamang and other gibbons are congeneric has continued furiously to this day [ANDERSON, 1878; ELLIOT, 1913; POCOCK, 1927; KLOSS, 1929; SCHULTZ, 1933; SIMONETTA, 1957; FRISCH, 1965; FOODEN, 1969; GROVES, 1972].

This study of the ecology and behaviour of free-ranging siamangs was initiated partly to help resolve this problem, in the belief that behavioural information is a necessary supplement to morphological (and cytological) data in producing a taxonomy which accurately reflects phylogenetic relationships. The other main reason was that, apart from a few species or subspecies of gibbon, the siamang was the only ape species which had not been systematically studied in its natural habitat. This is of special importance to physical anthropology as gibbons (and siamang) share certain features only

Fig. 1.1. Distribution of hylobatids in South-east Asia [constructed from FRISCH, 1965; FOODEN, 1969; GROVES, 1970; WILSON, W. and C., pers. comm.; personal observations].

with man among hominoids, such as monogamy and habitual orthogradism.

Whereas gibbons (lesser apes) occur in large numbers throughout south-east Asia – from Assam to Borneo, from China to Java (fig. 1.1) – and fossil siamang occurs in Java [HOOIJER, 1960], the siamang is today restricted to Sumatra and Malaya. The Malayan siamang was not described as a separate subspecies until the beginning of the 20th century [THOMAS, 1908] and it has since been shown to differ from its Sumatran counterpart in features, such as smaller body size [FRISCH, 1971]. Unlike most gibbons, the siamang is monomorphic in colour (frontispiece). This black ape is nearly twice the weight of gibbons at nearly 11 kg with a relatively broader chest; with shorter legs (relative to trunk), and much longer arms (relative to legs); with a higher face, and a large throat pouch; and with webbing (syndactyly) between toes II and III [SCHULTZ, 1933]. The siamang has 50 chromosomes, the concolor gibbon *(H. concolor)* of Indo-china 52, and all other gibbons 44 [BENIRSCHKE, 1969; HAMERTON *et al.*, 1963]. Because GROVES [1972] in his detailed analysis of gibbon material finds the siamang and concolor to be equally distinct from all those of the *lar* group (including *hoolock*, *agilis*, *moloch*, and so forth), he advocates the recognition of three subgenera within *Hylobates: Symphalangus*, *Nomascus*, and *Hylobates*, respectively.

Although the gibbon has been increasingly kept in captive or semi-captive situations in zoos and laboratories in tropical and temperate regions, and is assuming importance in biomedical research, little is known of the siamang except for some observations on the birth and growth of infants in San Diego and San Francisco zoos [HILL, 1967; SOO-HOO, 1966], and on a social group in Milwaukee Zoo [FOX, 1972]. Nothing is known of the detailed ecology and behaviour of free-ranging siamangs. Such information is essential for their future conservation in an increasingly threatened tropical rain forest environment.

The gibbon was the subject of one of the first primate field studies [CARPENTER, 1940]. It has for long been established that they are unusual among primates in their organisation into monogamous family groups in small, stable, defended territories. Suspensory behaviour predominates in travel and feeding on fruit; there is little sexual dimorphism. These observations have been confirmed for the white-handed gibbon in Malaya by ELLEFSON [1967], and extended to include the siamang as behaviourally and ecologically similar. But ELLEFSON suggested that the siamang would live in a larger territory, and be dominant over gibbons in areas of sympatry. He confirmed observations from Sumatra [CARPENTER, 1938] that the siamang also lived in family groups. Some additional information was provided by MCCLURE [1964], KHAN [1970], and KAWABE [1970].

But it was not known whether siamang and gibbons have a similar diet and social organisation within the family group, or whether they use and maintain territories in the same way – hence the present study. CARPENTER [1940] predicted that, compared with other gibbons, the siamang would show the greatest variation from the distinctive hylobatid behaviour patterns. Some preliminary conclusions on these differences have been drawn from a comparison of this study of the siamang with ELLEFSON's gibbon study [CHIVERS, 1972 a].

Aims and Scope

The main aim of this dissertation is to describe the ecology and behaviour of the siamang in detail, and in such a way as to facilitate a direct comparison with other rain forest mammals, especially primates. It seeks to describe (1) the way the siamang utilises its habitat, and (2) the social organisation of siamang (within and between groups). The emphasis is on quantitative description: behaviour patterns will only be described qualitatively where the siamang differs from the whitehanded gibbon (see CARPENTER [1940] and ELLEFSON [1967] for full accounts).

Following a general description of the Malay peninsula, the study areas, and study methods (see below), there will be a detailed analysis of ranging behaviour in terms of distance travelled, and proportion of territory utilised (Chapter 2). This includes a description of patterns of use, with seasonal and climatic variations. The relation beween the siamang and its habitat is further investigated by analyses of (1) vegetation – tree density and forest structure, (2) siamang diet in terms of tree part consumed, tree species, and food tree dispersion about the territory, and (3) feeding behaviour in terms of forest level, habit, posture, rate and movement (Chapter 3).

Daily activity patterns are presented for the group, and for each age and sex class as number of minutes/hour for each maintenance activity, and as number of bouts/hour. The characteristics and periodicity of these activity bouts, and the effects of weather and seasons are investigated (Chapter 4). The social organisation of the siamang family group is clearly indicated by the spacing of individuals, in terms of actual distance, synchronised activity, and close proximity (number in same tree). These relations are further elucidated by analyses of grooming, aggressive play, and sexual behaviour (Chapter 5).

Vocalisations are described in terms of individual and group calling patterns. This provides the basis for an analysis of calling behaviour, which varies in its characteristics between groups, and between hylobatid species. The relation between these frequent and loud calls and the spacing of groups is discussed with reference to the siamang and gibbon (Chapter 6). The determination of distribution, and estimates of population size, of these two species in the Malay Peninsula are largely based on calling behaviour (Chapter 7). In this way a detailed comparison of related species is initiated (the secondary aim of this study).

Finally there is a discussion of the position of the siamang in tropical rain forest ecology, and of its relations with other animals, especially primates. The siamang is reassessed in relation to other gibbons and the great apes in the light of this new evidence on its ecology and behaviour.

Malay Peninsula

This is the surviving land link between the Asian continent and Sundaland. In the early Palaeozoic there was a large land mass between present-day

Asia and Australia. Since the Mesozoic the continually changing land surfaces of Sundaland have been partially isolated from the neighbouring continents. Because of this isolation, the intermittent connection between islands, and the distinctive climate, the 'islands' of the Sunda Shelf – Borneo, Java, and Sumatra – have become an important centre of animal and plant evolution [MEDWAY, 1970]. The changes in land connections in relatively recent times have introduced an important dynamic aspect into gibbon evolution. It appears that gibbons are still actively speciating in isolation, with fascinating situations developing where sympatry occurs between populations that were thought to be reproductively isolated [MARSHALL *et al.*, 1972]. Thus this region provides a unique situation for the study of the mechanisms of hominoid phylogeny (fig. 1.1).

Intermittent volcanic activity over a long period has thrown up several arcs of mountain ranges running much of the length of the peninsula. All have provided important refuges for animals at times of high sea level. The peninsula is dominated by the main mountain range on the west, rising to just over 7,000 ft above sea level, and the Trengganu Highlands in the east (up to 5,000 ft).[1] They are partially linked along the Kelantan-Pahang border by Gunong Tahan (7,174 ft) in the Taman Negara (National Park). High temperatures and heavy rainfall have helped the intense weathering which these mountains have undergone. The final phase in the formation of the Peninsula has been widespread coastal sedimentation, with sea level varying to 300 ft above and below the present level [OOI JIN-BEE, 1963].

The Malay Peninsula, defined zoogeographically by CHASEN [1935] as that part of the peninsula south of latitude 10° N, contains the original Malay States, now known as West Malaysia, extending between latitudes 1° 20′ and 6° 40′, and between longitudes 99° 35′ and 104° 20′ E. Malaya is about 1,300 km long, and up to 500 km wide, the peninsula narrowing to 100 km at its junction with the Asian mainland – the Isthmus of Kra in peninsular Thailand (fig. 1.2). West Malaysia covers an area of 130,750 km^2, of which about three fifths (nearly 90,000 km^2) is still forested. Over half the land area is over 500 ft above sea level.

There is a dense network of rivers and streams draining the highlands, but no single river dominates the drainage as in other nearby countries;

1 Where feasible, metric units are given, but if there is no analytic advantage, or if certain units are likely to lose their meaning with many readers the more familiar units have been retained. Thus kilometres, hectares, and grams have been used for distance, areas and weights, respectively, but degrees Fahrenheit, feet, and inches have been retained for temperature, heights, and rainfall, respectively.

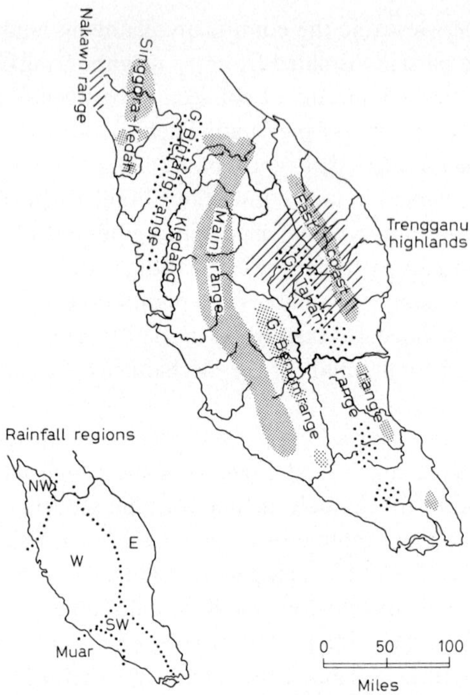

Fig. 1.2. The southern part of the Malay Peninsula, showing main rivers and mountain ranges, and contrasting rainfall regions [from Ooi Jin-Bee, 1963].

Sungei Pahang and S. Perak are only 700 and 440 km long, respectively. Although there is year-round precipitation, and no river course is ever completely dry, the torrential and localised nature of the rainfall results in rapid fluctuations in river levels. Uncontrolled forest clearance has aggravated the problems of erosion and flooding to serious proportions.

The Malay Peninsula has an equatorial climate modified by monsoonal and maritime factors. The dense vegetation increases air humidity and rainfall potential, and reduces ground temperature. Such seasonality as there is in Malaya is most marked on the east coast, because of the rain-laden northeast monsoon at the turn of the year. This is followed by a dry season, until May or June when the west coast receives the main force of the milder southwest monsoon. The centre of the country in the rain shadow of both mountain ranges is the driest, sometimes with less than 70 in of rain per year [Dale, 1959], whereas the annual total often approaches 200 in in Kedah and Perak, and rises above 150 in in the wetter parts of Kelantan and Trengganu. The number of rain days per year ranges between 120 and 230 days according to location and altitude [DALE, 1960]. There is a daily temperature range of about 15°F, with a mean annual temperature in lowland Malaya of about 80°F, whereas over 4,000 ft above sea level it may be as low as 65°F. There is a small drop in temperature during the monsoon periods. At Temerloh, over nearly 20 years, sunshine averaged 5.8 h/day (varying between 4 and 7 h according to the time of year); the sunniest hour was around noon. On Bukit Fraser there were only 4.8 h (varying from 3 to 6 h) with the sunniest hour in mid-morning [DALE, 1964].

The vegetation of the Malay Peninsula is Indo-Malaysian. It can be distinguished from the northern deciduous monsoon forests of India, Burma, Indo-China, and Thailand, and from the eastern evergreen monsoon rain-forests of the Philippines, Celebes, New Guinea, and north-west Australia [CORNER, 1952]. This central evergreen rain-forest of the Sunda Shelf can be divided into five climatic climax formations [WYATT-SMITH, 1953]. The forest canopy is dominated by trees of the family Dipterocarpaceae up to about 4,000 ft above sea level. Within this range there is an altitudinal zonation – correlated with floral change – into lowland, hill, and upper dipterocarp forests, with the main transition zones at about 1,000 and 2,500 ft above sea level, respectively. Above 4,000 ft are montane oak-forests, which are replaced at about 5,000 ft by montane ericaceous forests.

The study of siamang ecology and behaviour was made in lowland and hill dipterocarp forests which belong to the world formations of tropical lowland evergreen rain-forests. Hylobatids in Malaya, however, do occur in lower-montane evergreen rain-forest, and the siamang has been recorded in tropical upper-montane forest at 6,000 ft above sea level [MEDWAY, 1972a].

Study Areas

Following a two-month preliminary study at the University of Malaya Field Studies Centre in Ulu Gombak[2], 15 miles from Kuala Lumpur, west of the pass over the main range into Pahang, a 5-month country-wide survey was made. The aims were (1) to plot the distribution of hylobatid species, (2) to estimate the population size of each hylobatid species in West Malaysia, and (3) to find at least two contrasting areas suitable for prolonged observation of siamang family groups.

2 Malay names: Bukit = hill; Kuala = river or stream mouth; Ulu = head waters; Gunong = mountain; Sungei = river, stream; Orang asli = aboriginals.

At least two main study areas were required, because study of an elusive arboreal primate in only one area might yield few results. The other major disadvantage of concentrating observations in one area would be the inability to relate the detailed ecology and behaviour of this one population to the species as a whole. Thus two accessible areas were selected for intensive study; they showed contrasting climate, altitude, flora, fauna, and degrees of disturbance by man.

1. Ulu Sempam

This locality on the northern slopes of Bukit Fraser was chosen as one area. The hylobatid population was the most dense, and the terrain the least rugged, of all highland areas surveyed. The hill dipterocarp forest appeared to be the main siamang habitat. The camp was situated at 1,500 ft above sea level by Sungei Sempam 8 miles along the logging track from the village of Teras on the main road across the mountains from Raub on the east to Kuala Kubu Baharu on the west (fig. 1.3, plate I). The study was carried out on a ridge rising southwards from the camp to 2,400 ft above sea level (and thence on up to the summit of Bukit Fraser at 4,279 ft. An irregular network of trails was cut (both following, and at right angles to, the contours). Some trees were climbed, using 6-in nails, for observations of the forest canopy. On either side were two tributaries draining into S. Sempam – S. Yet on the east, and S. Man on the west (fig. 1.4, plate III).

It eventually transpired that parts of this forest were selectively logged at about 5-year intervals, so that it was possible to record the effects of this on the primates. Two logging camps, on opposite banks of S. Sempam 4 miles below the study camp, operated in this area. Because this is not State Forest Reserve their activities were not subject to the rigid control that is enforced elsewhere. Nevertheless, their selective logging produced an habitat apparently enriched for primates, because the removal of much of the main canopy of dipterocarps encouraged the growth of lower canopy species. These provided much food for the siamang and other primates.

Ulu Sempam is essentially typical of hill dipterocarp forest in the Malay Peninsula. There are always several species of *Shorea*, such as *S. curtisii* on the ridges and *S. leprosula*, and many other dipterocarps such as species of *Dipterocarpus* and *Anisoptera*. Also there are species of Anacardiaceae, such as *Melanorrhoea*, *Melanchyla*, and *Bouea*, and large numbers of the palm *Oncosperma horrida*. The logging camp usually operated along the river valley

Fig. 1.3. West Malaysia, showing the States and their boundaries, large towns, railways, main roads, and the siamang study areas (from Survey Department, Kuala Lumpur). R = Ulu Sempam, T = Kuala Lompat, G = Ulu Gombak.

and up into the hills in the west. But in 1964 they crossed S. Sempam and extracted timber from the lower reaches of S. Yet, while the other camp came over the hill to work the ridge east of S. Yet, crossing higher up to come down and extract timber from the study ridge (fig. 1.4, plate IV). Both operations were repeated at the beginning of 1970, near the end of the siamang study.

It is for physical reasons that timber felling is mainly confined to the valleys and ridge tops. The extraction of trees from ridge tops is supposedly restricted to prevent excessive soil erosion. Siamang were mainly found on

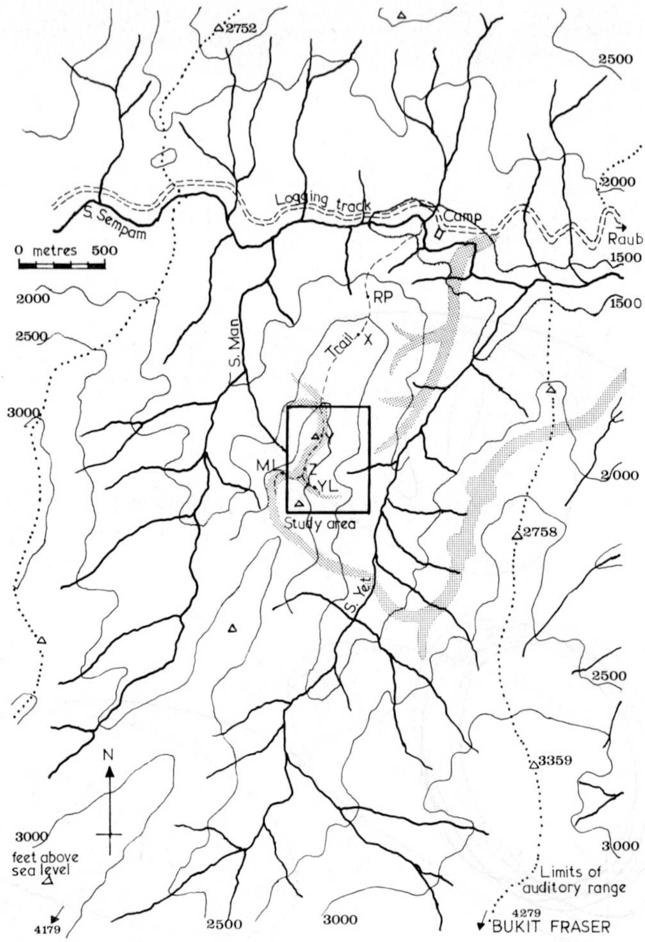

Fig. 1.4. Highland study area in Ulu Sempam, near Raub in west Pahang, showing camp on logging track, study ridge, rivers and hills. Also shown are limits of auditory range for calling siamang groups, and the direction of logging activities around study area in 1964 and 1970 (from 1-inch OS maps, Survey Department, Kuala Lumpur).

the hill slopes, and this cannot be explained solely by logging operations. It seems that nature and type of soil on the slopes may be more favourable for food plants than the drier ridge tops and the damper valley bottoms [SOEPADMO, personal commun.]. Although the siamang have adapted their locomotion to travel through bamboo, the excessive growth of such vegetation in logged areas may act as a barrier to movement.

Plate I. Ulu Sempam. Aerial view of siamang study area in hill dipterocarp forest. Extends from camp between logging track and Sungei Sempam (marked on left) southwards along ridge from 1,800 to 2,200 ft above sea level (near lower right).

Plate II. Kuala Lompat. Aerial views of siamang study area in lowland dipterocarp forest. Game Ranger Post at river confluence (above), and south-west corner of siamang territory near man-made clearing by Sungei Lompat (below, note landrover track).

Plate III. Ulu Sempam. Siamang study area ridge of hill dipterocarp forest between Sungei Yet (on left) and S. Man (over ridge to right), looking south from top of tree (*Shorea* sp., 110 ft) by camp on left bak of Sungei Sempam. Area shown is part of territory of siamanggroup RS 3.

Plate IV. (a) Hill dipterocarp forest at 2,200 ft above sea level on study ridge in Ulu Sempam along timber extraction track (RS2 territory). Note dense secondary, foliage and palms, *Oncosperma* sp. *(b)* Floor of lowland dipterocarp forest along trail used by Jelutong-tappers at Kuala Lompat in middle of siamang territory. Dense undergrowhth behind indicates tree fall in recent years.

Although many dipterocarps have been felled from the study area, they have not been totally removed from any part, and large relatively undisturbed patches still remain. But in most parts the lower canopy species have flourished – many *Ficus* species; and species of *Aglaia*, *Cynometra*, *Dialium*, *Dillenia*, *Durio; Glochidion*, *Macaranga*, *Antidesma*, *Mallotus*, *Xylopia*, *Diospyros*, *Litsea*, *Adinandra*. Many trees are covered with climbers which provide a variety of primate foods, especially young leaves. The most important are species of Leguminosae; there are also many fruiting rattans, especially the giant *Calamus urens*. The Ulu Sempam flora is summarised in Appendix II.

2. Kuala Lompat

The forest around the Game Ranger Post at Kuala Lompat at the west end of the Krau Game Reserve was selected as the other main study area. Although near the periphery of siamang distribution, and in lowland at the base of the central massif of Gunong Benom, this area appeared to have large numbers of hylobatids. It was readily accessible being eight miles west of the road along the Sungei Pahang from Temerloh to Jerantut along a track through two small villages (fig. 1.3., plate II).

The Game Ranger Post is at the confluence of Sungei Krau and Sungei Lompat about 150 ft above sea level, one mile away from the *orang asli* settlement of Penderas by S. Krau (fig. 1.5, plate II). The trails of these local people were cleared, and some extra ones cut. The forest is typical of lowland dipterocarp forest in the Malay Peninsula being much richer in species than Ulu Sempam. Although in hill forest there are greater numbers of certain species, in lowland forest there are generally smaller numbers of a much wider variety of species. Species of *Dipterocarpus* and *Shorea* are again well represented, with species of *Parkia*, *Intsia*, *Dyera*, *Sindora*, *Alstonia*, *Parinari*, *Ficus*, and the magnificent giants *Koompassia excelsa* and *malaccensis*. Conspicuous in the lower canopy are species of *Sloetia*, *Randia*, *Xylopia*, and *Ficus* (Appendix B).

As happens in mature forest tree falls are not infrequent, particularly of the Jelutong, *Dyera costulata*, which have been tapped for their latex this century for the manufacture of chewing-gum. These tree falls result in temporary localised patches of more open forest with dense secondary growth. Generally the shadier forest floor was clearer than in Ulu Sempam; woody climbers especially leguminous were widespread (plate V). Along lines of drainage, because of soil conditions, there are different floral associations. Poor aeration in these swampy areas, because of water cover at some times

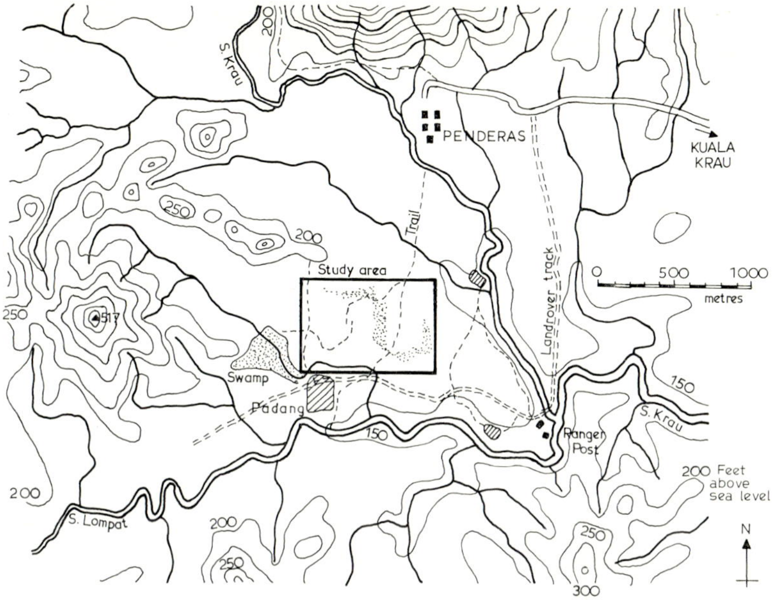

Fig. 1.5. Lowland study area of Kuala Lompat near Temerloh in central Pahang, eight miles west of Kuala Krau, showing Game Ranger Post at confluence of Sungei Krau and S. Lompat, *orang asli* village of Penderas, mainly lowland, man-made clearings (hatched), and swamps (stippled) (from 1-inch OS maps, Survey Department, Kuala Lumpur).

of the year, results in lower denser forest, with pandans, sedges, and small palms conspicuous on the forest floor (plate V). There are two common tree species – one of Rubiaceae, the other of Euphorbiaceae – and certain species of *Ficus*, and *Knema laurinum* (Myristicaceae). The forest margins show signs of disturbance being more open with smaller trees; in places there are small clearings (plate II). But there does seem to be a distinctive riverine habitat preferred by the long-tailed macaques, *Macaca fascicularis*. Men from the nearby village were often seen in the forest on their way to fish, harvest fruit, or to collect rattans.

3. Ulu Gombak

Observations were continued here, on the western slopes of the main range, for a few days each month throughout the two years. At lower altitudes it is very similar to the hill dipterocarp forest of Ulu Sempam, but the higher slopes are dominated by species of *Eugenia*, *Canarium*, and *Santiria* – kedongdong forest – indicative of dryness. The absence of logging from these higher ridges has preserved the very distinctive *Shorea curtisii* (plates VI, VII). The direction of slope of land in relation to the prevailing wind and rain may be of greater importance than being on the east or west slopes of the main range.

As at Ulu Sempam there are different floral associations around the rock streams where, in damp surroundings, there is a low dense growth dominated by species of *Saraca* and *Pometia*. The close proximity of Kuala Lumpur is increasingly threatening Ulu Gombak and its flora and fauna. Land development for cultivation is taking place in the lower parts of the valley, for tourism in the Genting Highlands, north of the pass over to Pahang, and timber is extracted from nearby valleys as well as from the lower parts of Ulu Gombak. This activity may be responsible to some extent for the high density of primates within the valley. This advantage for detailed study, however, is more than offset by the rugged terrain which precludes continuous observation of hylobatid groups from dawn to dusk. But excellent discontinuous observations may be made from the roadside. Now the valley is to be further disrupted by the construction of a highway and a tunnel through into Pahang.

Study Methods

Siamang groups were studied in detail in the two selected study areas from January 1969 until May 1970 inclusive. Observations were made in Ulu Sempam in the first half of each month on 173 days during 1,651 h in the forest (forest time), and in the second half of the month at Kuala Lompat on 190 days during 2,318 h, with the assistance of my wife. Siamang were under observation for 35.6% forest time in Ulu Sempam, and for 73.8% forest time at Kuala Lompat. There were 20 additional days when botanical activities and so forth were carried out in the two study areas, and 52 days (400 h) of roadside observation in Ulu Gombak. Although observation conditions were potentially better in Ulu Gombak, siamang could only be observed for 8.4% time (Appendix A).

The proportion of forest time during which the siamang were under observation is a reflection of the relative difficulties posed by the terrain and in the habituation of study groups in the two areas. At Kuala Lompat field work was relatively easy on the flat open forest floor with good visibility and

Plate V. Kuala Lompat. Pandans and buttressed trunks of common swamp species of Euphorbiaceae in a seasonally flooded part of forest (above). Habit of strangling fig L242, *Ficus stupenda*, on host *Koompassia malaccensis* – early stage, note aerial roots and straight pale branches (lower left). Strangling fig L287, *Ficus sumatrana*, on host *Dipterocarpus* sp. – later stage, note massive fig root and dense foliage of other climbers (lower right).

Plate VI. Morning mists over the Trengganu hills looking seawards (above); the hill forest of Kelantan looking west towards the Main Range (middle); and forest clearance for rubber planting in south Pahang (below).

Plate VII. Ulu Gombak. Looking north-west from road at 21st mile across territory of siamang group GS16C towards Gunong Bunga Buah (above). Aerial view of forest around road at 21st mile from north-west. Territories of GS6 above road, and of GS16C, GS16B, and GS16A from left to right below road (below).

primates that were used to the presence of man. No attempt was made to hide from the siamang, which showed excellent ability to detect the observer's presence. For detailed quantitative observations to be possible it was important that the siamang should come to accept the continual presence of the observer. Because of the exclusively arboreal habits of the siamang there was a greatly reduced risk of the observer's presence interfering with patterns of behaviour, after the initial habituation period.

Despite the many difficulties, this study of the behaviour of these quiet, elusive, arboreal primates was greatly facilitated by (1) small size of social groups, (2) the small area of forest used, (3) occasional loud group calls, (4) regularly used arboreal pathways, and (5) habitual patterns of activity with the small group acting in close concert.

1. Habituation

Thus before satisfactory observations could be made it was necessary that the observers should become well acquainted with the forest and the general behaviour of the siamang, especially their main pathways through the trees, and that the siamang should accept the presence of the observers. Difficulties were aggravated in Ulu Sempam, where the siamang were more shy than expected, presumably because of the recent logging activities. The rugged terrain and denser vegetation made it difficult to sustain the chase once the siamang group was located, so that it took much longer to learn the main arboreal pathways.

It appeared that the siamang were more curious than fearful. Thus, rather than flee a considerable distance when located, the group would, until habituated, rapidly move a short distance, and then sit quietly for a long time, presumably watching the observers. The basic problem in learning the siamang territory was to know when and where this happened. Persistent chasing was not very successful in Ulu Sempam, and after patrolling the area in which the siamang were known to be each day, the two observers would separate. Contact was maintained by two-way radio and/or whistle signals between the most likely vantage points in an attempt to locate the group. It was important that siamang be seen before the group saw the observer and moved out of view. It is difficult to sit still in the tropical rain forest, with all the activity that occurs therein – pleasant and unpleasant. A 'good book' markedly reduced the tendency for movement, so that an approaching group could be heard before it came into visual range, enabling the observer to be prepared.

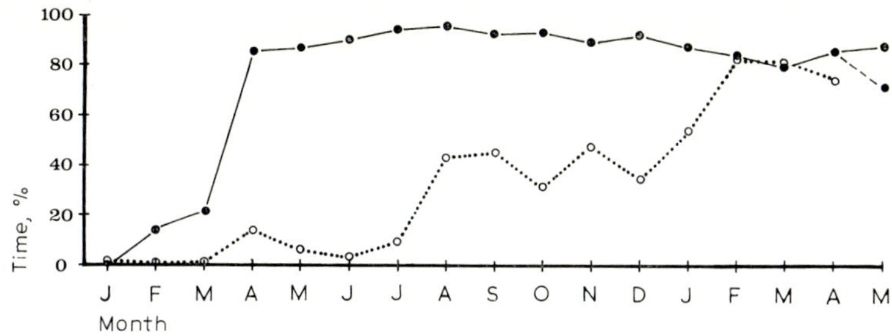

Fig. 1.6. Siamang observation time as a percentage of forest time for each monthly observation period in Ulu Sempam (open circle and dotted line) and at Kuala Lompat (closed circle and continuous line) illustrating contrasting patterns of habituation. The slight decline at the end of the study reflects an increase in activities other than observing siamang, e.g. plant collection. The dashed line shows the difference for the last month at Kuala Lompat.

From January to July observations of siamang at Ulu Sempam were limited to chance encounters and the discovery of fruiting trees. In August a large fig tree was fruiting, which was well placed for observation from the slopes above. As the siamang fed there at least once each day, it provided the first real opportunity for tracing the group arboreal pathways, and for siamang and observers to become accustomed to each others' presence. This marked the first phase of the breakthrough that had occurred at Kuala Lompat in April after only 30 days in the forest. It was not until the following February that the group accepted the observers' presence so that they could be observed from dawn until dusk. This delay was caused mainly by the untimely advent of the rainy season, and by less favourable ranging patterns of the study group.

Once observers and siamang were habituated there were few problems. As long as the group was located before leaving the sleeping tree, it could be followed and observed continuously throughout the day. Unless one knew precisely where to find them, the siamang could not usually be located during the day, because of their quiet habits. This, and habituation problems, did not make it feasible to study in any depth more than two groups during this 18-month period. The same energy and time expenditure would have been necessary to observe more than one group in one area, but there would not have been the great advantages of having information from different parts of the country.

From April 1969 for 14 months 85 – 95% of time in the forest was spent observing siamang at Kuala Lompat (fig.1.6). This level was achieved after

34 days (294 h) in the forest, and about 40 h of observation. The greater ranging activity of the siamang group in the dry season in many ways makes it a less favourable time for starting such a study (Chapter 2). At other times of the year the weather is not so favourable, and although the siamang may be easier to follow, they are more difficult to locate because of their reduced activity. In Ulu Sempam it was 139 days and 1,245 h in the field before contact was maintained with the siamang group for more than 80% forest time.

2. Data Collection

Observations were made to produce a quantitative description of the siamang habitat, and of individual behaviour and social organisation within and between groups, through an annual cycle. Although a vast amount of information has been accumulated, a fuller statistical analysis and interpretation must await the collection of further data over a longer time period. Thus the emphasis here is on description, and on the selection of sampling methods for the various parameters measured:

(1) Habitat. In each month, for each study area, the position and state (e.g. flowering, fruiting, leaf flush, leaf fall) of trees important to the siamang were recorded. Thus a botanical description (quantitative in part) of each study area could be produced. Trees were counted, classified according to height, and identified where possible along transects in the Krau Game Reserve. In plots 15 m wide, and 300 m long, 1.5 m^2 sheets of polythene were set in two rows of 20 at intervals of 15 m. This separation was devised so as to ensure accurate sampling of fruit fall from the trees above.

In addition to these botanical data for detailed comparison with siamang behaviour, the variety and numbers of other animals – especially primates – observed while following the siamang were recorded.

(2) Weather. Each 15-min interval from 0600 to 1900 h inclusive (day time) was scored for sunshine, cloud cover, or rainfall. (Each hour was scored as sunny, cloudy, wet, or variable.) A recording thermohygrograph was set in a Stephenson screen in a clearing to obtain a continuous record of temperature and relative humidity. Maximum and minimum thermometers were read each morning in the clearing, on the forest floor, 40 ft above the forest floor, and occasionally in the canopy at 100 ft. Rainfall was measured daily at 0700 h from a rain gauge set in the clearing.

(3) Calling. The time of first gibbon and siamang calls was noted each morning. The onset time and duration of subsequent calls were also noted. Where possible, groups would be identified by compass direction and estimated distance from the observer. A record was kept of the calling behaviour of each hylobatid group in the vicinity of the study area, and of the hours of the day in which each primate species was heard calling. This study was aided by using a Uher 4000-L Report tape recorder, thereby providing material for an analysis of vocalisations.

(4) Group behaviour. The times that group activities started and finished each day were noted. In one notebook the time, group location, direction and distance of movement with timed sequence of individual activity, tree number, and activity were recorded. At the end of each day this enabled a day range to be plotted, and group times spent in the major activities of rest, feed, and travel to be calculated. Data were available for an analysis of territory use, activity patterns, grooming, travel order, and so forth. Where possible behaviour was photographed with a Pentax Spotmatic 135 mm camera with a 200- or 400-mm telephoto lens, or with a Bolex K1 8-mm zoom lens cine camera.

(5) Individual behaviour. Each day one group member was selected as subject. In a second notebook the behaviour of this individual was recorded almost continuously – certainly at 1-min intervals – from dawn to dusk. This gave a more precise measure of time spent in each activity, and of social relations, which could be analysed in terms of age and sex. It provided a detailed measure of interactions, especially grooming, and of postures, locomotion, and feeding behaviour. Each animal was the subject for at least two consecutive days each month; each study group was observed for at least ten consecutive days each month (where possible – this was not achieved until near the end of the study in Ulu Sempam).

As the study developed it became possible to sample the major activity and location of each member of the group at 10-min intervals. In some months the distance of each animal from the subject was recorded at these intervals, to give a measure of spatial relations within the siamang group during the day, and in different activities. In trees where the group fed openly for long periods each group member was sampled at minute intervals for posture, handedness, and/or rates of feeding and, sometimes, frequency and distance of movement about food tree. Relative positions of group members within a tree were often recorded diagrammatically at regular intervals.

These observation methods will be amplified during the presentation and discussion of results in subsequent chapters.

A Siamang Day

As an introduction to the main text, and to facilitate a superficial interspecific comparison [see for example JAY, 1965; SCHALLER, 1965; JOLLY, 1966; ELLEFSON, 1967; KUMMER, 1968], an outline of the siamang daily activity cycle will now be presented.

Day dawns soon after 0600 h local time at these latitudes. At first light the siamang family group stirs; the adult male, followed by juvenile, moves in from his sleeping position in the terminal branches of one of the large emergent trees. The subadult, perhaps in a neighbouring tree, also stirs. After several minutes one siamang starts urinating; soon the air is full of falling excreta, as the group urinates and defaecates in concert. The female and infant, the last to move, eventually join in. If there is a tree in full fruit nearby, the male may lead the group to feed immediately; otherwise, they will rest further until after the time of sunrise – especially on cloudy days.

After the first feed, which may involve several trees, and last until about 0900h, the group occasionally moves up into an emergent tree to call in concert in response to a neighbouring group or some other stimulus. Although the male may start this calling, it is the female who directs the pattern of the chorus. After about 15 min the calling may be taken up elsewhere, and the group rests and grooms for about 30 min – sometimes together, sometimes in pairs – before the female continues to lead the group around the small territory. As they move from one food tree to the next, the male carries the infant, and the subadult follows behind.

Infrequently the group may meet a neighbouring group on their boundary. In this event, the female and young hide, and the males (including subadult if present) chase to and fro intermittently for 15 – 60 min. The family group then resumes the pattern of brief periods of travel alternating with long bouts of feeding. In the early afternoon, in the heat of the day, activity decreases as individuals rest and groom for more than an hour. If satiated, or if heavy rain threatens, the group will move up into a sleeping tree before 1600 h. But there may be a final feeding bout before they settle for the night – usually by 1700 h, sometimes after 1800 h on very active days.

The group is usually active for about 10.5 h each day, of which just over half the time is spent feeding. They will have travelled about 900 m around their small territory of about 25 ha. There are considerable fluctuations (short- and long-term) according to the stage of (1) the annual climatic cycle, (2) the longer floral cycles, (3) the 2- to 3-year reproductive cycle, which effect the ranging behaviour and patterns of social behaviour, respectively [CHIVERS, 1971a]. Comparison with observations on captive siamang confirms the following stages of maturation: (1) infant (two phases – at first dependent on female, then on male), from 0 to 3 years; (2) juvenile (independent young) from 3 to 5/6 years; (3) subadult (or adolescent) from 5/6 to 8/9 years, and (4) adult (physically *and* socially mature) over 8/9 years. (The composition of study groups will be described in Chapter 2 under Territory Size.)

Weather and Seasonality in Central Malaysia (1969 – 1970)

Seasons are not usually marked in the humid tropics. It has already been mentioned that the Malayan climate is characterised by an annual period of increased rainfall (often with serious flooding) during the north-east

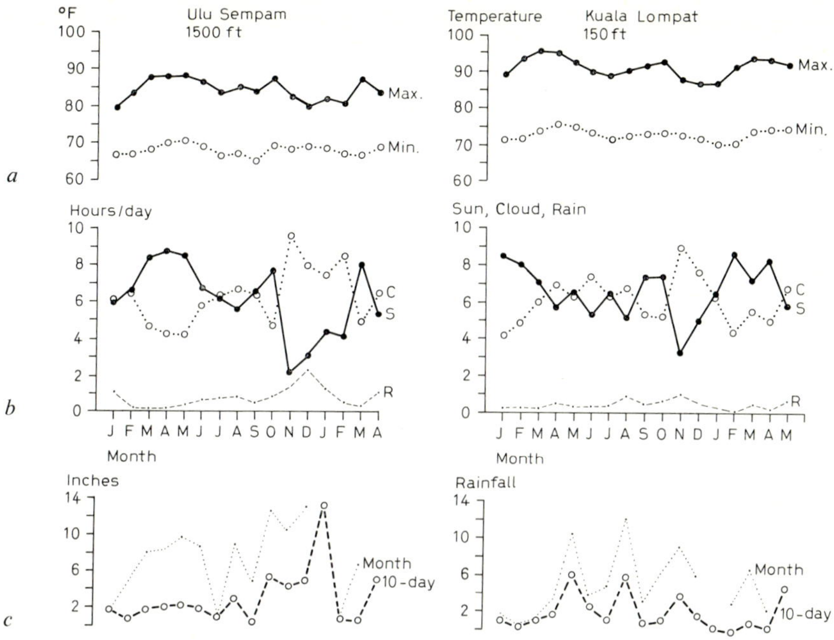

Fig. 1.7. Mean daily values in each month of (a) maximum and minimum temperatures (°F); (b) sun, cloud and rain, hours, and (c) rainfall, inches, in Ulu Sempam and at Kuala Lompat.

monsoon at the turn of the year, followed by a prolonged period of high temperatures and low rainfall. The timing and duration of these seasons varies from year to year; at other times the weather is less predictable.

In this study three criteria were used to define the limits of the wet and dry seasons: (a) mean daily maximum and minimum temperatures; (b) mean daily hours of sunshine and cloudcover, and (c) the amount of rainfall (number of inches for each 10-day observation period, and for each month). These criteria give slightly differing results (fig.1.7), which highlights the problem. But there are seasonal changes which exert a profound effect on the forest flora [MEDWAY, 1972b], and possibly thereby on the behaviour of the fauna.

Daily shade temperatures varied between 61 and 92°F in Ulu Sempam (1,500 ft), and 67 and 98°F at Kuala Lompat (150 ft above sea level). The highest temperatures were recorded in March and April, the lowest between July and September; this is contrary to the expectation of greatest extremes in the dry season, when the nights tend to be cloudless. Nevertheless, the wet season was characterised by the lowest range in daily temperatures, and the dry season by the greatest daily change.

In both years at Kuala Lompat the dry season started in mid-January and lasted until April; it was a little later and shorter in Ulu Sempam, especially in 1970 (fig.1.7). There was nearly 9 h of sunshine daily on average, and less than 2 in of rain in 10 days. The wet season started early in November 1969 in both areas (after a dry, sunny period) ending in January at Kuala Lompat, and early February 1970 in Ulu Sempam. The increase in cloud cover (up to 8 – 10 h/day) was as much a characteristic of the wet season as the increase in rainfall (up to 15 in in 10 days, with rain in 2 h of daylight). The weather was more variable in the intermediate season (fig.1.7; April to September at Kuala Lompat, for example). There was less rain than usual in 1969 in both areas [Dale, 1959], which helps to explain the fluctuations from month to month. The secondary rain season (north-west monsoon) was not distinct in 1969, but was developing in May 1970 as the study was being concluded.

The degree of change in some aspects of behaviour during the course of the study at Kuala Lompat can be more clearly demonstrated by grouping the monthly scores into three nearly equal periods: (a) May to August; (b) September to December, and (c) January to April inclusive (Chapters 2 – 6). The latter periods cover respectively the wettest, coolest, and cloudiest months of November and December – the wet season – and the driest, hottest, and sunniest months of February and March – the dry season. One month either way makes little difference to the seasonal patterns observed, and so there is no cause for concern in adding data from April 1969 and May 1970, in some cases, to the intermediate and dry seasons, respectively.

While it is clear that in some cases changes in behaviour are related to these weather changes, in others different factors are involved to varying degrees. Efforts have been made to investigate the various ways in which weather affects each aspect of behaviour: (1) directly from hour to hour, or day to day; such an effect may be cumulative and not obvious for 2 – 3 days, especially in a higher mammal able to be more independent of small environmental variables, and (2) indirectly, including the short-term effect on vegetation, after 20 days or more; longer term consistency composes seasons which are less marked and regular in the humid tropics. Annual cycles are the other main pattern imposed on the vegetation, which have probably become an evolutionary expectation to the forest fauna.

Chapter 2: Ranging Behaviour and Territory

Introduction

Prolonged observation reveals that most primate social groups restrict their activities to a limited area [DeVore, 1965]. This occupied area, which provides all necessary resources, is referred to as the home range for the resident group, but problems arise for the observer in determining the boundaries [Burt, 1943] and in describing differential utilisation within these limits [Altmann and Altmann, 1970]. In some cases all or part of this home range may be defended; the concept of territory among primates has been much discussed in recent years [Chivers, 1969; Bates, 1970; Wilson, 1972]. Debate has centered on one or more of the following diagnostic features: an area used by the resident group (1) habitually and uniformly, (2) exclusively from conspecifics being, (3) defended and/or (4) otherwise clearly defined by behaviour or topography. Kummer [1971] described territory as a 'field of repulsion... fixed in space' containing all the resources required by the inhabitants for permanent life. Unlike home range, which does not involve such marked repulsion, many of the resources have been 'created' through time by the resident group. Such a relationship between animals (with minimal conflict), and between animals and space, shows a high degree of stability.

Because the white-handed gibbon in Malaya fulfilled all the conditions outlined above, Ellefson [1967] applied the concept of territory in detail. Preliminary observations indicated a similarity between siamang and gibbon [McClure, 1964; Kawabe, 1970]. For the present, the area of tropical rain forest habitually used by a siamang family group will be referred to as a territory. The aim of this chapter is to determine the extent of territorial behaviour in siamang, and the ways in which this behavioural complex differs from that shown by the gibbon.

Following a description of the ranging patterns of two groups, and an analysis of how the limits and intensity of ranging varied at different times of year, and under contrasting weather conditions, there will be an investigation of the differential use of each part of the territory, in terms of hectare quadrats.

There will be a description of the dispersion of night sleeping trees, and of territorial maintenance with an assessment of territory size. In this way it will be shown to what extent a siamang group has exclusive use of its territory, and how it is defined by behaviour (including defense) and topography. In conclusion, the application of the term 'territory' to the siamang in particular, and primates in general, will be discussed.

Measures of Movement

Methods

Five measures of group daily movement have been used in this analysis:

1. Day range (DR) is the linear distance travelled by the siamang family group in single file progression between dawn and dusk.

2. Number of hectares (nH) is an areal expression of day range. The study area map was overlaid with a grid of 100 m unit length, so that each square represented an area of 10,000 m^2 (equals 1 ha). Each entry of a quadrat is scored even if it is a return visit on the same day.

3. Night position shift (NPS) is the distance between sleeping trees on consecutive nights.

4. Maximum radius (MR) is the distance between the night sleeping tree and the farthest point from it to which the group travels during the following day.

5. Maximum radius time (MRt) is the time taken from leaving the sleeping tree to reach the farthest point from the night position.

These parameters reflect different aspects of daily movement. Day range provides a measure of actual distance travelled, and from plots of day range the number of hectares covered daily can be counted. It may be the only way to record the day range in species where the spread of a moving group is considerable, and where no linear plot can be determined. Alternatively, the number of different quadrats visited daily provides an extra dimension, because it is not necessarily correlated with day range, and cumulative nH scores are of special value in an analysis of differential patterns of territory use [Altmann and Altmann, 1970; Struhsaker, 1967]. Night position shift and maximum radius shed further light on the pattern of day range, with particular reference to start, finish, and extent. MRt, with the number of hectares covered in each 3-hour period (see below), and the number of minutes of travel in each hour (Chapter 4), reveal the timing of movements about the territory.

Table 2.1. Overall measures of daily movement by siamang groups at Kuala Lompat and Ulu Sempam

	Kuala Lompat, TS1, 148 days			Ulu Sempam, RS2, 30 days		
	$\bar{x}$	M	R	$\bar{x}$	M	R
Day range, m	969	820	320–2,860	800	778	485–1,390
Night position shift	293	270	0–560	186	193	0–395
Maximum radius	361	340	110–730	260	260	120–400
MR time, min	422	438	99–687	362	352	20–636
Number of hectares	9.1	8	5–23	7.0	7	3–11
Percent territory	20	17	11–50	30	30	17–48
NPS:DR	0.30			0.23		
DR:NPS	3.31			4.31		
NPS:MR	0.82			0.71		
MR:NPS	1.23			1.41		

Together these measures provide sufficient information for an analysis of ranging patterns, the extent of ranging being balanced with the intensity of use of different parts of the territory. The means, medians[1], and ranges of each parameter are given in table 2.1 for 150 days of observation of the siamang group at Kuala Lompat (TS1), and for 30 days in Ulu Sempam (the group hereafter being referred to as RS2).

Results

Day range was nearly 1 km on average (usually 0.8 km, varying between 0.3 and 2.9 km at Kuala Lompat). In both study areas NPS was about 0.25 DR, and MR approached 1.5 NPS (table 2.1). On only about 12% days did NPS equal MR. TS1 used about one fifth of the 46 hectare quadrats each day, whereas RS2 covered one third of the smaller area of 23 ha in Ulu Sempam. Only one quadrat was revisited on the same day on average. It was usually 6–7 h from leaving the night position before the siamang group reached the farthest point in its day range. There was, however, considerable variation in all parameters from day to day, which must now be investigated.

1 The close similarity between medians and means confirms the usefulness of the latter, although comparisons are made by non-parametric twotailed tests as described by Siegel [1956]. Unless otherwise stated, all probabilities of less than 0.05 are regarded as significant departures from the null hypothesis, of less than 0.01 as highly significant. See Chivers [1972b] for a more detailed statistical analysis of the results in each chapter.

The 140 scores for each parameter, spread evenly over 14 months at Kuala Lompat, were tested by the Kruskal-Wallis one-way analysis of variance test. There was a highly significant difference (d.f. = 13, corrected for ties) between day ranges (H = 85.1, p = 0.001), maximum radii (H = 44.8, p = 0.001), and MRt (= 33.5, p = 0.01 – 0.001), but only a significant difference between night position shifts (H = 22.6, p = 0.02 – 0.05).

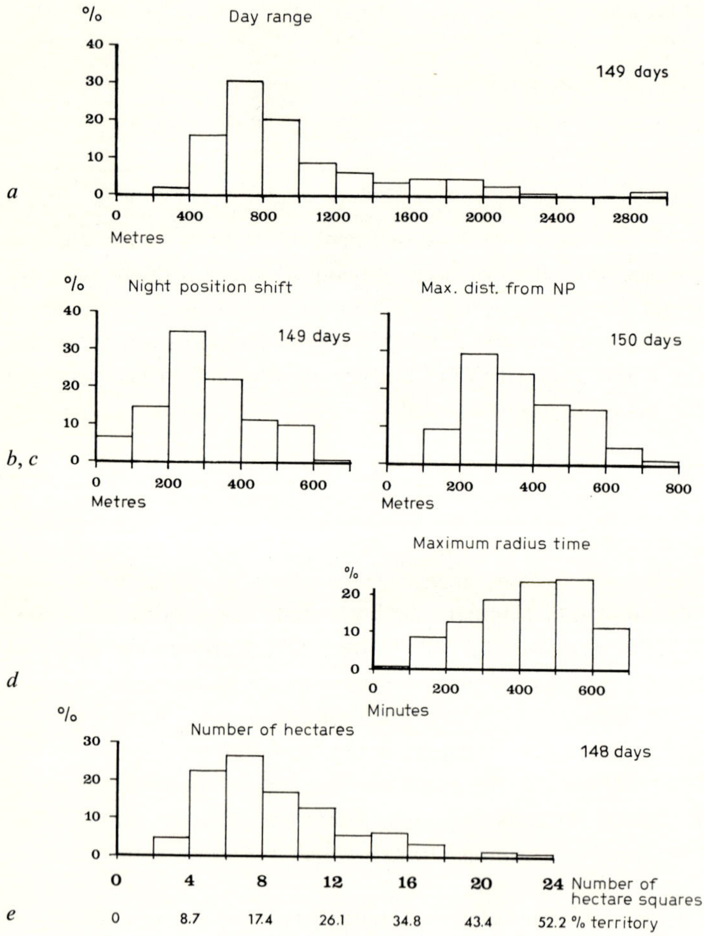

Fig. 2.1. Distribution of observations of (a) day range, (b) night position shift, (c) maximum radius, (d) maximum radius time, and (e) number of hectare quadrats for TS 1 at Kuala Lompat on 148 days from April 1969 to May 1970 inclusive.

1. Temporal Variation in Movement

Plotting the daily measures of movement overall in frequency histograms clearly demonstrates the pattern of scores for each parameter (fig. 2.1). Thus 30% of DRs were between 600 and 800 m, with the frequency approaching 50% in the 'wet season'. More than one third of NPSs were between 200 and 300 m, which was also the peak for MR. Nearly 50% of MRts fell between 400 and 600 min. Over 25% of day ranges covered 7 or 8 hectare quadrats. These patterns were more distinct in the 'wet season'. In the 'dry season' nearly 20% of DRs were between 1,800 and 2,000 m; 25% of NPSs and 40% of MRs were between 500 and 600 m; nearly 25% of MRts were between 300 and 400 min; and nearly 60% of day ranges covered 11 – 16 hectare quadrats.

DR and nH scores were doubled on average in the 1970 'dry season', so that DR was nearly five times NPS and three times MR. These increases in

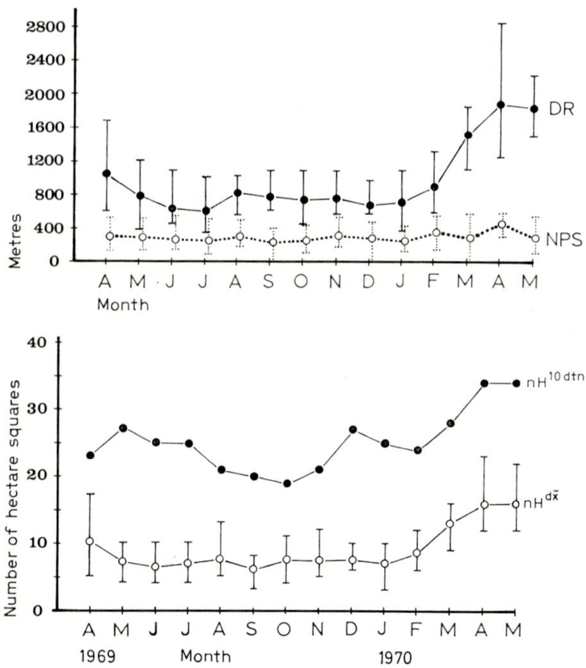

Fig. 2.2. Measures of movement for TS1 at Kuala Lompat (mean and range for each 'monthly' score). Day range, DR, and night position shift, NPS (above); number of hectares, $nH^{d\bar{x}}$ = daily mean, nH^{10dtn} = number of different squares covered in 10-day observation period (below).

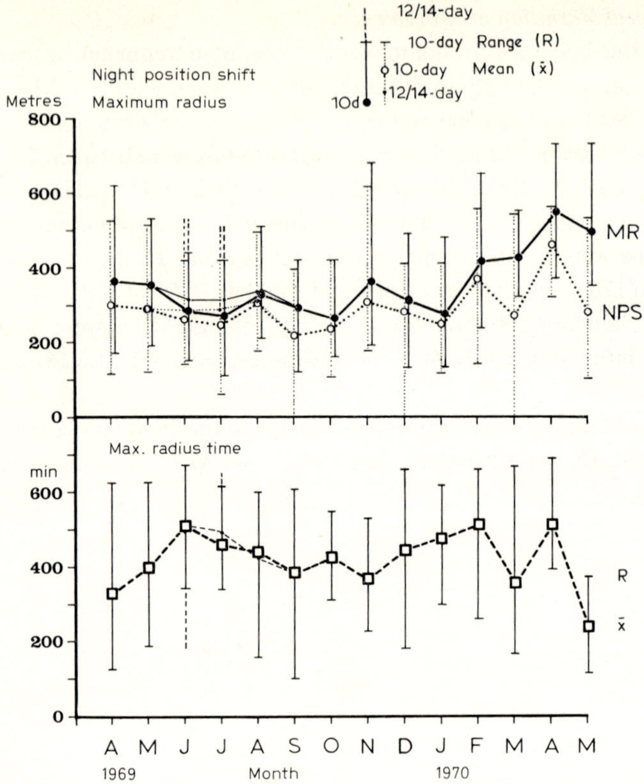

Fig. 2.2 (continued). Measures of movement for TS 1 at Kuala Lompat (mean and range for each 'monthly' score). Night position shift, NPS, and maximum radius, MR (above); maximum radius time, MRt (below).

all parameters (with a slight decrease in MRt) were highly significant for DR, MR, and nH, and significant for NPS. Although DRs fluctuated within relatively constant limits between June and January, there was a highly significant increase between February and March, which was sustained until the end of the study (fig. 2.2).

There was no corresponding increase in NPs because longer DRs tended to return the group to near the starting position, e.g. March; whereas a short direct DR may nearly equal NPS, and occasionally MR, e.g. December. Similarly, a longer DR does not necessarily mean that the MR was reached more quickly; MRt was inversely correlated with DR only when the same rate of group movement was achieved.

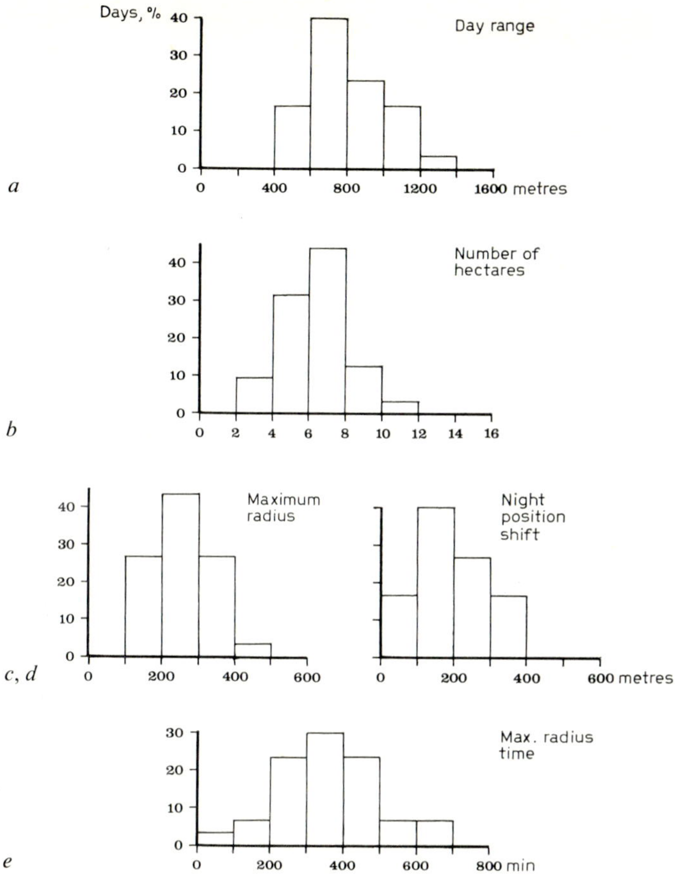

Fig. 2.3. Distribution of observations of (a) day range, (b) number of hectares, (c) maximum radius, (d) night position shift, and (e) maximum radius time on 30 days in Ulu Sempam (RS2) in 1970.

The calculation of Spearman rank correlation coefficients on the monthly medians for each measure in certain months shows MR and nH to be highly correlated with an increase in DR, whereas NPS shows a less significant increase, and the probability of MRt being negatively correlated is 0.20 – 0.10. A Kendall partial rank correlation between DR and MR, with the third variable NPS held constant indicated that DR and MR were independent of NPS ($r_{dm \cdot n} = 0.70$, $r_s = 0.71$). But a partial correlation of DR and NPS, with MR held constant, indicated the relative dependence of NPS on the other variables ($r_{dn \cdot m} = 0.12$, $r_s = 0.53$).

There was a similar constancy in the number of hectare quadrats covered daily until March 1970. But there was a greater fluctuation in the total number of different quadrats visited in each 10-day period, with increases in May to July and December to January inclusive (fig. 2.2). This reflects more extensive ranging than from August to November. Thus, although the hectare grid has no biological significance as such, it provides a useful analytical technique in the study of ranging patterns, as will be elaborated below.

In Ulu Sempam absolute values for RS 2 for each parameter were smaller, but larger relative to the small area covered (table 2.1). Until these differences in area are corrected for (see below) a more detailed comparison is not justified. 40% of DRs were between 600 and 800 m covering 7 or 8 ha; 40% of MRs were between 200 and 300 m and the NPS peak was at 100 – 200 m. On 30% of days MR was reached in 300 – 400 min (fig. 2.3).

2. Effect of Weather on Movement

The effect of weather on measures of movement was more precisely investigated by selecting days of weather extremes and analysing the ranging data on those days. There were 21 days when the air temperature in the shade reached 94 °F, and when the sun was shining for more than 80% of daylight hours. (Data collected from different levels of the forest indicated that readings

Table 2.2. Seasonal variation in daily measures of movement at Kuala Lompat (TS 1)

	'Season'		
	intermediate, April – August, $\bar{x}$ 50 days	wet, Sept. – Dec., $\bar{x}$ 40 days	dry, January – May, $\bar{x}$ 49 days
Day range, m	808	725	1,358
Night position shift	294	257	320
Maximum radius	334	305	440
Maximum radius time, min	435	404	422
Number of hectares			
Total	7.7	7.1	11.6
Different	6.8	6.4	10.9
NPS:DR	0.36	0.35	0.24
DR:NPS	2.75	2.82	4.24
NPS:MR	0.88	0.84	0.73
MR:NPS	1.14	1.19	1.38
DR:MR	2.45	2.38	3.09

taken in the clearing in the shade were a good approximation of the temperature in the main part of the canopy.) In contrast there were 17 days when the temperature did not rise above 85°F, and more than 80% of daylight hours were cloudy. All but two of these cloudy and wet days occurred during 1969, whereas ten of the hot and sunny days were in 1970.

First the hot and sunny days were compared between the dry season and the wet and intermediate seasons to see if the increase in ranging in the former can be explained directly in terms of weather. The means ($\bar{x}$), medians (M), and ranges (R) of the measures of movement on these days (table 2.3) show

Table 2.3. Variation in daily measures of group movement at Kuala Lompat (TS1) and Ulu Sempam (RS2) on days with contrasting weather

	Hot and sunny			Cloudy and wet		
	$\bar{x}$	M	R	$\bar{x}$	M	R
TS1	21 days (10 + 11)			17 days		
Day range, m	1,081	925	370–2,860	698	685	320–1,100
1969	849	850	370–1,675			
1970	1,336	1,235	710–2,860			
Night position shift	324	320	0–560	315	305	125–540
1969	299	320	105–440			
1970	324	365	0–560			
Maximum radius	377	350	160–620	352	360	150–540
1969	340	350	160–620			
1970	417	390	270–600			
Maximum radius, time min	473	522	209–687	453	445	225–615
1969	459	546	209–648			
1970	487	484	296–687			
Number of hectares	10.7	10.0	4–23	7.2	8	4–11
1969	9.3	9.0	4–17			
1970	12.3	11.5	8–23			
Hectares / 3 h (0600–1800 h)	4.0:3.4:3.3:2.4			2.8:2.9:2.5:1.8		
1969	4.0:2.5:3.5:1.8					
1970	4.1:4.3:3.1:3.0					
RS2	3 days			4 days		
DR, m	1,002	985	960–1,060	743	698	525–1,040
NPS	188	195	160–210	149	160	40–235
MR	277	250	180–400	235	260	120–300
MRt, min	358	348	268–458	304	284	220–429
nH	8.3	8	8–9	7.8	7	6–11
nH/3 h	3.0:1.3:2.0:2.0			2.8:2.5:1.3:0.8		

quite a change between dry days in the different seasons, but the samples were small and only the increase in DR in 1970 was significant (Mann-Whitney U test, $p = 0.05-$). There was no significant difference for any parameter when comparing hot and sunny, with cloudy and wet days in the intermediate and wet seasons. But comparing the whole sample of each weather type DR and nH were significantly increased on hot and sunny days with an increase in distance moved in the first 3 h, but NPS and MR were not. MRt was decreased on hot and sunny days, but not significantly. The small sample for RS2 shows similar trends.

Thus, the apparent increase in ranging on hot and sunny days is mainly a function of the increased ranging in the dry season, which is not influenced directly by the weather. The main explanation must lie elsewhere, but could certainly involve climatic seasonality indirectly through its effect on the phenology of the flora.

Ranging Patterns

The outer limits of ranging during the study (= cumulative day range limits) were mapped out for TS1 (fig. 2.4) and for RS2 (fig. 2.5) with plots of the main arboreal pathways used. Also shown are the hectare quadrats (identified numerically), and, for RS2, the small area of overlap with adjacent groups. To investigate major movements and differential use the territory of TS1 was divided into three sections, that of RS2 into two. Following this gross sectional analysis, there will be a detailed investigation of differential ranging in terms of number of visits to, time spent in, and rate of coverage of hectare quadrats in each month, season and overall.

Plots of consecutive day ranges for RS2 in the last three months in Ulu Sempam can be compared with proportion of time spent in, and number of visits to, each hectare quadrat (fig. 2.6). The same information is presented for each month at Kuala Lompat in Appendix C, and summarised by seasons and overall in figure 2.7. In the latter, 10-day ranging limits in each month are plotted, rather than actual day ranges. It is these which give the greatest insight into the territorial boundary, particularly of a group with no immediate neighbours – which was the situation for TS1. It is clear that although ranging is confined to a small part of the forest, and that only part of the territory is intensively used, this part varies at different times of the year but almost all quadrats were visited at least briefly each season.

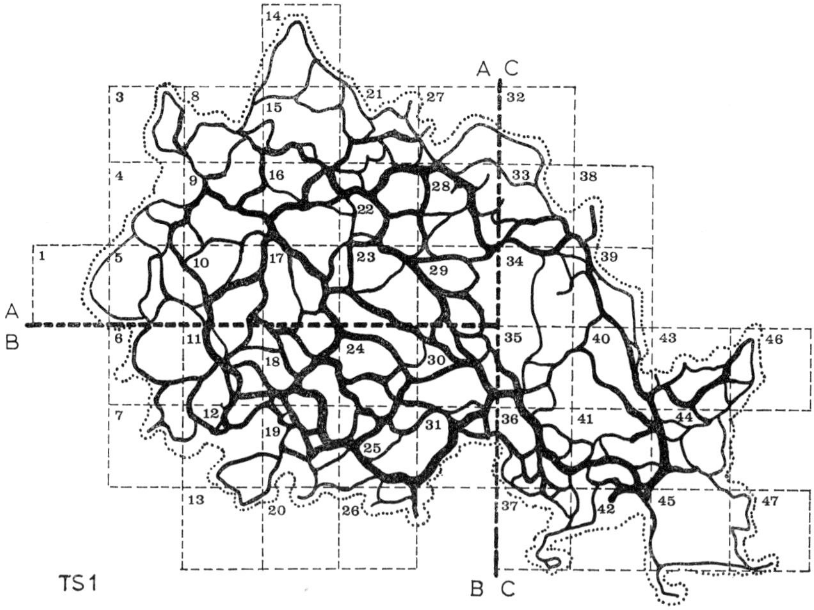

Fig. 2.4. TS1 territory at Kuala Lompat, showing cumulative day range limits, main arboreal pathways, grid of hectare quadrats, and sections A, B, and C.

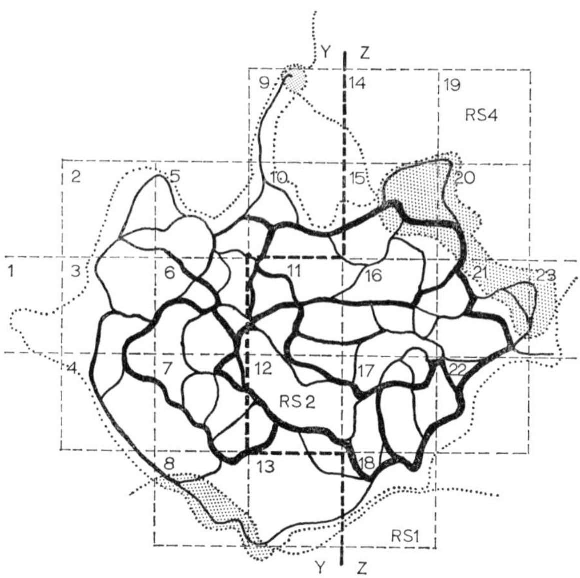

Fig. 2.5. RS2 territory in Ulu Sempam, showing cumulative day range limits, main arboreal pathways, grid of hectare quadrats, sections Y and Z, and areas of overlap with groups RS1 and RS4.

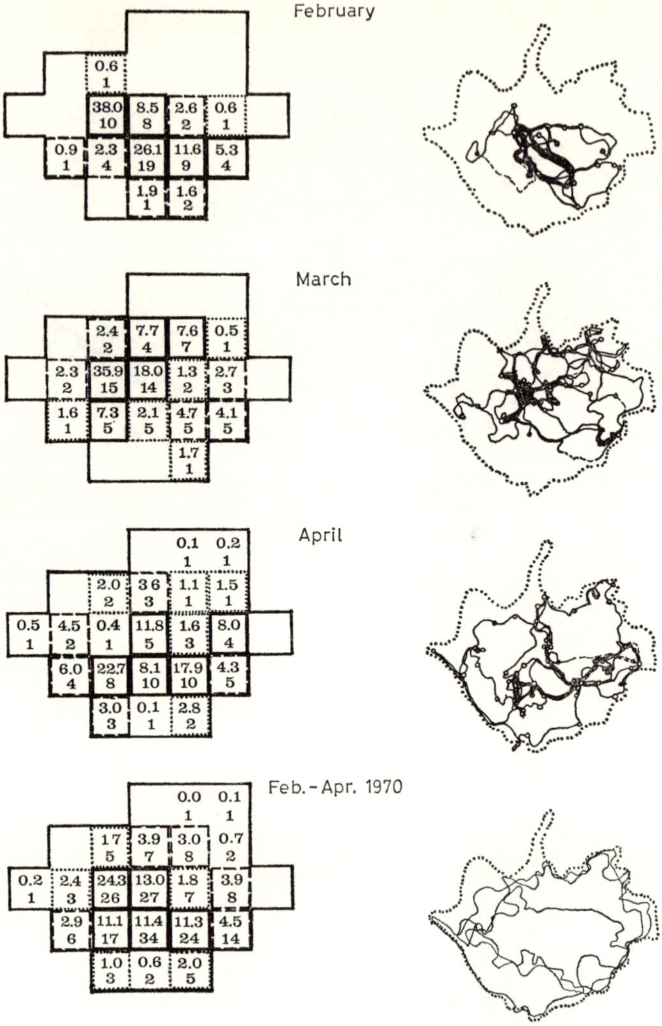

Fig. 2.6. Pattern of territory use by RS2 in Ulu Sempam showing number of visits and percent daily activity time in each hectare quadrat, and plots of day range (cumulative limits for each month overall).

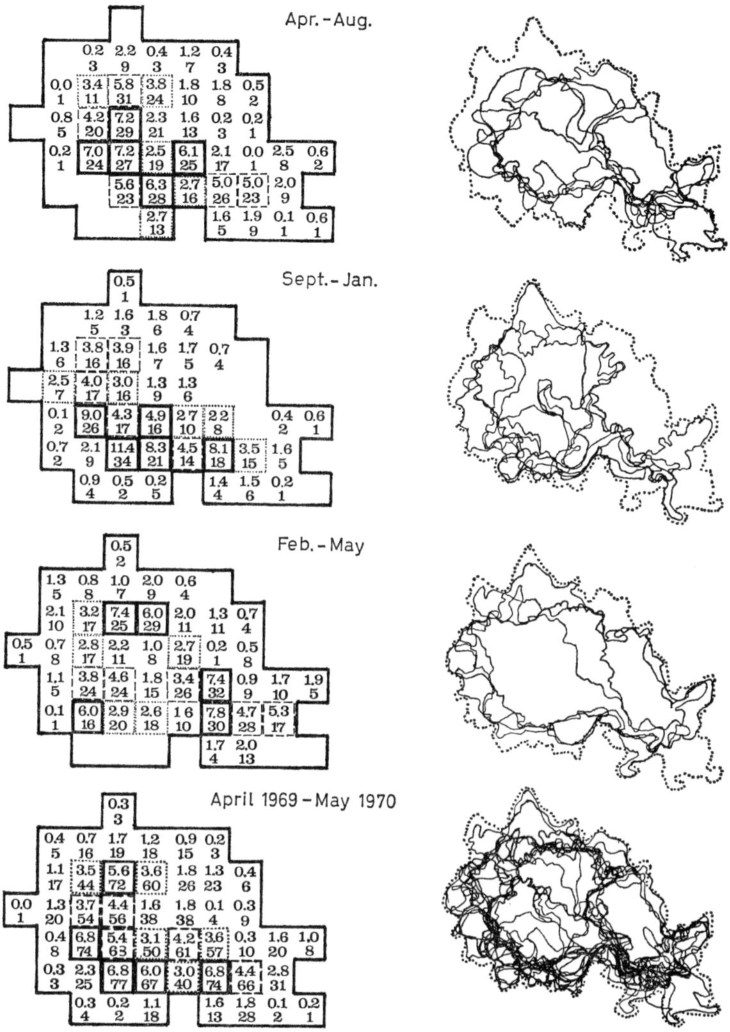

Fig. 2.7. Pattern of territory use by TS1 at Kuala Lompat showing number of visits and percent daily activity time in each hectare quadrat, and cumulative limits of day range in each month for each season and overall.

Table 2.4. Use of sections of territory by TS1

	Number of visits per ten days				Average interval between visits, days			Percent time		
	A	B	C	t	A	B	C	A	B	C
April	9	2	5	16	1.1	*5.0*	2.0	*68*	20	12
May	7	6	4	17	1.4	1.7	*2.5*	*42*	39	16
June	4	9	3	15	*2.5*	2.0	1.7	19	34	*47*
July	9	4	2	15	1.1	2.5	*5.0*	*48*	40	12
August	3	9	4	16	*3.3*	1.1	2.5	10	*63*	28
Mean	6.4	5.2	4.2		1.6	1.9	2.4	35	40	25
September	1	8	6	15	*10.0*	1.3	1.7	3	*63*	34
October	7	7	1	14	1.4	1.4	*10+*	49	49	2
November	7	10	2	19	1.4	1.0	*5.0*	28	*63*	8
December	5	5	7	17	2.0	2.0	1.4	24	21	*54*
Mean	5.0	7.5	3.8		2.0	1.3	2.7	30	50	20
January	6	7	1	13	1.7	1.4	*10+*	47	*52*	1
February	5	8	7	20	*2.0*	1.3	1.4	17	38	*44*
March	10	8	2	20	1.0	1.3	*5.0*	*66*	29	5
April	10	9	10	29	1.0	*1.1*	1.0	23	29	*47*
May	7	6	10	23	1.4	*1.7*	1.0	39	13	*49*
Mean	7.6	7.6	5.8		1.3	1.3	1.7	36	28	36

Forty-six squares: A = 17 (37%); B = 13 (28%); C = 16 (35%).
Italicised figures in Average interval column denote highest score = time to circle territory.
Italicised figures in Percent time column indicate highest score each month.

Sectional Analysis

TS1's territory was partitioned into three, partly on ecological grounds. Section C was separated from A and B by a swampy area running North to South. The canopy was continuous but lower, and the siamang group usually used one of only two main arboreal pathways into C from sections A and B (fig. 2.4). Thus, movement into section C was a distinctive feature of the ranging pattern. Section A and B were less distinct, but they denoted two main areas of activity – in and around hectares 16 and 19, respectively.

The following were calculated: (1) the proportion of time, and (2) number of visits per 10 days, (3) the mean interval between visits to each section in each month, and (4) the mean time to circle the territory each month (i.e. to visit all three sections and return to starting section). An alternative measure of time to circle territory is to take the maximum mean interval between visits to the same section (3), but this does not require that each other section be visited in the interval. Thus, not only can one indicate the main emphasis in monthly and seasonal ranging, but the rate of movement about the territory can be calculated (table 2.4).

Ranging was concentrated in section A in April, July, and March, in section B in August, September, and November, and in section C in June, December, April, and May. There was no significance to these patterns in the intermediate and dry seasons, but the concentration in section B in the wet season was highly significant (χ^2 test: d.f. = 2; χ^2 = 16.28; p = 0.001 –). Decrease in number of visits to a section was often countered by an increase in time spent in that section or in rate of movement about it.

In the intermediate and wet seasons sections A and B were visited every 1–2 days on average by TS1, and section C every 2–4 days (table 2.4). In the dry season each section was visited every other day at least, often every day. Thus, in April and May 1970 TS1 circled its territory every day, as indicated by the very long day ranges, but in October and January, when section C was not visited during the 10-day period, the circling time was much longer. In 10 days TS1 consistently visited 50–80% quadrats in section A. There was greater variability in quadrat visits in section C with 5–65% being visited in different months.

There was also considerable variation in the time spent in each section each month, even for section B (13–63%, table 2.4). There was a high correlation between sectional visit frequency and time spent in sections C and A (r_S = 0.87 and 0.75, p = 0.001 – and 0.01 –0.001, respectively), but a low correlation for section B (r_S = 0.47, p = 0.10 – 0.05).

RS2 visited each section, at least briefly, every day. Section Y covered the higher ridges; section Z the lower ridges and gullies. Thus RS2 effectively circled its smaller territory every 2–3 days; important food trees being situated probably both on the ridge tops, and down by the *Saraca* streams.

Quadrat Analysis

In each month (10 days) the number of visits to, and proportion of daily activity time (Chapter 4) spent in, each hectare quadrat were calculated (Appendix C). Thus each set of data were ranked, and a relatively stable pattern emerged (fig. 2.8). In each month TS1 spent between 48 and 67% time

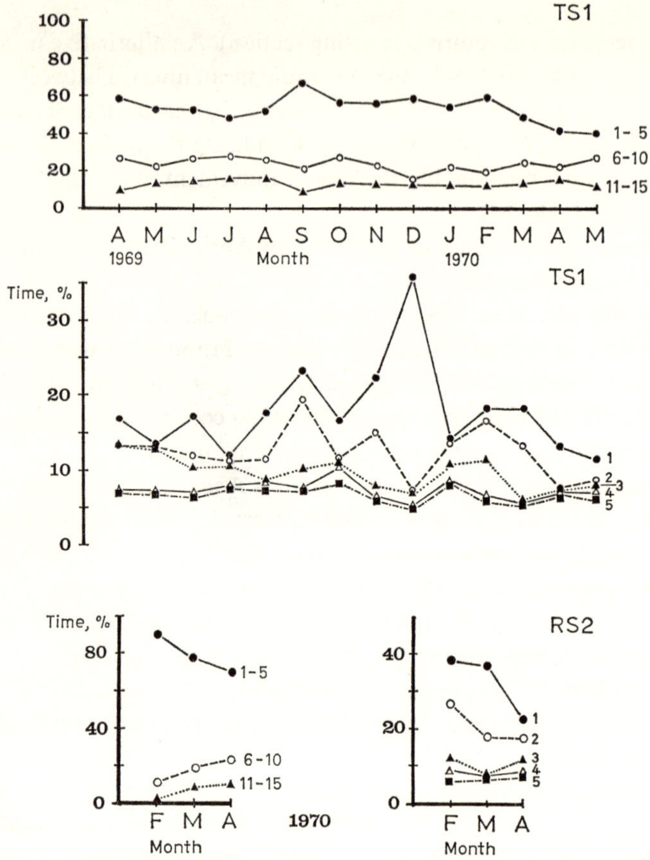

Fig. 2.8. Monthly patterns of territory use: percent of time group spent in quadrats ranked 1–5, 6–10, 11–15, and percent of time spent in quadrats ranked 1, 2, 3, 4 and 5 at Kuala Lompat, and Ulu Sempam.

in only 5 hectare quadrats (11% territory), although this score fell to 40% when ranging was at a maximum. Ten hectare quadrats accounted for 72–88% time, declining as low as 63% at the end of the 1970 dry season. RS2 also spent 70–90% time in nearly one quarter of the territory. The only variations that occured in this pattern emerge on considering hectares individually. Although the proportion of time spent each month in the quadrats ranked 3rd, 4th and 5th remained relatively constant, there were marked fluctuations for the two in which most time was spent (usually reciprocated within the top five), especially in December 1969.

No satisfactory way has yet been found to test the significance of these time differences. Testing the difference in the number of visits between suc-

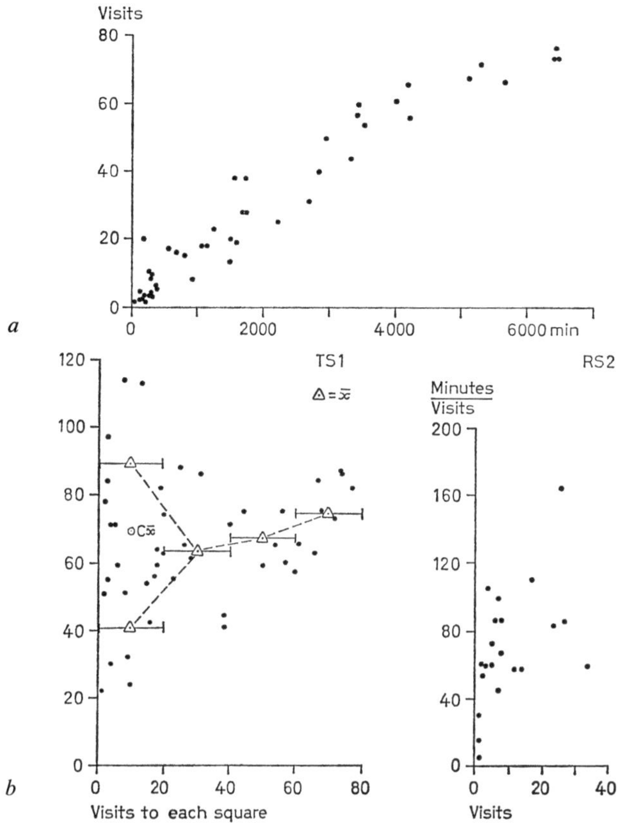

Fig. 2.9. Bivariate plots of (a) total number of visits and total time for each hectare quadrat of TS1 territory, and (b) total number of visits and mean visit duration for each hectare quadrat of TS1 and RS2 territories.

cessive months to the squares ranked 1 – 10 in turn by the Kolmogorov-Smirnov two-sample test produced no significant difference. The same was true for a seasonal treatment of the scores by χ^2 tests. But this does not necessarily mean that the time differences were not significant, for increased visit frequency to a square corresponded with more time per square, i.e. longer visits (fig. 2.9). Even considering in turn by the same method the ten squares visited most there was no significance for numbers 19, 11, 36, 18, 25, 41, and 30, but a significant difference for 16 (ranked 4th) and 35 (ranked 10th), with a highly significant difference for 22 (ranked 9th).

Measures of frequency of visit to each hectare ranked by season (rather than by month) and tested by the χ^2 test confirmed that there was usually no

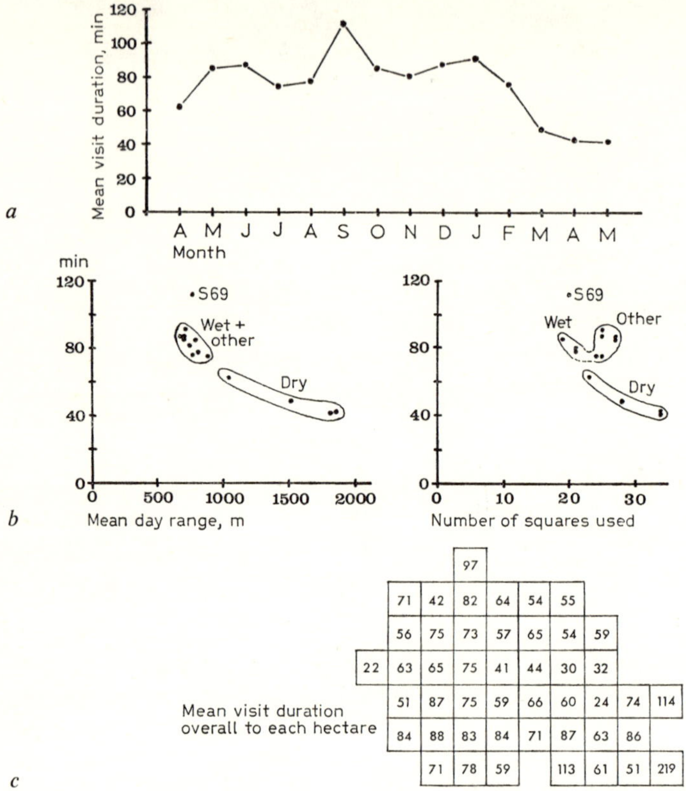

Fig. 2.10. (a) Monthly pattern of mean visit duration to hectare quadrats, (b) bivariate plots of mean visit duration (day range and number of hectares visited – monthly means), and (c) mean visit duration (overall) to each hectare quadrat of TS1 territory at Kuala Lompat.

significant variation between the number of visits to squares of each rank at different times of the year. But the decrease in visit frequency in the wet season (and increase in the dry season) was significantly different for quadrats 36, 41, and 16, and highly significant for 30, 22, and 35 (ranked 8th, 9th, and 10th, respectively). There was no significant fluctuation in visit frequency to the most used squares 19, 11, 18, and 25 (ranked 1st, 2nd, 5th, and 6th, respectively).

Thus, some of the quadrats visited most overall figure prominently in most months. But generally the small intensively used area shifts significantly from one month to the next, although the temporal pattern stays remarkably similar. This was most effectively demonstrated by calculating Kendall's coefficient of concordance for the two situations. While there was a highly significant concordance between the seasonal ranks of the ten squares in which most time was spent each month, there was no concordance between the seasonal rankings of visit frequency for the 10 identified squares used most overall (cf. W = 0.86 and 0.13, respectively).

Mention has already been made of the tendency for 'visits' to last longer in those squares most frequented. But of squares visited less often some were characterised by visits of long duration (mean visit duration = total time/total number of visits), whereas others were visited only briefly (fig. 2.9). Mean visit duration was calculated for all squares visited each month (fig. 2.10), as well as for each one overall (cf. Appendix C). There was a marked decrease in visit duration towards the end of the 1970 dry season, when day range increased significantly. In the intermediate and wet seasons the scores were relatively constant, except for a marked increase in visit duration in September 1969. In that month 42% time was spent in two squares – 19 and 31.

Extent and Rate of Ranging

Although the siamang group spent 77% of time on average each month in 22% of the territory (fig. 2.8), during this period TS1 covered between 40 – 60%, and RS2 60 – 90%, of their territories in terms of quadrats visited. Thus, in a period of ten days a small part of the territory was used intensively, although at least half the area was visited if only briefly. Over a longer period, such as the seasonal intervals containing 40 – 60 days data, ten squares accounted for only 50 – 60% activity time, and 80% of time was spread over nearly 40 % of the territory (almost double the 10-day figure), with 80 % of visits to nearly half the territory (the ten most used squares each month, in terms of visits and time are identified in Appendix C).

The cumulative proportion of time spent in each hectare quadrat (ranked) overall as a proportion of the territory size for TS1 and RS2 were similar (fig. 2.11). TS1 spent even more time (54%) in the most visited (50%) quarter of the territory (table 2.5). Time and visit frequency were closely correlated[2] (r_S = 0.93 – 0.97), but the latter was almost invariably a lower proportion of the whole sample. Curves plotted for specific months further illustrate these ranging patterns. In February RS2 was restricted in its movements, reverting

2 The value of comparing these to some extent dependent samples is to search for such associations that are not correlated as expected.

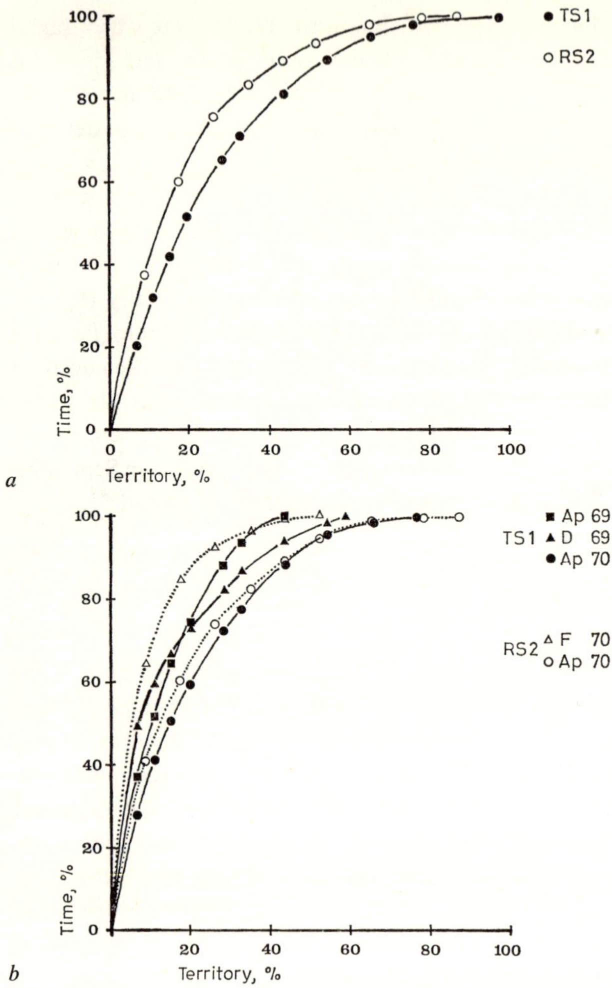

Fig. 2.11. Cumulative proportions of territory (ranked hectare quadrats) and observation time for (a) TS1 and RS2 overall, and (b) TS1 in April and December 1969 and April 1970, and RS2 in February and April.

to a more normal ranging pattern in April. The latter closely paralleled that for TS1 in the same month at the end of the 1970 dry season.

Thus the important variables were the size of area covered, and the intensity of use of each part of that area. In the dry season a much higher proportion of the territory was covered, and more evenly, than in the wet season. In the former, 80% of time was spent in 40% of the area, whereas 65% of time was spent in 15% of the territory in the wet season.

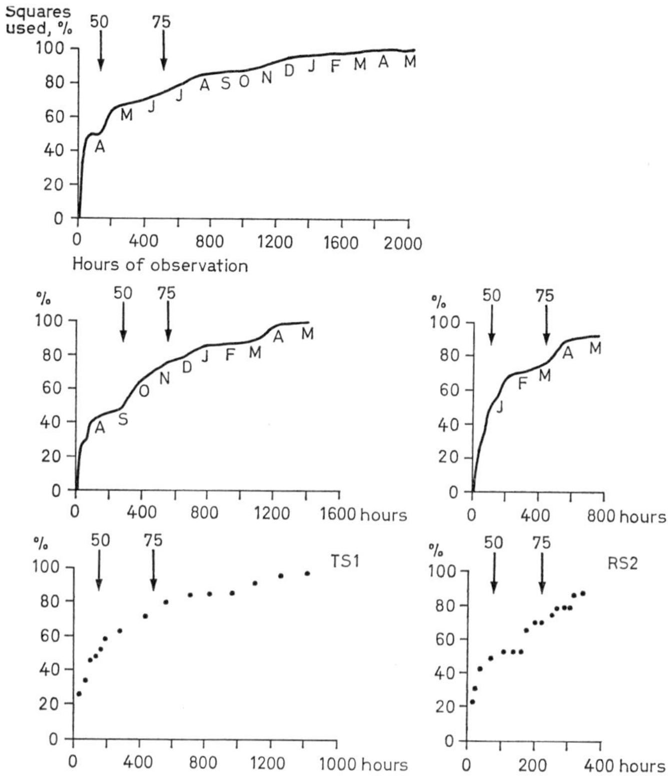

Fig. 2.12. Cumulative proportion of territory covered with increasing observation time of TS1 in different seasons – April 1969 to May 1970; August 1969 to May 1970; January to May 1970; TS1 overall 0–1500 h, and RS2 from February to April 1970 inclusive indicating time to observe 50 and 75% territory, and successive months of study.

On the assumption[3] that 12 months observation revealed territorial limits, the proportion of the area covered after each month was calculated (fig. 2.12). This was repeated from the start of the wet and dry seasons, which showed that whereas 80% of the territory had been observed after 600 h in the intermediate season, it was 700 h in the wet, and 500 h in the dry season.

3 Subsequent observation by ALDRICH-BLAKE [personal commun.] and CHIVERS *et al.* [in prep.] has shown such an assumption to be not wholly valid. It has thereby provoked further thoughts on the nature and mechanisms of territoriality in siamang (see Discussion, and paper in preparation).

Table 2.5. Rate of use of hectare squares

	Number of hectare squares entered[1]						Mean daily total of different quadrats
	0600–0900	0900–1200	1200–1500	1500–1800	mean daily total	n	
TS 1							
April	5.5	2.0	1.8	0.8	10.1		8.5
May	2.4	2.3	2.1	0.4	7.3		6.3
June	2.6	1.6	1.5	0.7	6.4		5.7
July	2.8	1.9	1.5	0.8	7.0		6.4
August	3.1	2.0	1.5	1.1	7.7		6.8
Mean	3.3	1.9	1.7	0.8	7.7	49	6.8
September	2.9	1.4	1.3	0.5	6.1		5.3
October	2.9	2.1	1.4	1.1	7.5		7.0
November	2.9	2.5	1.9	0.2	7.5		6.8
December	2.9	1.9	1.8	0.9	7.5		6.5
Mean	2.9	2.0	1.6	0.7	7.1	50	6.4
January	2.2	2.3	1.6	0.9	7.0		6.3
February	3.1	2.9	1.6	1.3	8.7		8.2
March	4.2	4.8	2.4	1.7	13.1		11.4
April	5.2	4.9	3.7	2.6	15.9		15.0
May	6.8	3.2	3.6	2.4	16.0		13.9
Mean	4.2	3.6	2.6	1.7	11.6	39	10.9
Overall mean	3.5	2.6	2.0	1.0	9.1	138	8.1
RS 2							
February	2.3	2.0	2.3	0.3	7.0	9	5.4
March	2.5	1.5	2.4	1.0	7.0	12	5.8
April	2.3	2.4	1.5	0.8	7.0	11	6.2
Mean	2.4	1.9	1.9	0.8	7.0	32	5.7

1 Ten-day mean.

In the latter, 50% of the territory was observed after 100 h; it was 300 h before a similar proportion was seen in the wet season. Finally, the mean curves (constructed from the three seasonal patterns) are shown for the 150 days at Kuala Lompat (TS1), and the 30 days in Ulu Sempam (RS2) (fig. 2.12). The fluctuations are important (most clearly shown by RS2), because they represent a ranging pattern of intensive use of certain parts of the territory for several days, before moving on to repeat the cycle. (This is over-emphasised because of adding temporally distinct observation periods.)

Finally, reference must be made to the rate of movement through the hectare quadrats daily. The daily pattern is remarkably constant, only the absolute scores varying seasonally (table 2.5). In all seasons more than half the daily movement in terms of nH takes place before midday, with the major part during the first 3 h. Because the siamang have effectively moved up out of the sphere of daily activity for the night (see below), the first quadrat utilised in the following day range was counted, even if it was the last one entered on the previous evening. Return visits to a quadrat were less frequent on days in the wet season.

It has already been shown that there was a highly significant difference in area covered on days of contrasting weather, and in different seasons (also by χ^2 contingency test, $\chi^2 = 15.60$ for sun/cloud, 6.94 for sun in different seasons), and between TS1 and RS2 ($\chi^2 = 37.1$). But there was no significant difference between the daily pattern under these different conditions; they were very similar (χ^2 k-sample test, $\chi^2 = 2.34-4.85$), especially comparing mornings with afternoons ($\chi^2 = 0.001-0.84$). These results have been presented now to prepare for a discussion of territorial maintenance, but such daily patterns of activity will be amplified in Chapter 4.

Night Sleeping Trees

Although observations were made mainly on the daily behaviour of the siamang – from the start to cessation of daily activities (Chapter 4) – data were collected on some aspects of nocturnal behaviour. Thus, the number and dispersion of trees in which the siamang group passed the night were determined. The group was usually dispersed among the terminal branches of one tree high above the main forest canopy. But sometimes they were lower than this, and one or more siamang might be in an adjacent tree. TS1 spent 28% of nights in the 29% of sleeping locations involving more than one tree. There appears to be no preference for this less frequent situation. Individuals were often closer, however, sleeping in adjacent trees, than in the terminal branches (opposite sides) of the same tree (Chapter 5).

TS1 slept on 165 nights in one of 34 trees or groups of trees. These appeared to be situated around one or more of 7–8 groups of feeding trees frequently used in the late afternoon (Chapter 3). 60% of the 47 trees involved were identified; only 5 of two species *(Ficus consociata* and *Koompassia excelsa)* provided food at any time during the study. These identified trees represented only nine species (table 2.6). Their frequency does not correlate

Table 2.6. Identity and distribution of night sleeping trees

Tree species	TS 1	
	% trees	% nights
Koompassia excelsa[1]	15	18
Shorea leprosula	13	8
Intsia palembanica[1]	9	16
Koompassia malaccensis	6	5
Sindora coriacea	6	8
Ficus spp.	4	5
Cynometra malaccensis	4	10
Dipterocarpus grandiflorus	2	9
Anisoptera laevis	2	4

	TS 1			RS 2	
Section	A	B	C	Y	Z
Percent territory	37	28	35	48	52
Percent night trees	44	27	29	50	50
Percent nights	35	33	32	20	69
Percent day time	34	40	26	48	52

1 Strong preference of the high open habit of *K. excelsa*, and the lower closed habit of *Intsia* was noted.

with the distribution frequency of the same species throughout the territory (r_S = 0.35, and see Chapter 3).

52% of nights were spent in 6 (18%) of the following tree groups:

Tree group number	Tree species
1	*Koompassia excelsa*
2	*Dipterocarpus grandiflorus*
3	*Intsia palembanica, Ficus consociata, K. malaccensis*
9	*Cynometra malaccensis, K. excelsa*
17	2 *K. excelsa, Intsia palembanica*, sp. indet.
21	*Intsia palembanica*

The location of all night sleeping trees, and number of nights used is shown in figure 2.13, with identity (where known) in figure 3.2. NT-17 represents the widest dispersion of individuals observed overnight, although no more than two of the four trees were used on any one night. 76% nights were spent in 15 of the 34 tree groups.

Night sleeping trees were dispersed evenly throughout sections A, B, and C (χ^2 = 0.76), but there was a significant decrease from expected in nights

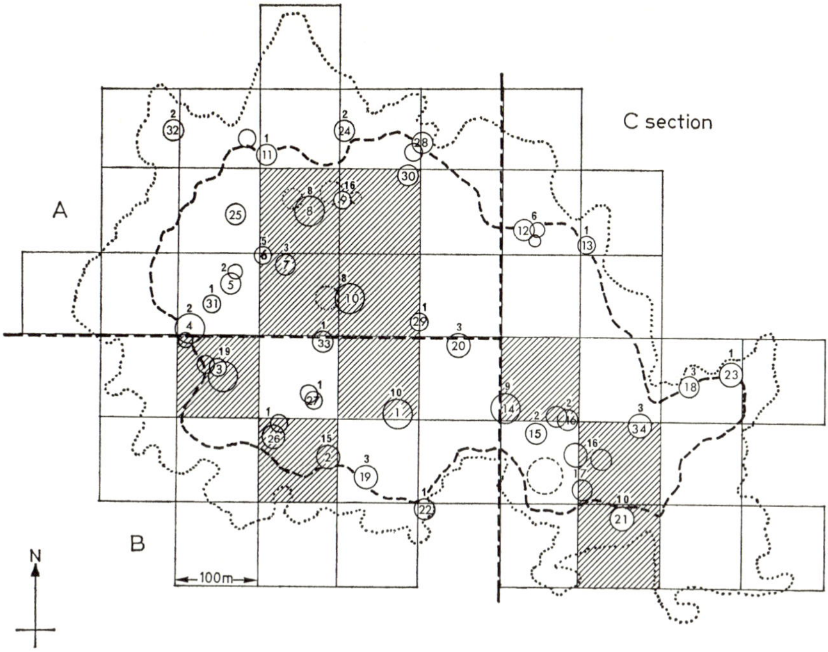

Fig. 2.13. Territory of TS1 at Kuala Lompat showing location and identity of night sleeping trees (circles), with number of nights in each, the ten quadrats most used for sleeping, cumulative day range (...), and usual ranging limits (— — —), with sectional boundaries.

spent in section C ($\chi^2 = 5.93$, $p = 0.05$). Thus the increase in ranging in section B was also reflected by night positions. But in terms of hectare quadrats, the ten most frequented sleeping quadrats did not correlate with the ten most used during the day, either for visit frequency or time ($r_S = 0.50$ and 0.39, respectively). But the ten most used sleeping hectares were centrally situated within the territory (fig. 2.13).

In Ulu Sempam sleeping trees were evenly dispersed between sections Y and Z – ridges and gullies – and so was daily activity (table 2.6). But the great majority of nights were spent downhill (fig. 2.14). Thus, siamang tended to sleep away from the immediate vicinity of daily activity, although at Kuala Lompat 71% trees and 83% nights were spent in the central 43% of the territory (i.e. those quadrats not forming the territorial boundary). The same trees were used consecutively on less than 6% nights. To what extent these reflect responses to actual or potential disturbance has still to be determined.

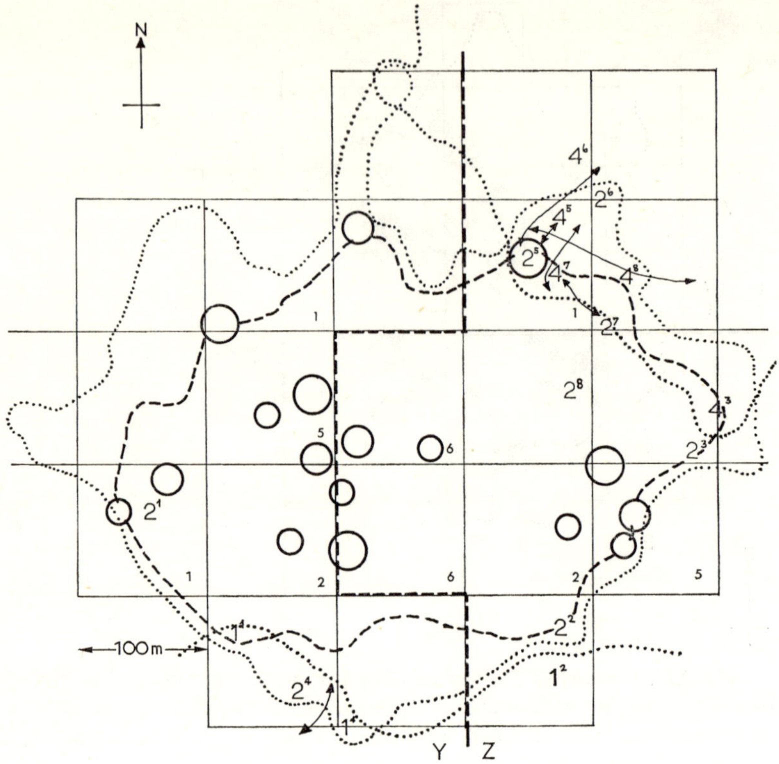

6 Number of nights in each hectare

1^4 RS1 start position and interaction number

⟵⟶ Chase direction

Fig. 2.14. Territory of RS2 in Ulu Sempam showing location of night sleeping trees (circles), and frequency of use, location and movements of groups involved in territorial encounters, cumulative day range (...), and usual ranging limits (— — —).

Maintenance of Territory

The siamang group appears to maintain or defend its territory by (a) regular ranging near the boundary, (b) calling at and/or chasing neighbouring groups actually or about to intrude, and (c) loud group calls in the morning. Ranging patterns have already been described (see especially fig. 2.4, 2.5 and Appendix C), and calling behaviour will be described in Chapter 6. This section will mainly refer to interactions between groups across territorial boundaries. TS1 had no siamang group territories adjacent to it (a situation that has now changed [SHEILA HUNT, personal commun.]), so such observations are limited to RS2.

For the present territorial boundary will be equated with the cumulative limits of ranging. RS2 shared 30% of its boundary with RS4 to the north-east, and RS1 to the south; possibly part of the west boundary was shared with RS7b. During 586 h of observation RS2 was only involved in seven inter-group disputes. The mildest involved RS1 calling after coming round the hill and seeing RS2 feeding in a fig tree 100 m away. RS2 showed no response other than glancing frequently in the direction of RS1 while continuing to feed. Both groups subsequently moved apart.

Three times two groups came up onto a ridge from opposite sides simultaneously. There followed intense calling and vigorous movement to and fro from adjacent high trees for about 15 min, followed by eventual withdrawal into respective territories (often after at least one group had rested for as much as 30 min). Similar behaviour was observed several times in Ulu Gombak. On three occasions (once with RS1, twice with RS2) this calling was followed by vigorous chases to and fro across the boundary in the lower levels of the canopy. These involved males (adult and sub-adult) while the female and young hid, and lasted for 15, 54, and 11 min, respectively. For almost half the time the males sat watching each other and the surrounding forest, then one would intrude and be chased out. One day the males of RS2 (two) and RS4 (one) chased each other up and down the gully that divided their territories for 67 min with no preceding calling having occurred (fig. 2.14).

Apart from the sound of crashing foliage, there were no other sounds, such as calls, to indicate what was taking place. Thus until it was possible to follow RS2 from dawn to dusk some such interactions may have been missed. In the last 29 days only four territorial disputes took place (14% days), and RS2 called on 24% days. TS1 called very infrequently until the 1970 dry season; their earlier silence was perhaps a function of the absence of immediate neighbours, although their calling has not increased appreciably now that the subadult male is establishing a group and territory adjacent to them [ALDRICH-BLAKE, personal commun., 1971]. The increase in calling appeared to have reproductive rather than territorial significance (Chapter 5).

However, TS1's regular pattern of ranging, whereby they covered the same pathways month after month (fig. 2.7), their watchfulness in the forest around them – especially near the periphery – and the notice they took of calling groups across Sungei Lompat to the south, and to the west, were indicative of active maintenance of their territory for exclusive use. After ten days observation in March 1970 a tape recording of TS1's group calls was played back to the group following a calling bout. The subadult male moved lower towards the observer, before following the rest of the group on about the day range. On the three following days the group returned to that place and called.

The most pertinent information on territorial maintenance concerns changes in behaviour when the resident group is near the boundary of the territory. Calling and chasing have been described, but they did not occur as frequently as one might expect in a territorial species. It could be that the integrity of the territory was being maintained in more subtle ways – by the frequency and quality of movements to the periphery. The group might approach the boundary more frequently or regularly than expected, either more briefly and warily or for longer periods, with conspicuous vigilance (or sentinel) behaviour.

Thus the ranging data were reanalysed to see whether: (1) the peripheral quadrats were recognised as quickly as the central ones; (2) the more frequented central quadrats were visited less regularly than the peripheral ones, and (3) whether the proportions of time spent feeding varied between central and peripheral. In some cases, for TS1, the 12 incompletely covered quadrats were omitted from the analysis, so that 19 peripheral and 15 central quadrats were compared (rather than 23 of each). Similarly, in Ulu Sempam the behaviour in six central quadrats was compared with that in nine peripheral ones.

At Kuala Lompat 96% of the central quadrats (the core of the territory) were visited in the first 20 days of observation, but only 38% of peripheral quadrats had been visited. It was 100 days before 90% of the latter were visited (fig. 2.15, cf. fig. 2.12). In Ulu Sempam RS2 visited 90% of central, and 63% of peripheral quadrats in 20 days; all peripheral quadrats were visited in 40 days – this could indicate increased territorial behaviour in the presence of neighbouring groups.

TS1 visited central quadrats more often than peripheral ones (mean scores of 50.1 and 27.5, respectively) for slightly shorter periods (65.2, cf. 73.3 min on average). RS2 visited central squares nearly three times more often than peripheral ones, although visits to the latter were much shorter (60.9, cf. 93.8 min). But it must be remembered that a visit to a peripheral square often meant two visits to a central one (including the return journey), and that in a small hilly territory it is not always necessary to range right up to the boundary to defend the territory. There were several occasions when RS2 was believed to be watching for intruders from the ridge in the centre of

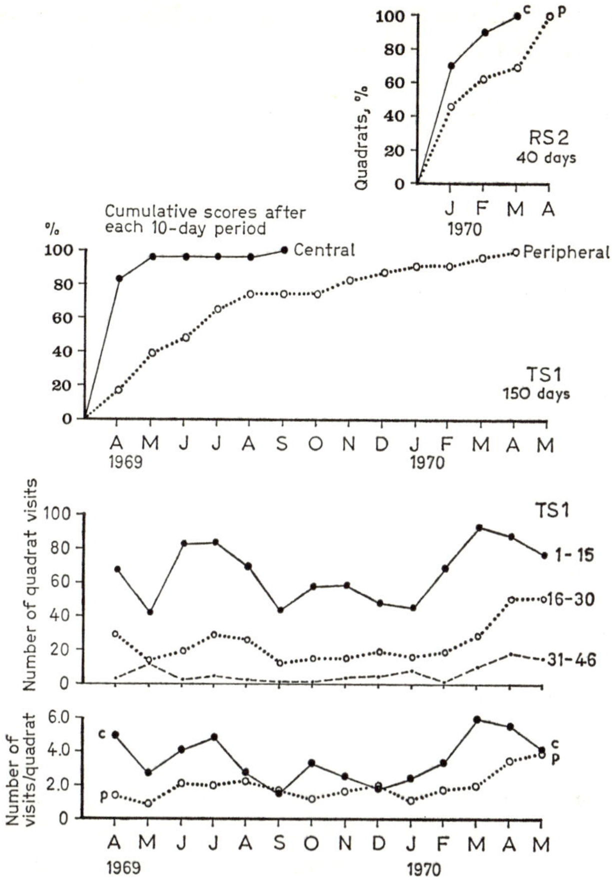

a

b

Fig. 2.15. (a) Cumulative proportion of central and peripheral quadrats visited after each 10-day period at Kuala Lompat (TS1) and in Ulu Sempam (RS2) and (b) number of visits to quadrats ranked by ranging time 1 – 15, 16 – 30, and 31 – 46, and number of visits/quadrat in the centre and periphery of TS1's territory.

its territory. At least once this resulted in a quick movement to the boundary to clash with RS1. Similar 'sentinel' behaviour was seen on most days at Kuala Lompat in 1972 when TS1 was ranging near the territory of the new group, TS1A.

While there was little difference in the proportion of time spent feeding in peripheral and central squares at Kuala Lompat, in Ulu Sempam RS2 spent 35% of time in central squares feeding, and only 31% in peripheral squares. This is more clearly demonstrated in terms of the number of visits to food trees as a proportion of the number of visits to each square – 1.35 in central quadrats (a similar level to TS1 throughout its territory), and only 0.73 in peripheral quadrats (Chapter 3).

From one month to the next TS1 was more consistent in its frequency of visiting peripheral quadrats – 1–2 visits/quadrat in each 10-day period (fig. 2.15), cf. 2–5 visits/central quadrat. A similar pattern emerges if quadrats ranked by total time spent are divided into three groups (fig. 2.15, cf. fig. 2.8) – the central quadrats were in fact these in which most time was spent (fig. 2.7), and in which most feeding occurred (fig. 3.16).

Thus, although there were fewer, shorter visits to peripheral quadrats, with a lower proportion of feeding visits and time, the evidence collected so far on changes in behaviour near the territorial boundary is quantitatively inconclusive. Nevertheless, the lack of overlap between adjacent groups, the frequent calling, occasional inter-group conflicts, increased vigilance near the periphery, and intensive ranging indicate that siamang effectively obtain exclusive use of the same tract of forest from one year to the next.

Territory Size

The cumulative day ranges of TS1 covered an area of 34.2 ha. Those of RS2 covered only 14.7 ha. TS1 was composed of five individuals (adult male, adult female, subadult male, juvenile male, and infant female), RS2 of four (adult male and female, subadult male, infant male, see Chapter 5). In Ulu Sempam the territories of RS1 (male, female, subadult female, and infant), RS4 (male, female, juvenile male, and infant), and RS3 (male, female, subadult male, and juvenile) were estimated to be 15, 18, and 25 ha, respectively. The maximum distance across TS1's territory was 1,000 m, the minimum 600 m; similar figures for RS2 were 600 and 500 m, respectively.

In Ulu Gombak near the University Field Studies Centre (15th mile) GS1 and GS2 (both four individuals) had territories estimated (from cumulative sightings) to be 30 and 25 ha, respectively. At the top of the valley (21st mile) GS16C (male, female, juvenile male and infant) lived in about 23 ha, and GS6 across the road in about 27 ha (male, female, subadult male, juvenile). Adjacent to GS16C was GS16B (male, female, subadult female, and subadult male from S16B temporarily) in about 21 ha. The male and female down-valley from them (S16A) had a territory of about 17 ha.

Thus, siamang groups lived in territories of about 23 ha on average, in these three locations in central Malaya.

Discussion

One of the first descriptions of territorial behaviour was by HOWARD [1920], who suggested its main functions were to assist pair formation and ensure adequate food for the young. Because this concept cannot be defined as readily as most other range concepts – because of the behavioural as well as spatial components – there has been much controversy in its development. HINDE [1956] used the definition of territory as 'any defended area' [NOBLE, 1939], in his review of the biological significance of bird territories. He illustrated the diversity of territorial behaviour among birds by categorising territories for discussion purposes as breeding areas (with/without food), as nesting areas (colonial birds), and as pairing or mating areas.

There are two main components that HINDE draws attention to, apart from self-advertisement – the restriction of some or all behaviour to a particular area, and defense of that area. The various possible functions of territory now considered include the advantages of familiarity with an area, the regulation of population density, helping pair bond formation and maintenance, specialisation of aggressive behaviour for defence of nest site, food supply maintained, and so forth. Among birds, therefore, territory appears to be a complex concept applied specifically in certain clearly defined situations, which are often transient in the life of a species (absent altogether in others). They may recur, however, at regular intervals, e.g. territories for nesting, feeding, mating.

Since the study of primate behaviour is a young discipline it is sensible to refer closely to an established one, such as bird ethology. It is obviously advantageous from all points of view to be able to apply concepts across major animal taxa. This has probably been the motivation behind the application of behavioural concepts, such as territory, to primates. In some cases this has been unsuccessful because the approach has not been rigid enough, in others the situation did not lend itself to this classification. In fact, there are few primates which fulfil the condition of defending a clearly defined area, or even of defending any area. What is most often defended is the space immediately around the social group, with no topographical reference [MASON, 1968].

It is the relatively permanent occupancy of an area by a social group, that appears to be the main characteristic of primate societies. Few examples of nomadism or seasonal migration have yet been described in primates. Each study shows the social group confining its activities to a specific area within the habitat [DEVORE, 1965; ALTMANN, 1967; JAY, 1968]. Although there is some movement of immature animals between groups, the social group has otherwise a high degree of stability in its dynamic equilibrium with the environment. What varies between species, and even populations of the same species, is the degree of ranging intensity about the 'home', and the degree of intolerance shown to the proximity of conspecifics alien to the social group. Attempts to apply general ethological concepts to these situations appears to obscure their fundamental properties.

In primates the home range is a 'home' in reality. This could be more effectively demonstrated by invoking a concept such as 'domicile' [MEDWAY, personal commun.]. Such an area is difficult to define precisely in many species, especially as short-term studies reveal an inconstancy in ranging patterns. This contrasts with the greater stability within a wider area revealed over a longer period. But the limits of this domicile are not as important as the fact that such a stable 'home' exists. If one accepts this to be the fundamental property of a primate social group's relation with the environment, the next step is to define the variation that occurs within this-to measure the stability of ranging limits, the degree of overlap with adjacent groups, and the pattern of range use.

The majority of primate species show considerable tolerance to adjacent groups of conspecifics, as long as sufficient distance is maintained at any given time. The limits of ranging are not clearly defined, and the domicile of one group overlaps with that of its neighbours [ALTMANN and ALTMANN, 1970]. Usually mutual avoidance (facilitated by certain behaviour patterns) ensures that group cohesion is increased by its exclusiveness. But aggressive encounters can have the same result; group space is thereby preserved, whether or not one chooses to invoke the concept of 'mobile territory' as an explanation. Avoidance is also facilitated if ranging is concentrated in a part of the domicile away from the area of overlap – the core area [KAUFMANN, 1962]. In some species the core area is defended, conspecifics being excluded and the occupants rarely leaving this clearly defined area. Ecologically territory and core area are identical, the former being distinguished by the behaviour involved in creating and maintaining the resources [MASON, 1968].

Precise, prolonged observation is required to determine whether a group is 'defending' the space around it, or an area of habitat. The critical information concerns the location of inter-group disputes, the constancy of short-term range limits, the degree of uniformity of ranging patterns and changes in behaviour near the periphery. Within the framework proposed this distinction is subtle, but it nevertheless reflects real differences in group characteristics of behaviour evolved from individual behaviour patterns (which are determined by both heredity and environment). These are likely to assume greater importance under increased habitat pressures [CHIVERS, 1969]. They could reflect a clear dichotomy within a 'domicile' concept.

The white-handed gibbon [ELLEFSON, 1967, 1968] represents an extreme among primates in the intensity with which it uniformly covers a small area of forest, and with which it excludes conspecifics from this area of about 40 ha. Although twice the biomass of the gibbon, the siamang family group confines its activity to about half the area of the gibbon territory [CHIVERS, 1972a]. ALDRICH-BLAKE [personal commun.] confirmed that TS1, nearly three years after observations started, was still ranging about the same 34 ha. In 20 days in 1971 over two months they covered 18 ha almost completely within the original 34, using many of the same food and sleeping trees, and arboreal pathways. They were ranging about the same territory at the start of 1973 [MACKINNON, personal commun.].

The siamang group is, therefore, familiar with all the resources of the area, the pair bond is maintained (proved for TS1 over 5 years), the young are reared with considerable success – judging by the large number of immature animals in most groups surveyed (Chapter 7) – sufficient food is constantly available, and certain behaviour patterns have been developed (or ritualised) to defend these resources. Regular movement about the territory (as convention would name it), or defended, restricted domicile, could ensure that intruding conspecifics can be detected and chased out. That this area is much larger than a bird territory with restricted visibility may be a problem in effective territorial behaviour. But visibility from the forest canopy is better than one would imagine. Furthermore, intrusion is most likely in search of preferred foods [ELLEFSON, 1967]. The resident group appears to know these sources through their regular movements, and through the activity of frugivorous birds and small mammals, which provide visual and auditory cues over a considerable distance. Thus they will be ranging in the vicinity of these food resources, and intrusion will be less possible.

There remains the problem of determining the limits of the territory. So far it has been equated with the daily limits of day ranges, which is all that one can do if no neighbours are present (TS1). This absence of close neighbours could reflect the efficiency of territoriality in siamang, i.e. the dispersion of groups by regular loud calls in the breeding period and/or fruiting season. The great distance between groups would then permit range expansion without territorial encounters at times of food scarcity (see also Chapters 3 and 6). Observations of RS2 revealed that there was a small degree of overlap between the cumulative day ranges of adjacent groups, and that one group, or the other, was more wary (i.e. more alert, mobile and/or vigilant) in each part of these zones (fig. 2.7). Observations of interactions between groups suggested that one might regard a boundary as running through the middle of these zones. It appeared that no arboreal pathway was used by equal confidence by each group. But further study is required before this problem can be resolved. At Kuala Lompat TS1 had major arboreal pathways around the territory within about 100 m of the day range limits (fig. 2.6). That these are regularly patrolled is illustrated by the coincidence of monthly ranging limits around this region (fig. 2.9). It can be argued that this represents the boundary of the territory, but it is just as possible that from these pathways they are effectively maintaining a territory that includes most of the limits of ranging.

These problems must await further study for resolution; the changed situation at Kuala Lompat will facilitate this. The main difference between siamang and gibbons in their territorial behaviour is the much lower intensity displayed by the former in a smaller area [Ellefson, 1967; Chivers, 1972 a]. Calling by the siamang was less frequent (Chapter 6), as were border disputes, which were also less intense than described for the white-handed gibbon. The reasons for this can be more clearly perceived following a description of feeding behaviour and the dispersion of food trees (Chapter 3). But it appeared that the siamang showed a greater awareness of the limits of its territory, which reduced the need for such frequent and active defence.

The siamang ranging pattern combined an intensive use of a small part of the territory, with ranging about a major part of the territory (at least briefly) in a relatively short period. Although they usually spent less than 10% day time over 10 days in any one hectare, they spent two thirds time in one tenth of the territory, and yet covered nearly two thirds of the area. The former can be interpreted as food-getting behaviour, the latter as territorial and food searching behaviour. This was more clearly demonstrated with regard to territorial behaviour in the smaller territory in Ulu Sempam, although the dichotomy was clearer at Kuala Lompat.

More conclusive proof (in quantitative terms) of territorial behaviour in siamang must await further investigation; difficulties are caused by the multifactorial nature of the problem. Certainly there is a lower level of territorial behaviour than in white-handed gibbons. One explanation could be that the concentration of suitable food in a small area reduces contact between adjacent groups. Or this could represent phylogenetic dietary adaptations, whereby exclusive use of a defended area has become less important. The third alternative concerns the possible facultative nature of the behaviour, so that it increases when resources are scarce. There was some evidence from both areas that territorial behaviour increased in the breeding period – in terms of ranging (fig. 2.6 and 2.7) and calling. It may be that this is the way in which siamang periodically redefine their territorial boundaries, subsequently reinforcing them by group calls and the occasional encounter.

Ten days were chosen as the length of each observation period, partly because two such periods were required in each month, but also because the patterns for each month seem to have been established after six to seven days. While it was advantageous to have observations of standard duration at regular intervals, it is difficult to infer about ranging over a longer period. Increasing observations by four days (or 48 daylight hours) as in June and July indicated that after a certain routine has been followed for several days, ranging shifts slowly into a new pattern. This pattern can be discerned more clearly after 14 days, but it seems unlikely that more than two thirds of the territory will be covered in any one month (at least at Kuala Lompat) except in exceptional months (75% in April and May 1970). Signs that the group was only partially habituated at the beginning of the study of TS1 (longer day ranges) were subsequently shown to be part of an annual cycle of increased ranging at the end of dry season.

Description of other aspects of siamang behaviour (Chapters 3, 4 and 5) should help to reveal the extent to which the increased ranging in the 1970 dry season can be equated with an annual climatic cycle, affecting the forest flora, or a 2- to 3-year reproductive cycle affecting the social relations within the group. So far it has been shown that the seasonal increase is not specifically associated with weather, and that it appears to be an annual event of varying intensity.

Summary

The siamang group at Kuala Lompat was followed from dawn to dusk on at least ten consecutive days in 14 months; the group in Ulu Sempam was followed when possible. From plots of day range, it was possible to calculate the number of hectare quadrats covered daily, the distance between night trees, the farthest distance reached from the previous night sleeping tree, and the time taken to reach this point. Further information on the extent and intensity of ranging was provided by an analysis of the number of visits to, and time spent in, each section (gross measure) and each hectare quadrat of the territory.

Thus TS1 restricted their activity to 35 ha, and RS2 to 15 ha. Each day

TS1 travelled 969 m on average, with a 293-meter shift in night position, and a maximum radius of 361 m (reaching this point after about 7 h); RS2 maintained relatively high scores in their smaller territory, especially with a day range of 800 m on average. Although two thirds time were spent in one fifth of the area, nearly two thirds of the territory were visited in only ten days. Thus concentrated activity in a small part of the territory was combined with frequent sorties through the remainder – especially by RS2 in Ulu Sempam.

Ranging increased in hot and sunny weather, especially when this was sustained for a long period as in the dry season during the first months of 1970; there was a striking increase in all aspects of ranging behaviour for TS1 at this time. But other environmental and behavioural variables were obviously involved in this. Generally there were few significant changes in ranging behaviour from one month to the next, and rarely within months. It required at least 12 months observation (at regular intervals) to determine the territorial boundaries, although the core area was defined in the first month.

Plotting the position of night sleeping trees, and calculating the frequency with which each was used, indicated a preference for high trees of certain species (often with contrasting leaf habit, e.g. *Koompassia excelsa* and *Intsia palembanica*). This preference was extended to specific trees, although the same tree was rarely used on consecutive nights; the location of night trees was not correlated with daily ranging activity, in terms of hectare quadrats.

Territories were maintained by regular ranging near the boundary, by irregular calling, and by infrequent conflicts with neighbouring groups. In these ways the resident group gained exclusive use of the clearly defined tract of habitat (of about 23 ha) in which their requirements were concentrated.

Chapter 3: Diet and Feeding Behaviour

Introduction

The literature on the behaviour of free-ranging primates has indicated, until recently, a reluctance to investigate food as one of the basic determinants of social organisation and behaviour. While there are certain species-specific characteristics of social organisation, which do not depend directly upon the external physical environment, the occurrence of considerable intra-specific variations in behaviour has recently been demonstrated (e.g. *Presbytis entellus*, JAY [1965]; SUGIYAMA [1967]; YOSHIBA [1968]; RIPLEY [1970] and *Papio anubis*, DE VORE and HALL [1965]; ALTMANN and ALTMANN [1970]). Thus genetic components of behaviour can be modified by environmental factors, such as food, to allow a relatively rapid adaptation to local conditions with the production of quite profound sociological changes.

Insufficient attention has been given to the details of, and relations between (1) feeding behaviour, (2) diet, and (3) nutrition in primates, and of determining the significance of feeding as a group activity to the individual. But a start has been made in remedying this situation in Panama, Gabon, and Ceylon [HLADIK, 1967, 1968; AMERASINGHE *et al.*, 1971]. This study recognises the fundamental importance of food to siamang behaviour. It attempts to initiate a programme of research, which will elucidate the complex temporal and spatial effects on the behaviour of groups and individuals of variation in (1) the mode, timing and intensity of feeding, (2) diet, in terms of food category, tree identity, and quantity and dispersion of food, and (3) the nutritional value of food intake.

Siamangs were feeding for a major part of the day, even if one excludes the time spent moving about the forest in search of food. The other conspicuous aspect of siamang feeding behaviour was that the group almost invariably fed together in the same tree, or group of trees. Usually in the morning, individuals moved around a tree of open habit, such as a fig, harvesting dispersed food (fruit or new leaf shoots); whereas in the afternoon feeding was usually less active, with individuals more stationary in dense foliage, where there was an abundance of young vine leaves. Feeding was very much an individual activity with a low tolerance for the close proximity of others (Chapter 5). The adults appeared to occupy the most favourable parts of the food tree initially.

Feeding was a continuous process; once a siamang started picking and ingesting food, it was unlikely to stop for some time (Chapter 4). Movement about the food tree, the extent to which siamangs sat or suspended themselves while feeding, with which hand they picked food, and rates of feeding (from which can be calculated wet weight of food consumed) will be described at the end of the chapter. The main analysis will be on the siamang diet and the dispersion and identity of food trees throughout the territory. The patterns of use of these food sources (monthly and seasonal) by the siamang groups will be compared with their ranging patterns (Chapter 2).

Vegetation Analysis

Two main methods were used in this preliminary survey. At the end of 1968 a government department survey team cut a series of transects through the study area, while investigating the possibility of flooding the area between the two rivers for rice paddy. One transect ran due west from the Game Ranger Post, the others went north-south at intervals of 150 m. They proved invaluable for moving about the area after the siamang, as reference points, and for this study of the vegetation.

The trees with trunks within 10 m on either side of 2,100 m of north-south transects were counted, classified, and identified (where possible) – a sample area of 4.2 ha, which was 12.3% of TS1's territory. Undergrowth of less than 6 in trunk diameter and/or 30 ft in height were not counted, as this represented the lower size limit for siamang use. Trees in the main canopy layer (80 – 120 ft) were classified as medium height; the emergents (120 – 250 ft) as high; and the under-story (30 – 80 ft) as low trees. At a density of 290 trees/ha there were 25 high (8.5%), 83 medium (28.6%) and 182 low (62.8%) trees/ha (fig. 3.1). Transects (actual and theoretical) were numbered on both axes to provide a grid for references; being 150 m apart, they should not be confused with the hectare quadrats (fig. 3.2).

The main emergent tree species contributing to the higher layer at Kuala Lompat were *Dipterocarpus grandiflorus*, *Shorea leprosula*, *Shorea* spp., *Alstonia* sp., *Dyera costulata*, *Parkia speciosa*, *Anisoptera laevis*, with some *Ficus* spp., and most conspicuously *Sindora coriacea*, and the two species of *Koompassia*. Most of these species were well represented in the main canopy

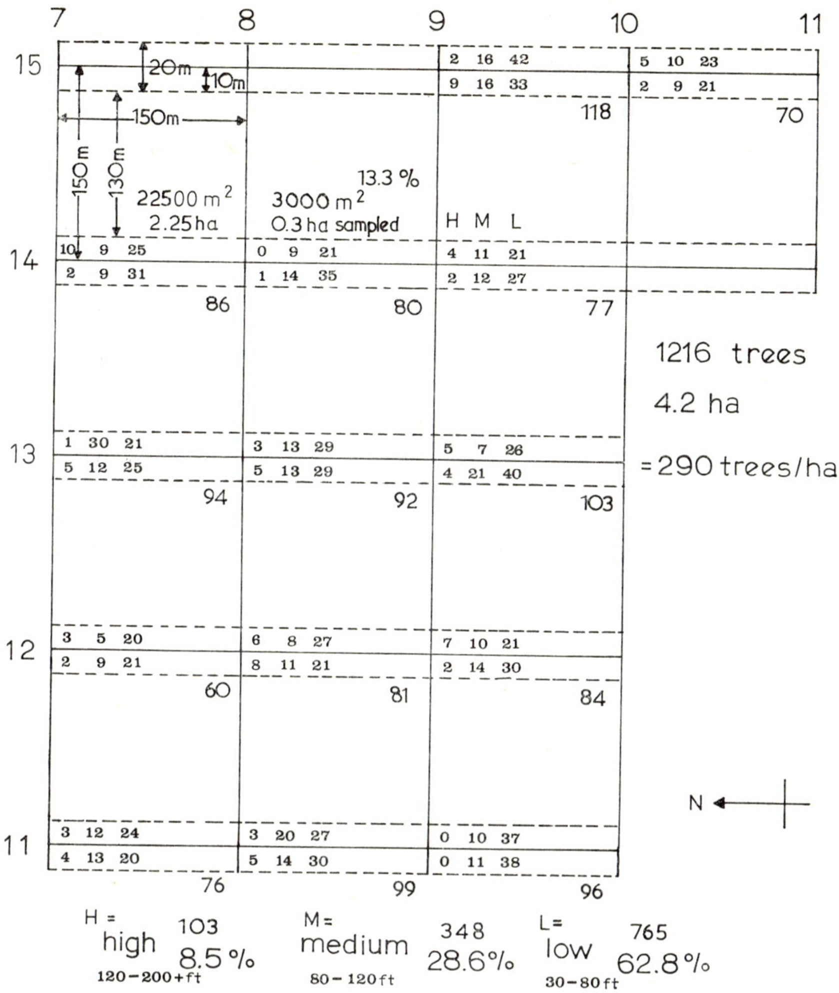

Fig. 3.1. Tree density and forest structure in siamang territory at Kuala Lompat. The number of high, medium, and low trees (in that order) are shown for each 150 × 10 m forest strip, with the total number of trees in each 150 × 20 m strip (below and to right). Also shown are: the density of trees from overall total and area sampled; dimensions of transects and 150 × 150 m quadrats, and characteristics of high, medium, and low trees, with total number and proportion of each.

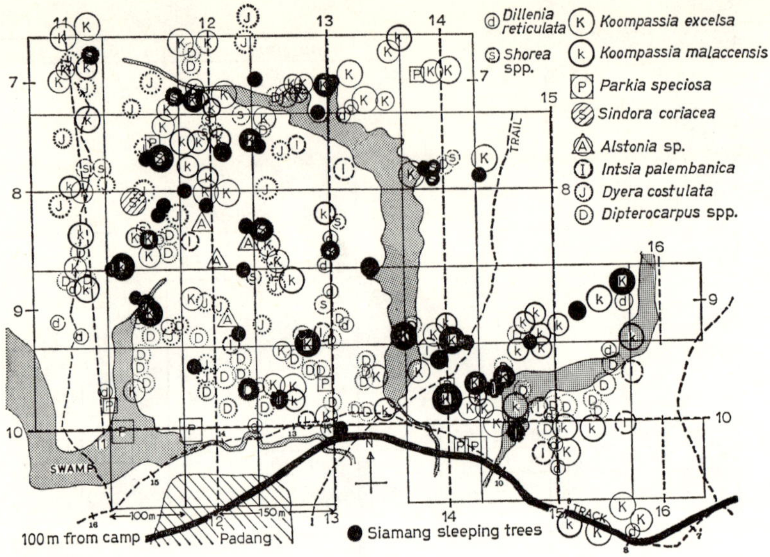

Fig. 3.2. Distribution of dominant tree species in siamang territory at Kuala Lompat, showing species of ten genera, transects with grid references, hectare quadrats, swamps, clearing, landrover track, trails, and siamang night sleeping trees.

Table 3.1. Density of dominant tree species in siamang territory at Kuala Lompat

Tree species	Sample area, 4.2 ha			Territory (TS1)	
	n	n/ha	n/t[1]	n	n/ha
Sloetia elongata	16	3.8	130	74	2.2
Ficus spp.	13	3.1	106	65	1.9
Koompassia excelsa	12	2.9	99	55	1.6
Dipterocarpus grandiflorus	11	2.6	89	46	1.3
Shorea leprosula	11	2.6	89	16	0.5
Dillenia reticulata	10	2.4	82	20	0.6
Koompassia malaccensis	10	2.4	82	32	0.9
Dyera costulata	5	1.2	41	22	0.6
Sindora coriacea	4	1.0	34	7	0.2
Intsia palembanica	2	0.5	17	44	0.4
Parkia speciosa	2	0.5	17	8	0.2
Dialium platysepalum	2	0.5	17	6	0.2
Alstonia sp.	1	0.2	8	4	0.1
Endospermum malaccense	1	0.2	8	28	0.8

1 Estimate from count in sample area of number of trees of each species in siamang territory of 34.2 ha.

at medium height along with species of *Eugenia*, *Parinari*, *Dillenia*, *Dialium*, and *Intsia*. Species of *Xylopia*, *Diospyros*, *Endospermum*, and *Sloetia* were also present, but they were mainly characteristic of the low level along with *Randia scortechinii*, and two species of *Rubiaceae*.

No measure has yet been made of the amount of foliage in each layer, although it is obvious that the main bulk is in the middle layer of the canopy, with the least in the emergents. The trees of the dominant species in the forest canopy were counted in this sample area, which enabled their density to be estimated throughout the siamang territory (table 3.1). This facilitated a comparison with the second method of vegetation study.

This involved plotting the position of each specimen of the dominant tree species in the 34 ha of the siamang territory (fig. 3.2). These species were conspicuous in their relatively high frequency (see Discussion), their major contribution to the forest canopy, and their identity. Each species was widely spread throughout the area, although *Parkia speciosa* occurred mainly near the river, and *Dillenia reticulata* in damper parts. But there were frequent clusterings of trees of the same species.

The discrepancy between the two methods for measuring the density of these tree species can be explained in two ways (table 3.1). Either the sample area was too small to allow an accurate estimate for a larger area of forest, because of the uneven distribution of some species through the area – not all were over-estimates. Or not all trees of the dominant species were counted in the siamang territory; it is difficult to count all specimens of even a few conspicuous tree species in 30 ha or more of forest, even after nearly 200 days in that forest. The most likely source of error was not counting young trees of those species; there appears to be a bias favouring the lines of arboreal pathways of siamang, and the terrestrial pathways of observers. Both sources of error appear to be contributory, and it is likely that the true figures lie between the two estimates. But the two methods show a very close correlation between trees ranked for abundance; this highlights their main value.

It appeared that near the river there was a lower tree density (but more undergrowth), a lower and more open canopy, with a dense low layer, which was a less suitable habitat for siamang. Thus the sampling method was repeated along the east-west transect for 1,350 m back to the Ranger Post (table 3.2). Dividing the transect into west, central, and east parts showed that there was a significant decrease in the number of trees ($\chi^2 = 6.78$, $p = 0.05 - 0.02$), but no difference in forest structure (pattern of high, medium, and low trees, $\chi^2 = 1.83$). But the western part of the transect is close to Sungei Lompat, so that a riverine component is involved there also. Comparing an equivalent

Table 3.2. Tree density along the east – west transect at Kuala Lompat

a. Distribution of trees by height class

Reference point	Tree type H	M	L	H	M	L	total		H+M:L
11 – 10	2	16	52				70		
12 – 10	7	23	51				81		
13 – 10	7	14	55	16	53	158	76	227	0.44
14 – 10	4	23	50				77		
15 – 10	8	19	47				74		
16 – 10	5	8	41	17	50	138	54	205	0.49
17 – 10	5	11	38				54		
18 – 10	6	17	45				68		
19 – 10	0	5	39	11	33	122	44	166	0.31
Total				44	136	418		598	
Percent				7.4	22.7	69.8			
	221 trees/ha						2.7 ha		

H = high; M = medium; L = low.

b. Density of dominant species

Species	Number	Number/ha
Koompassia excelsa	9	3.3
K. malaccensis	4	1.5
Dipterocarpus grandiflorus	6	2.2
Parkia speciosa	2	0.7
Intsia palembanica	3	1.1
Dillenia reticulata	8	3.0
Shorea leprosula	2	0.7

area (0.9 ha) from the centre of TS1's territory with this western part produced a highly significant increase in the number of trees in the former ($\chi^2 = 9.45$), but, again, there was no significant difference in forest structure ($\chi^2 = 1.03$), i.e. the increase in tree density involved trees of all heights.

Similar results were obtained when comparing TS1's territory (sample of 4.2 ha) with the east-west transect (2.7 ha – $\chi^2 = 9.34$ and 3.27, respectively). Comparing the central segment of the siamang territory with the east end of the east-west transect (near Sungei Krau) resulted in a highly significant difference in tree density ($\chi^2 = 25.62$), and a significant difference in forest structure ($\chi^2 = 7.29$); the riverine forest had fewer trees, particularly of high and medium height.

Thus although there was considerable variation within the siamang territory, there was a greater difference between the density of 290 trees/ha therein, and the riverine density of 221 trees/ha. Forest structure was significantly different, at least compared with some parts of the siamang territory. Overall there was a ratio of high to medium to low trees of about 1:3.3:7.4, whereas near the river the ratio was 1:3.1:9.5. There were about 16 high, 50 medium and 155 low trees/ha near the river, and 25 high, 83 medium, and 182 low trees in each hectare of the siamang territory; an estimated total of 9,910 in all, of which only 5.4% were fed in by the siamang on 150 days over 14 months. An analysis of the distribution and density of these food trees will follow a description of siamang diet. The pattern of use of different forest levels by the siamang groups in both study areas is described in Chapter 4.

The fruit trap experiment was not a success. In Ulu Sempam some of the plastic sheets were taken by *orang asli* collecting rattan in the forest. At Kuala Lompat part of the series was demolished accidentally by game rangers in the course of the construction of a landrover track. But while it did give a useful measure of fruit fall, and confirm the description by MEDWAY [1972b] of an annual fruiting season in Ulu Gombak for all three areas, it did not provide a useful measure of fruit availability to an arboreal animal. When the siamang did eat fruit, they were often unripe, which was usually a long time before they fell. It did contribute to a description of the forest flora, particularly of the difference between edible and apparently inedible fruit. But this method has greatest application to the diet of terrestrial animals.

Diet

In this study siamang food was categorised as (1) fruit, (2) leaves – mature, new, stems, and shoots, (3) flowers and buds, and (4) insects. The proportion of each food category in the siamang diet was measured in terms of (1) the number of food trees, (2) the number of feeding visits, and (3) the feeding time (in minutes) for each day, month, season, and overall for the two siamang groups. From this was calculated (4) the duration of feeding visits. This variety of measures was used because of the possibility that they varied independently. This section will be followed by an analysis of the dispersion of the tree species and trees, and the distribution of feeding time and visits about the hectare quadrats of the territory.

Fruit ranged from the tiny berries of *Carallia* to large seeded fruit of *Maranthes* and *Gnetum*, and some very large figs (*Ficus auriantacea*, plate X). New leaves were distinguished from mature ones by their smaller size and/or different colour (usually a pale green or some shade of red, e.g. *Sloetia elon-*

gata, plate XII). In some cases specific parts of the leaves were consumed, such as apex, base, or blade. Sometimes the stem (and midrib), usually of a new leaf, was eaten (e.g. *Koompassia excelsa*, or *Xylopia caudata*, plate IX). Shoots were recognised for some species (such as *Parinari oblongifolia*, and some *Ficus* spp.), where the young leaf was produced as a tightly furled succulent spike. It is planned to investigate the nutritional value of these categories in subsequent field work.

It was often difficult to distinguish small flowers from buds at a distance, and, in particular, between leaf and flower buds. Apart from caterpillars which seasonally infested trees of *Endospermum malaccense* (plate IX), the only other animal matter seen to be eaten was termites. The siamang would sit by a large branch (often the vertical trunk) and wipe termites out of their marching column into the hairs on the back of a hand from which they would be extracted with lips and tongue. This involved similar movements to those used sometimes for ingesting water (see below). Thus insect consumption could be measured in the same ways as other food categories.

Food Categories

1. Food Trees

At Kuala Lompat over 148 days, in 14 months, 535 food trees were located for TS1; over 22 days in 1970 in Ulu Sempam RS2 fed in 140 trees. TS1 spent 354 min feeding in 13 trees each day on average, whereas RS2 fed for 276 min in 12 trees. The number of trees was counted as (a) the total number used each day, and (b) as the number of different trees used over 10 days – expressed as a daily mean (fig. 3.3). Thus each month about one third of food trees were revisited on subsequent days. Similar data are presented for fig trees (irrespective of food category eaten) because of the preponderance of this genus in siamang diet (see below). The daily number of food trees equalled the number of feeding visits each day, except on the few occasions when trees were visited more than once in the same day.

One quarter of the food trees at Kuala Lompat yielded fruit, and two thirds young leaves, with small scores for flowers, buds, and insects; about 13% of trees were figs, *Ficus* spp. The proportions for RS2 were almost identical (table 3.3). The difference between RS2 scores for 1969 and 1970 (with apparently more fruit than leaves consumed in 1969) indicates the bias towards more conspicuous fruiting trees that can occur when observation is not continuous from dawn to dusk.

Plate VIII. Specimens from siamang food trees: (a) *Antidesma coriaceum*, L049; (b) *Sarcotheca griffithi*, L335; (c) *Bouea oppositifolia*, L405; (d) *Calophyllum* sp., L406; (e) *Baccaurea motleyana;* (f) *Vitis* sp., S118.

Plate IX. Specimens from siamang food trees (continued): (a) *Gnetum* sp., L325; (b) *Uncaria* sp., L020; (c) *Aglaia* sp., S073; (d) *Xylopia caudata*, L410; (e) *Knema laurina*, L324; (f) *Combretum* sp., L315; (g) Derris sp., L110; and (h) *Endospermum malaccense*, L370, note signs of caterpillar activity,

Plate X. Specimens from siamang food trees (continued): (a) *Ficus bracteata*, L350; (b) *F. annulata*, L126; (c) *F. auriantacea*, L350A; (d) *F. heteropleura*, S121; (e) *F. sumatrana*, S025; (f) *F. stupenda*, L041.

Plate XI. Feeding behaviour: (a, b) adult female (GS16C) sits (c) hangs, eating fruit of *Aglaia* sp.; (d) adult female (TS1) hangs eating new leaves of climber *Connarus* sp.

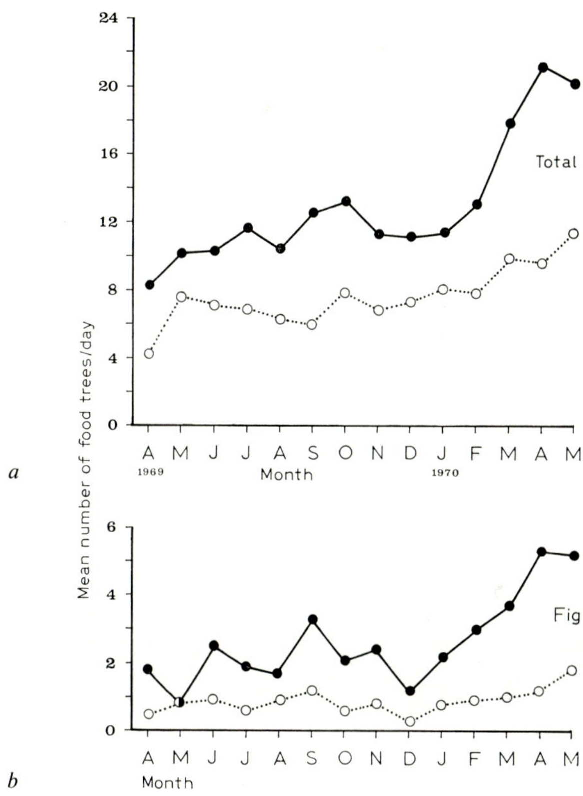

Fig. 3.3. Monthly pattern of mean number of food trees visited each day (closed circle, continuous line) and mean number of different food trees visited daily (open circle, dotted line) for (a) all food trees and (b) fig trees at Kuala Lompat (TS1).

Table 3.3. Overall proportions of each food type in siamang food trees

Group	Period	Food, % food trees				
		fruit	leaves	flowers	insects	figs
TS1	1969 May – August	18	62	16	1	12 (9)
	1969 September – December	25	69	4	1	10 (5)
	1970 January – May	22	65	7	5	13 (8)
Total		26	67	8	5	13 (7)
RS2	1969	54	44	2	0	22
	1970	31	59	9	3	17
Total		40	53	6	2	14
	1970 February	25	75	0	0	15
	1970 March	20	74	2	4	14
	1970 April	40	51	6	2	21
Total		29	65	4	3	17

Numbers in parentheses represent fig trees in fruit as percentages of total.

2. Feeding Visits

In contrast to the number of different trees utilised in each month, there was considerable variation in the number of trees of each category used daily, i.e. in the number of feeding visits. This is illustrated by plotting each food category in terms of (a) mean number of feeding visits, and (b) mean number of different trees visited each day (calculated from a 10-day total, see above). Thus the increase in daily feeding visits in the 1970 dry season (fig. 3.4) can partly be explained by repeated visits to the same trees – especially to those in fruit – rather than only by a significant increase in the numbers of food trees utilised.

The 'seasonal' differences in the numbers of trees visited for leaves, flowers, and insects were highly significant, but only significant for fruit. But the actual differences were small, especially for leaves, which highlights the limited biological significance of the χ^2 test with large samples. In Ulu Sempam there was no significant difference between months in the number of fruiting food trees, and in those with young leaves. Leaf intake was increased during the 'flush' at the beginning of both wet and dry seasons; the peak of flower consumption was soon after the dry season (during the north-west monsoon), and for insects at the end of the dry season, when fruit consumption was also markedly increased (tables 3.4 and 3.5).

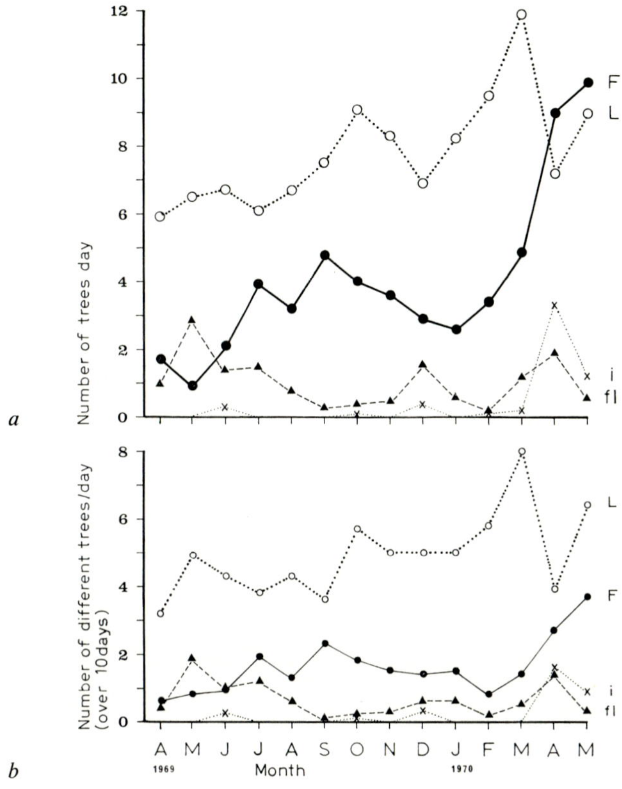

Fig. 3.4. Monthly pattern of (a) mean number of food trees visited each day, and (b) mean number of different food trees visited daily (calculated from total number of different food trees used in each observation period), for each food category eaten by TS1 at Kuala Lompat: fruit (F), leaves (L), buds and flowers (fl), insects (i).

Table 3.4. Seasonal variations in food-type proportions at Kuala Lompat

Food	'Season'							
	intermediate, 49 days May – Aug. 1969		wet, 40 days Sept. – Dec. 1969		dry, 49 days Jan. – May 1970		total, 138 days	
	visits/ day	time/ day, min	visits/ day	time/ day, min	visits/ day	time/ day, min	visits/ day	time/ day, min
Fruit	2.6	84	3.8	125	5.9	132	4.1	111
Leaves, young	6.5	206	7.9	202	9.1	202	7.8	206
Flowers, buds	1.6	50	0.6	21	0.8	18	1.0	30
Insects	0.1	0	0.1	2	0.9	13	0.4	5
Total	10.7	346	12.1	355	16.7	365	13.3	354
Figs	1.7	57	2.3	106	3.8	97	2.8	85
	visits/ day, %	time/ day %	visits/ day, %	time/ day, %	visits/ day, %	time/ day, %	visits/ day, %	time/ day, %
Fruit	24	25	33	34	35	36	31	32
Leaves, young	61	61	64	59	54	56	59	58
Flowers, buds	15	15	5	6	5	5	8	9
Insects	1	0	1	1	6	4	3	2
Figs	16	17	18	30	23	27	21	24
	visit duration, min		visit duration, min		visit duration, min		visit duration, min	
Fruit	32.2		31.9		22.4		27.2	
Leaves, young	31.9		26.8		22.2		27.4	
Flowers, buds	32.1		33.8		21.5		29.5	
Insects	6.0		20.8		13.8		13.9	
Total	31.6		29.5		21.8		26.6	
Figs	33.1		46.4		25.2		30.8	

3. Feeding Time

There was even more variation between months when measuring feeding on different foods in terms of time: (a) daily time for each category, and (b) daily time as a proportion of daily total feeding time (tables 3.4 and 3.5, seasonal, and fig. 3.5, monthly), but these measures are the most precise of those obtained. The difference between the monthly patterns of these two measures confirms the expectation that feeding time for each category tends

Table 3.5. Proportions of different food types eaten by siamang in Ulu Sempam

Food	Month, 1970							
	Febr., 6 days		March, 8 days		April, 8 days		total, 22 days	
	visits/ day	time/ day, min	visits/ day	time/ day, min	visits/ day	time/ day, min	visits/ day	time/ day, min
Fruit	4.3	140	5.4	106	6.5	149	5.5	131
Leaves, young	6.2	154	7.9	171	4.5	95	6.2	139
Flowers	0	0	0.1	7	0.4	6	0.2	5
Insects	0	0	0.3	5	0.1	1	0.1	2
Total	10.5	294	13.6	289	11.5	251	12.0	276
Figs	5.5	158	5.0	114	2.5	79	4.2	113
	visits/ day, %	time/ day, %	visits/ day, %	time/ day, %	visits/ day, %	time/ day, %	visits/ day, %	time/ day, %
Fruit	41	48	39	37	57	59	46	47
Leaves, young	59	52	58	59	39	38	51	50
Flowers	0	0	1	2	3	3	2	2
Insects	0	0	2	2	1	0	1	1
Figs	52	54	37	40	22	32	35	41
	visit duration, min		visit duration, min		visit duration, min		visit duration, min	
Fruit	32.2		19.7		22.9		23.8	
Leaves, young	25.0		21.7		21.2		22.4	
Flowers	0		(56)		17		26.8	
Insects	0		19		(5)		14.3	
Total	28.0		21.2		21.8		23.0	
Figs	28.7		22.9		31.7		26.8	

to vary less as a proportion of daily feeding time than as absolute values. Total group feeding time each day increased steadily through the seasons (Chapter 4).

Time spent eating fruit paralleled this increase (from 80 to 130 min daily on average), whereas leaves continued to occupy about 200 min (60%) of group feeding time each day. Flowers only made a significant contribution to siamang diet from May to July, and in December and March; insects in April 1970 only. RS2 spent less time feeding each day, but consumed a higher proportion of fruit (even when compared with TS1's 'dry season' peak).

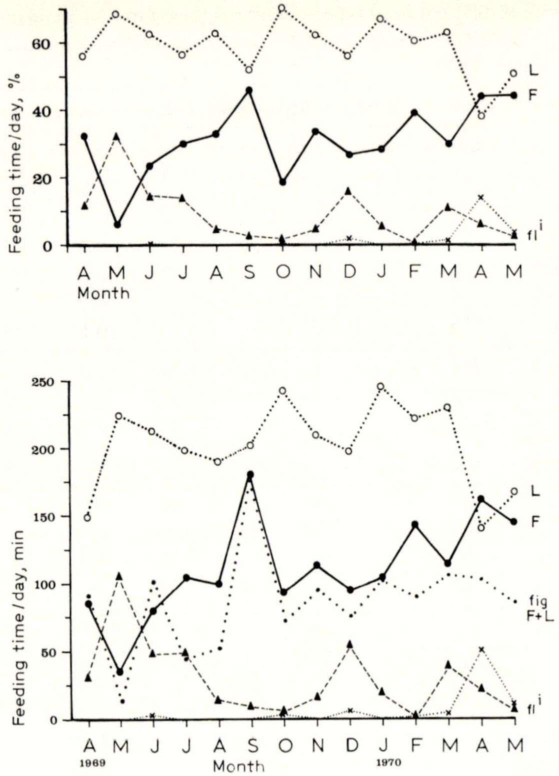

a

b

Fig. 3.5. Monthly variation in (a) percent of daily feeding time, and (b) number of minutes feeding each day, for each food category eaten by TS1 at Kuala Lompat: fruit (F), leaves (L), buds and flowers (fl), insects (i), and figs.

Further insight into these fluctuations in feeding time is gained by an analysis of day-to-day variations at Kuala Lompat (fig. 3.6). Group daily feeding time ranged from 188 to 477 min, with a mean 10-day range of 150 min. Between 5 and 32 trees were fed in daily, with a mean monthly variation of about 9 trees. Feeding time on fruit ranged from 0 to 315 min daily, with variations within months averaging at 152 min. For leaves scores ranged from 47 to 355 min, with a greater average variation from day to day of 183 min. Scores for figs closely approximated those for fruit. All these fluctuations in feeding time were highly significant by Kruskal-Wallis one-way analysis of variance tests, corrected for ties: (1) total, $H = 39.7$, $p = 0.001 -$; (2) fruit, $H = 31.8$, $p = 0.01 - 0.001$; (3) leaves, $H = 34.7$, $p = 0.01 - 0.001$, and (4) figs, $H = 43.9$, $p = 0.001 -$ (d.f. = 13).

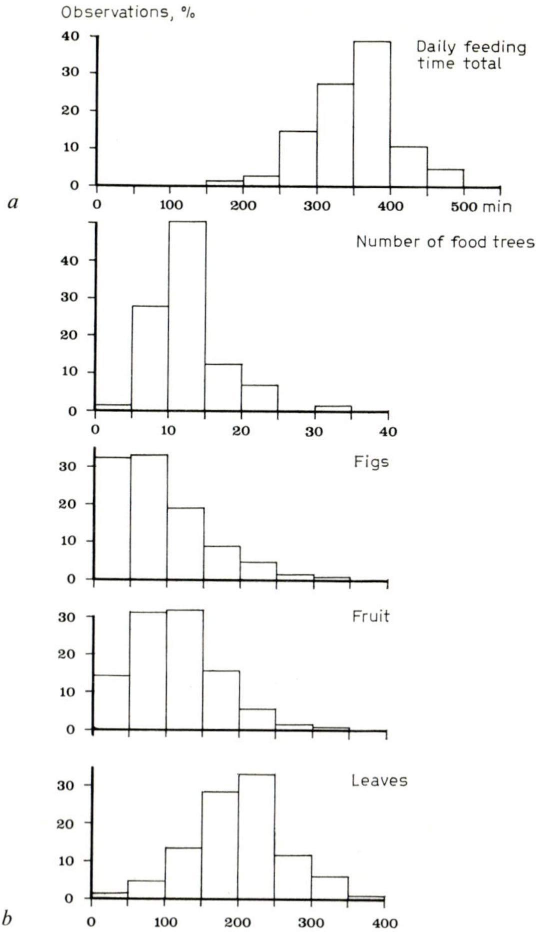

Fig. 3.6. Distribution of 148 daily scores for TS1 at Kuala Lompat for (a) total feeding time, and (b) number of food trees – figs (fruit and leaves), fruit, and leaves.

Comparing the distribution of daily scores of feeding time between 'seasons' produced a significant difference in fruit eating only between 'intermediate' and 'dry seasons'. No significant differences were found between 'seasons' for leaf-eating (Kolmogorov-Smirnov two-sample test). Comparisons between months (fig. 3.5) revealed a highly significant increase in fruit-eating in September (compared with October and August in turn by the Mann-Whitney U test). Leaf-eating was significantly increased in May and October 1969, and in January 1970 – but not in December (compared with the low scores of April 1970).

Table 3.6. Seasonal variations in proportions of different leaf parts in diet at Kuala Lompat

Leaf food	'Season'							
	intermediate		wet		dry		total	
	visits/ day	time/ day, min	visits/ day	time/ day, min	visits/ day	time/ day, min	visits/ day	time/ day, min
Mature	0	1	0.2	1	0.1	1	0.1	1
Young	5.8	179	7.0	179	7.6	160	6.8	172
Shoots	0.2	4	0.7	30	0.4	9	0.4	13
Stems	0.5	21	0.2	3	0.7	32	0.6	20
	visits/ day, %	time/ day, %	visits/ day, %	time/ day, %	visits/ day, %	time/ day, %	visits/ day, %	time/ day, %
Mature	0.4	0.4	0.2	0.2	0.6	0.3	0.4	0.3
Young	54.0	53.0	56.0	50.0	46.0	44.0	51.1	48.8
Shoots	1.5	1.3	5.8	8.3	2.2	2.5	3.0	3.8
Stems	5.0	6.1	1.4	0.8	6.0	8.9	4.5	5.6
	visit duration, min		visit duration, min		visit duration, min		visit duration, min	
Mature	32		(29)		11		18.6	
Young	31		26		21		25.4	
Shoots	26		41		25		33.4	
Stems	39		17		32		33.1	

The bulk of leaf consumption was of young leaves (51% of visits, and 49% of time daily, for TS1). Mature leaves were occasionally eaten, with stem consumption rising to a maximum between the dry and intermediate seasons; shoots were most consumed in the wet season (table 3.6). These seasonal changes were highly significant (χ^2 test). The pattern of leaf consumption was different between the two study groups – RS2 ate no shoots in April, but ate them for 30% of time in February.

RS2 fed in a higher proportion of fig trees (17%, cf. 13% for TS1). They visited 4.2 each day (35% of daily feeding visits, cf. 20% for TS1), which accounted for 41% of feeding time (113 min on average). In each month at least 5 (sometimes 11–13, and once 16) of the 64 fig trees in TS1's territory had fruit or young leaves; six were fruiting each month on average. It was only for fig trees that proportions changed significantly according to the different measure – trees, visits, or time. For the main categories they were almost identical.

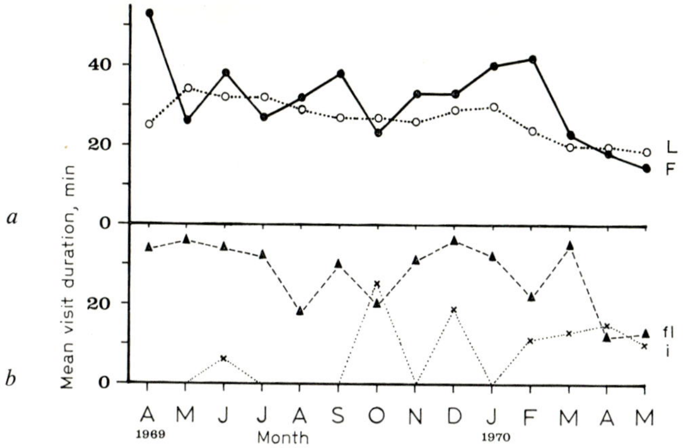

Fig. 3.7. Monthly variation in mean visit duration to food trees of each category at Kuala Lompat (TS1): (a) fruit (F), leaves (L), and (b) buds and flowers (fl), insects (i).

4. Feeding Visit Duration

The mean durations of feeding visits were calculated for each observation period from the total time spent feeding, and the total number of visits, to all food trees, and to trees of each category. Although feeding visits were recorded ranging from a few minutes to more than 2 h, on average most lasted for 20–40 min; the overall mean in both study areas was around 25 min (tables 3.4 and 3.5). At Kuala Lompat visits were longer in the 'intermediate season', and shorter in the 'dry season'. There was little difference in visit duration for different food categories, except for the much shorter visits to feed on insects; the fewer visits to flowering trees tended to be longer.

This parameter masks the bimodal pattern to fruiting trees, indicated by the fluctuations in the monthly plot (fig. 3.7). To some trees, such as fruiting figs, visits could be very long (up to 3 h); to others, such as the smaller trees, they were often very short. The pattern was less variable for leaf-eating, although mature leaves were consumed for a shorter duration, and stems and shoots (less often than young leaves) for longer (table 3.6).

As has already been suggested, there was a fairly close relation between the number of visits to food trees each day, and the amount of time spent feeding in trees of each food category. This was most distinct for flowers and buds, and more variable for fruit (see above). Furthermore, the increase in visit frequency to fruit and leaf trees was associated with a decrease in the duration of each visit (fig. 3.8). This pattern was less distinct for flowering and budding trees, but in some cases it appears that increased visit frequency

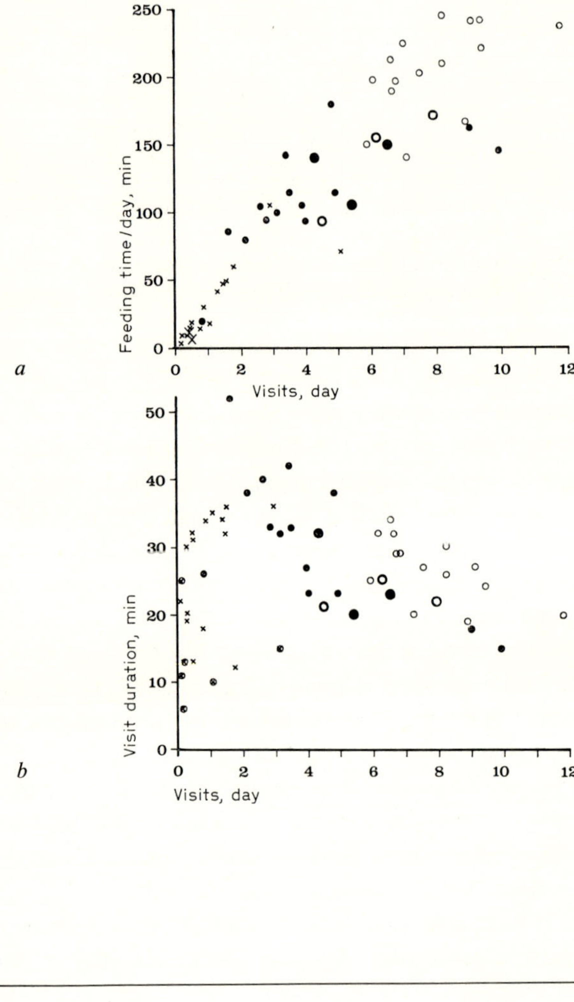

Fig. 3.8. Bivariate plots of monthly means for TS1 and RS2 for each food category: (a) number of visits to food trees and feeding time each day, and (b) number of visits and visit duration. Fruit = ●, leaves = ○, flowers and buds = × , and insects = ⊗ (symbols enlarged for RS2).

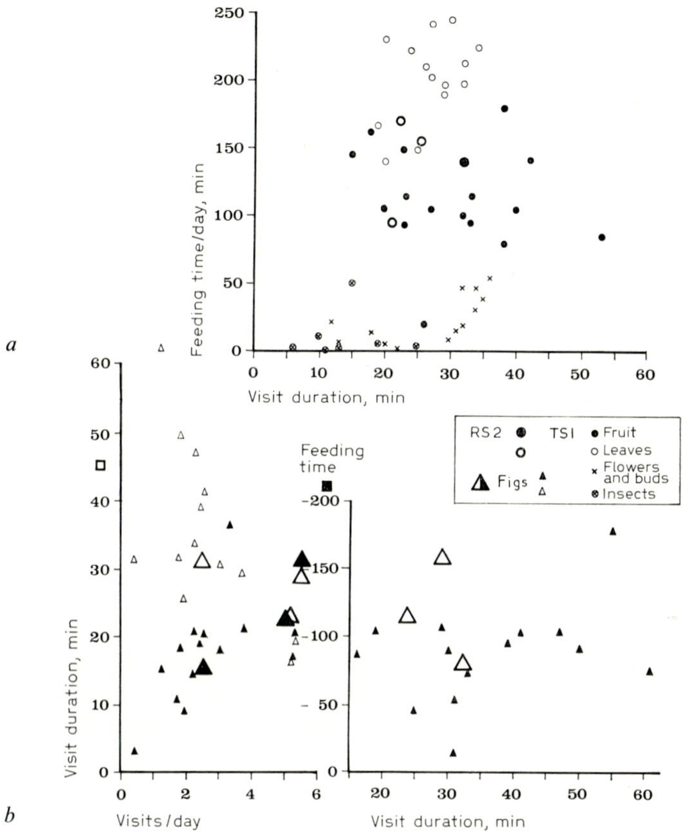

Fig. 3.8 (continued). Bivariate plots of monthly means for TS1 and RS2 for each food category (and figs): (a) visit duration and daily feeding time, and (b) for figs (fruit and leaves) – visit duration and number of visits each day, and visit duration and daily feeding time.

resulted in longer visits. A relation between feeding time and visit duration was less obvious, partly because scores for each category were not very variable from month to month. The same patterns applied to feeding in fig trees, and to the limited data for RS2.

Food Tree Identity

1. Kuala Lompat

Apart from at least 13 species of climbing or strangling fig, there were 21 species of 18 tree genera (usually one of each), which figured prominently in siamang diet (Appendix I). Their prominence could be measured in terms of number of trees used, high intensity of use at certain seasons, or continual light use.[1] Just over one third of siamang food trees remain unidentified. Most of the climbers involved appear to belong to relatively few genera, such as *Combretum*, *Connarus*, *Derris*, *Gnetum*, *Uncaria* (plate IX), and unidentified leguminous climbers. Thus it may be that certain of these less conspicuous climbing species may be of greater importance to siamang than is indicated in the following analysis, which was oriented towards the more prominent free-growing tree species.

The importance of species of *Ficus* (plate X) in the siamang diet has already been described. The only other major food source was a small tree, *Sloetia elongata*, which accounted for 11.4% food trees, and 13.4% feeding visits. This species was most important between March and July, when it flowered in the form of 'catkins', and at such other times as new leaves were available (table 3.7). *Endospermum malaccense* was visited for two months at the end of the dry season, and then only intensively in those years when there was an abundance of larvae of an unidentified Lepidopteran. It was on these that they fed, although they also consumed some of the tree's fruit (plate IX).

1 The main problem has been that siamang spend much time in trees bearing one or more species of climber (or vine). First one had to determine whether host or climber was being eaten, which was more difficult when more than one climber was involved. Secondly, the greatest problem was to collect that specimen. Even the tree-climbers hired for the purpose could not climb some of the large food trees to collect the correct specimen, but sometimes a young seedling of the same species was found on the forest floor nearby. Thirdly, many of the vines eaten (young leaves) were not seen in flower or fruit, which made specific identification of such specimens that were collected almost impossible. On the opposite bank of Sungei Pahang, Poore [1968] identified 375 tree species of 139 genera in 23 ha of forest.

Table 3.7. Pattern of use of main food tree species by siamang at Kuala Lompat (TS1) – 15.0 food trees/ha

Tree species	Number of trees, 35 ha	Density, trees/ha	Food trees, % (n = 525)	Feed visits, % (n = 1,912)	Months, % (n = 18)	Month – number of trees visited, main food obtained													
						1969									1970				
						Apr.	May	June	July	Aug.	Sept.	Oct.	Nov.	Dec.	Jan.	Feb.	Mar.	Apr.	May
Ficus spp. (13+)	64	1.8	12.2	20.0	100	5^{1}	7^{1}	13^{1}	9^{1}	11^{2}	12^{1}	6^{3}	8^{2}	3^{1}	8^{1}	9^{1}	10^{1}	$12^{1,3}$	16^{1}
Sloetia elongata	60	1.7	11.4	13.4	100	6^{2}	20^{4}	7^{4}	21^{4}	8^{2}	2^{2}	7^{2}	9^{2}	7^{2}	5^{2}	11^{2}	17^{2}	17^{4}	20^{4}
Endospermum malaccense	31	0.9	5.9	2.2	29			2^{5}										$22^{1,5}$	10^{5}
Randia scortechinii	13	0.4	2.5	1.6	50				1^{1}	3^{1}	3^{1}	5^{1}	6^{1}	3^{1}	2^{1}				
Koompassia excelsa	10	0.3	1.9	2.2	57	2^{2}	1^{6}	1^{4}	7^{1}	1						3^{2}	5^{6}		2
Nauclea sp.	10	0.3	1.9	0.8	14					1^{1}	7^{1}								
Wiedelandia sp.	8	0.2	1.5	2.0	43					1^{1}	2^{1}	7^{1}	5^{1}	2^{1}	1^{1}				
Carallia brachiata	8	0.2	1.5	1.5	43		2^{7}		2^{4}	1^{4}		4^{1}	1^{1}		1^{2}				
Sarcotheca griffithi and S. monophylla	8	0.2	1.5	3.2	43		1^{2}						1^{2}	1^{2}			5^{4}	6^{1}	4^{1}
Gnetum macrostachys and sp.indet.	7	0.2	1.3	1.3	43		3^{1}	1^{1}					1^{2}	1^{2}			1^{1}		6^{1}
Xylopia caudata and X. malayana	6	0.2	1.1	1.3	64				1^{4}	1^{2}	1^{2}		1^{2}		1^{4}	2^{2}	$3^{2,4}$	1^{2}	1^{6}
Knema laurinum	6	0.2	1.1	0.8	29									1^{1}			3^{1}	3^{1}	2^{1}
Eugenia sp.	4	0.1	0.8	0.7	14	1^{2}										4^{1}			
Aglaia sp.	4	0.1	0.8	1.2	43	1^{2}	3^{2}			1^{2}			1^{2}		1^{4}			3^{2}	
Parinari oblongifolia	4	0.1	0.8	0.8	43			1^{3}				3^{3}		1^{3}	1^{3}		1^{3}		1^{3}
Maranthes corynbosum	3	0.1	0.6	0.8	21					1^{1}	2^{1}	1^{1}							
Calophyllum sp.	3	0.1	0.6	0.4	14		1^{2}												3^{1}
Diospyros sp.	3	0.1	0.6	0.3	14			1^{2}			2^{2}								
Bouea oppositifolia	3	0.1	0.6	0.6	14													1^{1}	3^{1}
Total *Ficus*+21spp.	255	7.3	48.7	56.4		15	35	25	41	29	31	33	32	19	21	29	44	65	62
Other identified trees	78	2.2	14.9	15.2		36	51	25	43	38	52	42	47	25	30	43	44	68	61
Unidentified trees	192	5.5	36.6	28.4											Percent of monthly total				

1 Fruit. 2 New leaves. 3 Leaf shoots. 4 Flowers. 5 Insects. 6 Leaf stems. 7 Buds.

Plate XII. Feeding behaviour (continued): (a) adult female, and (b) adult male (TS1) hang eating new leaves of *Sloetia elongata;* (c, d) adult female (TS1) hangs eating fruit of *Ficus virens.*

Plate XIII. Feeding behaviour (continued): (a) Juvenile male (TS1) hangs eating fruit of large swamp tree, *Nauclea* sp.; (b) subadult male (TS1) hangs eating fruit of *Ficus* sp.; (c) subadult male (TS1) sits drinking from water-hole in *Koompassia excelsa;* (d) adult male (TS1) sits drinking from water-hole in *Sloetia elongata.*

Table 3.8. Pattern of use of main food tree species by siamang in Ulu Sempam (RS2) – 9.5 food trees/ha

Tree species	Trees (n = 142)		Density, trees/ha	1970, 22 days			Month – number of trees visited, main food obtained												
							1969									1970			
	number	%		Food trees, % (n = 85)	Feed visits, % (n = 264)	Time, % (n = 6,077)	Apr.	May	June	July	Aug.	Sept.	Oct.	Nov.	Dec.	Jan.	Feb.	Mar.	Apr.
Ficus spp. (13+)	24	16.9	1.5	17.7	35.2	41.1	2[1]	1[1]			2[1,2]	1[1]	4[1]	2[1,2]	3[1]	2[1]	4[1,3]	7[1,3]	11[1]
Cynometra inaequifolia	7	4.9	0.4	4.7	6.4	7.4								1[1]?	1[4]		1[3]	5[3]	
Connarus spp. indet. (2)	6	4.2	0.3	3.5	3.0	4.4						2[1]			1[1]	1[4]	1[4]	2[3]	2[2]
Aglaia spp. indet. (2)	5	3.5	0.3	2.4	3.8	4.5	2[1]				1[4]			1[5]		1[4]	1[1]	1[4]	1[1]
Gnetum sp.	4	2.8	0.3	2.4	0.8	0.3					1[1]	2[1]	2[1]	1[1]				1[4]	2[4]
Wiedelandia sp.	4	2.8	0.3	0										2[1]	1[1]	2[1]			
Calamus urens	3	2.1	0.2	3.5	3.0	1.3												3[1]	
Bouea oppositifolia	2	3.4	0.1	2.4	2.7	2.2													2[1]
Memecylon sp.	2	1.4	0.1	2.4	2.4	1.2											1[6]	2[1]	2[1]
Total																			
Ficus + 10 species	57	40.1	4.1	38.8	57.1	62.4	4	1			4	5	6	7	6	7	8	21	20
Other identified trees	18	12.7	1.3				100	33			44	63	75	35	43	33	40	45	44
Unidentified	67	47.2	4.8												Percent of monthly total				

1 Fruit. 2 Leaf stems. 3 Leaf shoots. 4 New leaves. 5 Flowers. 6 Buds.

The fruit of *Randia scortechinii* was eaten between July and January – mainly the large paired walnut-shaped seeds inside the tough skin. In August and September the small trees of a species of *Nauclea* (plate XIII) in the swamps produced a very large fruit – inflorescence covering a grossly swollen receptacle (superficially like 'breadfruit' in appearance). A second species of Rubiaceae, *Wiedelandia* sp., produced many small red berries in small trees between August and January. At least two climbing species of *Gnetum* fruited between March and May; also, *Knema laurinum* (a relative of the nutmeg) in the swampy areas with a seed covered in red aril (plate IX). The siamang ate the small purple flowers of *Sarcotheca griffithi* in March and the juicy fruit in April and May (1970 only). There were also juicy green fruit in trees of *Bouea oppositifolia* from April onwards (plate VIII).

Koompassia excelsa had an annual leaf flush near the end of the dry season (earlier in 1970), followed by flowering and the production of small-winged fruit. The siamang ate the very young leaves, the stems of the leaves as they matured, and finally the flowers and, to a lesser extent, the fruit. The same timing occurred with the young leaves and stems of *Shorea leprosula*. *Carallia brachiata* flowered in July and August, and fruited (very small berries) in October and November. The species of *Eugenia* fruited annually at the beginning of the dry season; the siamang scraped the tough green skin off the medium-sized fruit to chew the fibrous flesh. *Maranthes corynbosum* (Rosaceae) produced a purple drupe, with sticky white latex, annually in about August.

Thus some tree species were regular in their fruiting habits, others much less so; but most exhibited different seasonal cycles. This resulted in a continuous availability of fruit. Similarly, the cyclical variability, as well as a lack of synchrony, in leaf production in species of *Xylopia*, *Diospyros*, and *Parinari*, as well as *Ficus*, ensured a continuous supply of presumably nutritious young leaves for the siamang.

2. Ulu Sempam

Difficulties of observation prevented such a full analysis for RS2. The main differences were the importance of the leaf shoots of *Cynometra inaequifolia* in the dry season. Different species of *Aglaia* were present producing medium-sized fruit, and a species of *Gnetum* with an elongated fruit, large seed, and sticky brown latex (consumed) was more abundant. The fruit of the giant rattan, *Calamus urens*, and of a species of *Memecylon* were also eaten when available. The small red berries of *Wiedelandia* and the green drupes of *Bouea oppositifolia* were available as at Kuala Lompat (table 3.8). Although information is incomplete for the earlier months it is clear that sufficient food was continually available (see daily activity patterns, Chapter 4).

In both study areas between one quarter and two thirds of food trees were accounted for by the principal species described. Overall, in Ulu Sempam 40% trees, 57% tree visits, and 62% feeding time were to these trees of 9 genera. At Kuala Lompat 49% trees, and 56% visits were to feeding trees of 19 genera. In both areas more than half these food trees were identified. The dispersion of food trees of each genus and species illustrates the extensive distribution of *Ficus*. In Ulu Sempam *Cynometra* was concentrated on one steep slope, with *Gnetum* and *Connarus* nearby (fig. 3.9). At Kuala Lompat there were concentrations of *Sloetia*, *Endospermum*, *Randia*, and *Wiedelandia*, with *Knema* and *Nauclea* in areas liable to flooding (fig. 3.10 cf. fig. 3.2).

Principal Food Trees

There now follows a description of the pattern of food resource utilisation whereby a small proportion of food trees were involved in a much higher proportion of group feeding than would be expected by chance. At Kuala Lompat more than one third of food trees were visited only once in 148 days over 14 months; nearly 80% were visited five times or less (fig. 3.11). This indicates the diversity in siamang diet; the long tail to the curve shows the great importance of 7% food trees, which accounted for 30% feeding visits, and over 40% feeding time (fig. 3.12). More than 50% food trees were visited in only one observation period, but only 5% were visited in half or more months of the study.

In 22 days in Ulu Sempam more than 40% trees were visited only once. Nearly 90% food trees were visited five times or less; nearly two thirds were fed in for less than 50 min (fig. 3.13). But 12% food trees accounted for 45% visits and 55% feeding time.

These principal food trees were classified as: (1) *primary food source* – trees visited for a long time almost every day of an observation period (maybe more than one in 12 months), and about which day ranges were orientated during that period; (2) *secondary food source* – trees visited frequently for feeding in one or more observation periods; (3) *tertiary food source* – trees visited several times in several months.

Thus primary sources have a fundamental effect on ranging behaviour, secondary sources have a moderate short-term significance, whereas tertiary food sources are used least intensively of principal food trees, but over a longer period. All other food trees may be regarded as occasional food sources (with the potential to assume greater short-term importance). The location of each food tree is indicated in figures 3.14 and 3.15 for Ulu Sempam and Kuala Lompat, respectively.

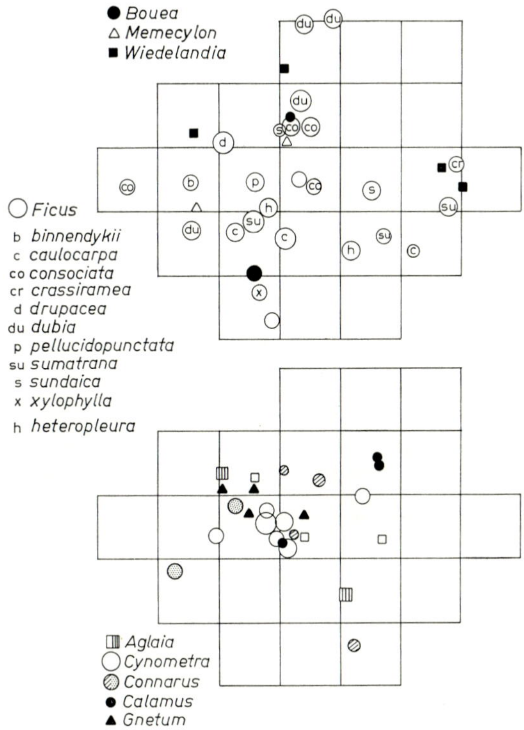

Fig. 3.9. Ulu Sempam – location of food trees of genera making greatest contribution to siamang (RS2) diet: *Ficus, Bouea, Memecylon, Wiedelandia* (above); *Aglaia, Cynometra, Connarus, Calamus, Gnetum* (below).

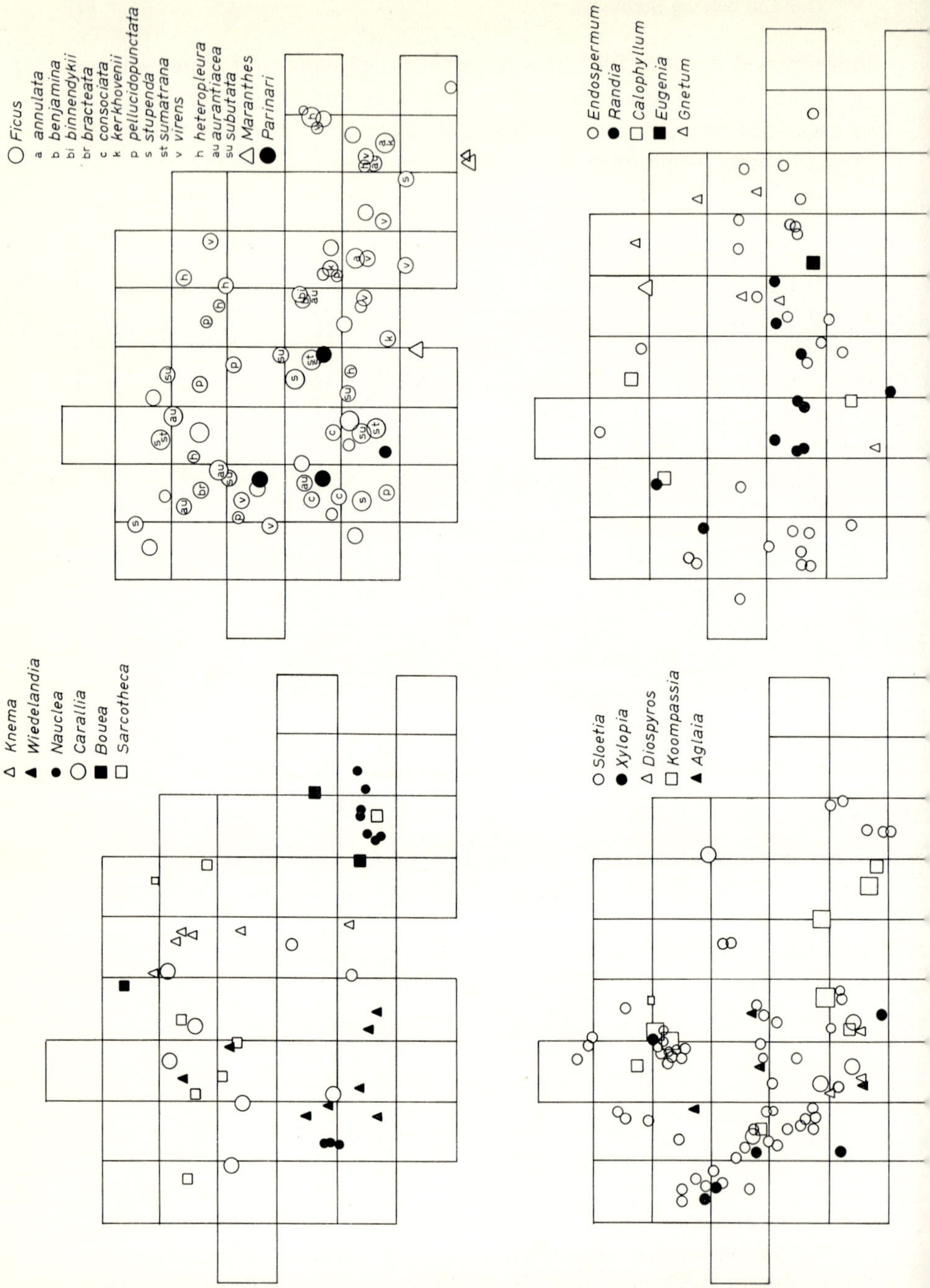

Fig. 3.10. Kuala Lompat – location of food trees of principal genera in siamang (TS1) diet.

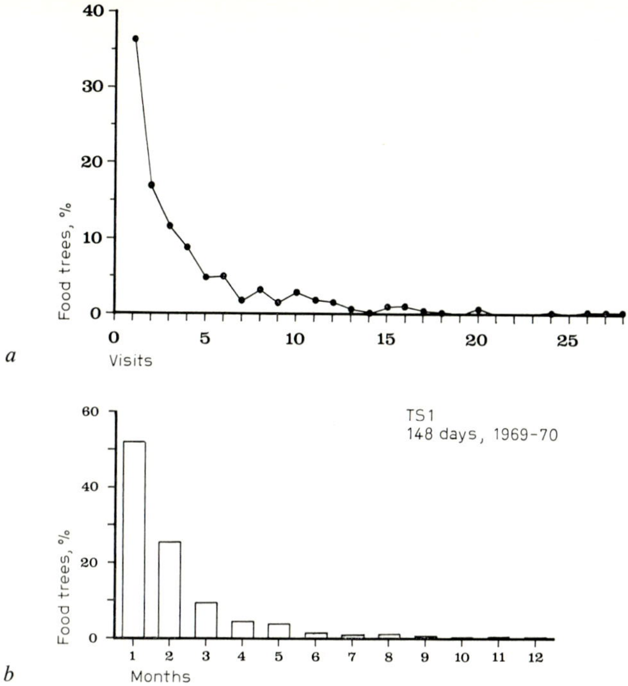

Fig. 3.11. (a) Frequency distribution of number of visits to each food tree by TS1 at Kuala Lompat, and (b) proportion of food trees visited for 1 – 12 months by TS1 at Kuala Lompat.

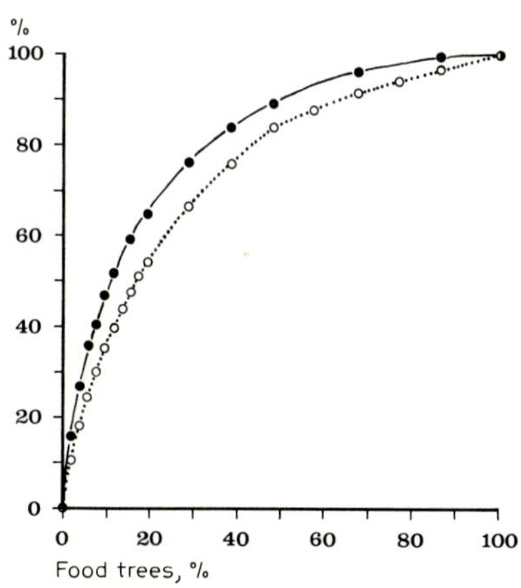

Fig. 3.12. Cumulative proportions of food trees (ranked), feeding visits (○), and feeding time (●) for TS1 at Kuala Lompat.

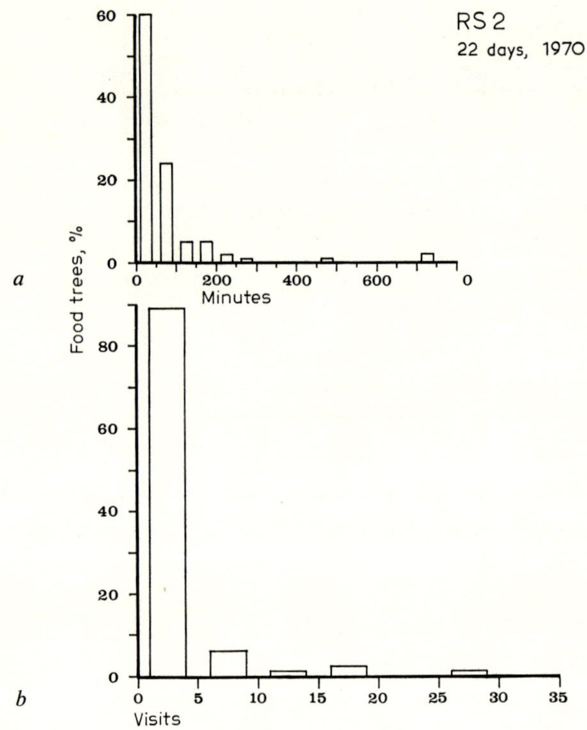

Fig. 3.13. Frequency distribution of (a) feeding time and (b) number of feed visits for each food tree in Ulu Sempam (RS2).

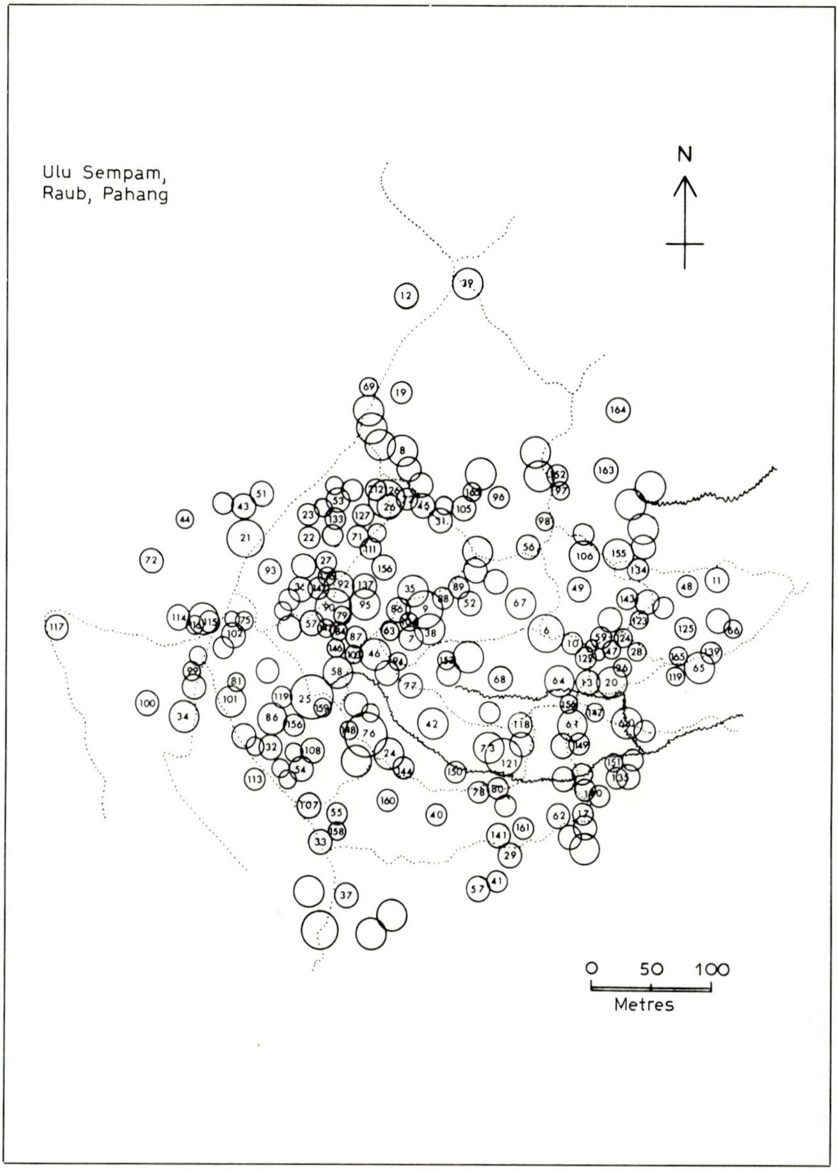

Fig. 3.14. Ulu Sempam – location, relative size, and identification number of each food tree visited by RS2 (1969–70). Also shown are trails, streams, and conspicuous trees along arboreal pathways.

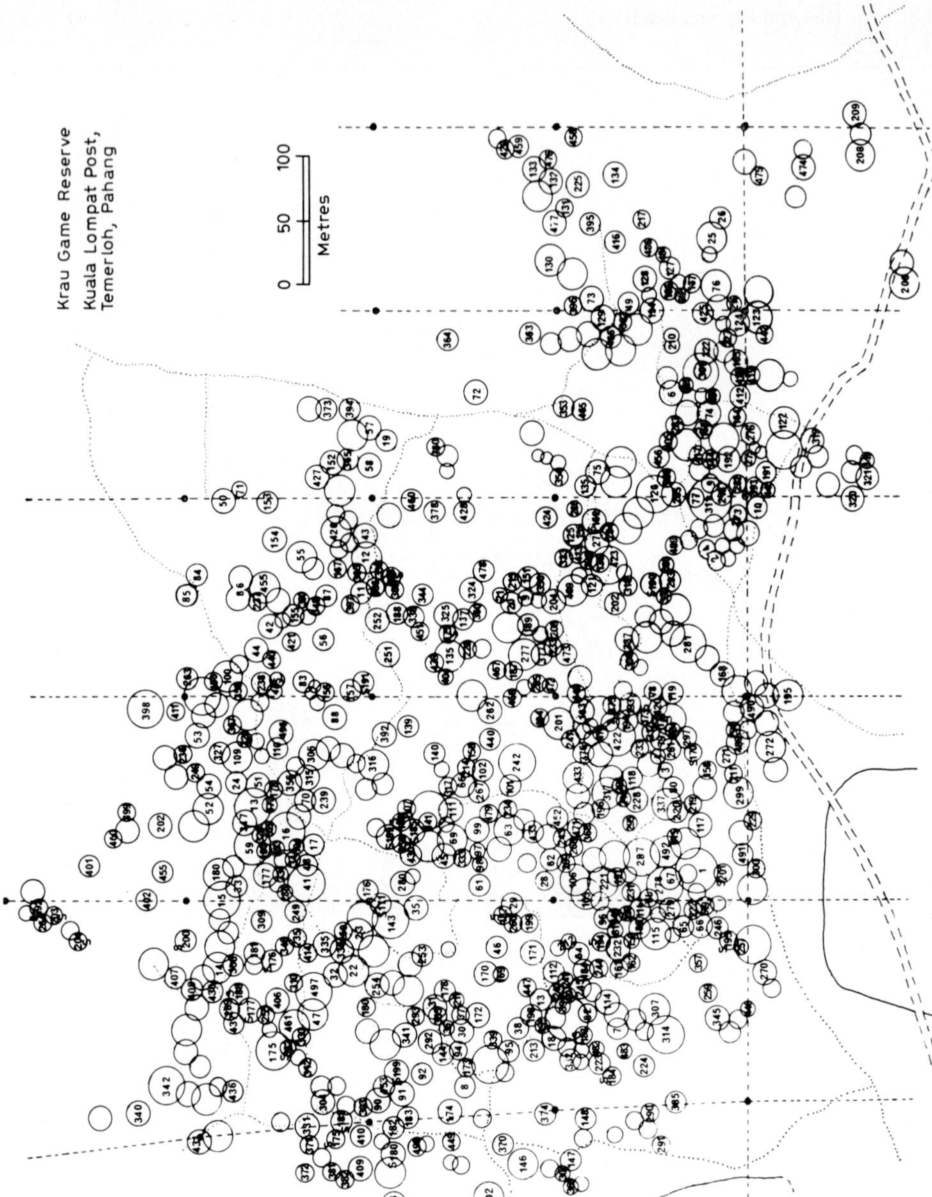

Fig. 3.15. Kuala Lompat – location, relative size, and identification number of each food tree visited by TS1. Also shown are trails, transects, and conspicuous trees along arboreal pathways.

1. Kuala Lompat

The monthly pattern of use of principal food trees at Kuala Lompat emphasises the importance of figs to siamang diet and ranging behaviour (table 3.9). The other major feature was their continuous overall availability throughout the year. Some species, especially *Ficus virens* (small pink fruit) were not synchronised throughout the area, and each tree fruited at about 6-month intervals. Thus it does not require many trees of this species alone to provide a continuous supply of figs. The same applied to a lesser extent to leaf shoot production by specimens of *F. subutata*. The fruit of *F. sumatrana* and *consociata* were somewhat larger, ripening red and yellow. The former appeared annually, the latter probably biennially. The red/yellow fruit of *stupenda*, and the buff-coloured fruit (L171, L217 unidentified) were even larger (about 2 cm diameter). The largest fruit (up to 5 cm across) were those of the climber *aurantiacea*, which fruited after each dry season. Unlike most other species, which ripened rapidly, ripe fruits were available over a longer period. This was also true for the other climbing species eaten by siamang – *heteropleura* with stalked fruit, especially important in Ulu Sempam. A less used food source was the purplish fruit (1 cm diameter) of the fairly common *pellucido-punctata*.

Table 3.9. Monthly pattern of feeding in principal food trees at Kuala Lompat indicating tree number, tree identity (where known) and food category

Month	Food tree				
	primary		secondary		tertiary
April	015	*Ficus stupenda*	032	*Aglaia* sp.	
	F	*Ficus sumatrana*	nL		
			017	*Sloetia elongata*	
			030	nL	
May	fl	*Sloetia elongata*	063	*Shorea*	064
	nL		st	*leprosula*	nvL
	048, 057, 062, 080+5		016	*Koompassia*	
			st	*excelsa*	
June	074	*Ficus virens*	116	*Ficus* sp.	064
	F		F		nvL
	126	*Ficus annulata*	127	*Ficus virens*	125
	nL		F		nvL
			151	*Ficus*	136
			F	*binnendykii*	nvL
			118	*Koompassia*	
			fl	*excelsa*	

Table 3.9 (continuation)

Month	Food tree		
	primary	secondary	tertiary
July	126 *Ficus annulata*	177 *Carallia*	175 *Ficus aurantiacea*
	F	189 *brachiata*	nL
	191 *Dracontomelum*	+3 F	277
	F *mangifera*	170 *Sloetia elongata*	nvL
		+10 nL, fl	
		192 *Koompassia*	
		excelsa	
		+5 nL, st, F	
		171 *Ficus* sp.	
		F	
August	195 *Maranthes*	001 *Ficus sumatrana*	077 nvL
	+2 F *corynbosum*	F	202 nvL
		168 *Ficus*	271 nvL *Combretum* sp.
		st *kerkhovenii*	and/or *Gnetum* sp.
		203 *Ficus virens*	
		F	
		281 nvL	
September	001 *Ficus sumatrana*	187 *Randia*	136 nvL
	F	+2 F *scortechinii*	284 nvL
	203 *Ficus virens*	210 *Nauclea* sp.	318 nvL
	F	+7 F	488 nvL
	278 *Ficus subutata*	219 *Wiedelandia* sp.	
	sh		
October	228 *Ficus subutata*	052 *Ficus* sp.	109 *Ficus* sp.
	254 sh	sh	
		233 *Ficus*	213 nvL
		nL *heteropleura*	
		248 *Wiedelandia* sp.	
		240 +4 F	
		232 *Carallia*	
		253 F*brachiata*	
		S176 *Sapium baccatum*	
		F	
November	018 *Ficus consociata*	066 *Aglaia* sp.	047 nvL
	F	nL	304 nvL
		242 *Ficus stupenda*	175 nvL *Ficus* sp.?
		F	168 nvL *Ficus* sp.
		240 *Wiedelandia* sp.	
		+5 F	
		S180 nL	
December	265 *Ficus virens*	S190 *Knema laurinum*	077 nvL
	F	F	
	274 *Leguminosae*	275 nvL	
	fl, nvL	283 nvL	

Table 3.9 (continuation)

Month	Food tree				
	primary		secondary		tertiary
January	201	*Ficus sumatrana*	015	*Ficus stupenda*	039 nvL or *Sloetia?*
	F		F		064 nvL
	233	*Ficus*	163	*Parinari*	181 nvL
	F	*heteropleura*		*oblongifolia*	
February	027	*Eugenia* sp.	233	*Ficus*	096 vb
	+3	F	F	*heteropleura*	275 nvL
	314	*Ficus stupenda*	312	*Sloetia elongata*	276 nvL
	F		nL		275 nvL
	311	*Koompassia*			318 nvL
	nL, st	*excelsa*			
March	315	*Ficus*	S171	*Sarcotheca*	316 nvL
	F	*pellucido-punct.*	335	*griffithi*	430 nvL
	287	*Ficus* sp.	+2	fl	
	F		341	*Ficus virens*	
	324	F	F		
	325vF	*Gnetum* sp.	334	*Sloetia elongata*	
			nL	+16	
April	i	*Endospermum*	O63	*Shorea*	123 *Ficus stupenda*
	(+F)	*malaccense*	st	*leprosula*	sh
	20+	trees	224	*Xylopia*	041 *Ficus* sp.
	049	*Antidesma*	nL	*caudata*	F
	F	*coriacea*	127	*Ficus virens*	
	356	*Sarcotheca*	st	074, 265 F	
	365	*griffithi*	314	*Ficus stupenda*	
	322, S171 F		F		
			221	*Ficus consociata*	
			F		
			012	*Ficus*	
			st	*heteropleura*	
			S169	*Ficus*	
			st	*kerkhovenii*	
May	126	*Ficus annulata*	072	*Gnetum* sp.	
	F		391,	398 vF	
	074	*Ficus virens*	049	*Antidesma*	
	F		F	*coriacea*	
	166	*Ficus*	396	*Bouea*	
	F	*heteropleura*	405F	*oppositifolia*	
	409	*Xylopia candata*	S169	*Ficus*	
	st		st	*kerkhovenii*	

Food categories: F = fruit; fl = flowers; i = insects; nL = new leaves; nvL = new vine leaves; sh = leaf shoots; st = leaf stems; vb = vine buds; vF = vine fruit.

An analysis of the overall use of principal food sources overlooks the different patterns of utilisation, so that primary, secondary, and tertiary food sources are intermingled. Food trees were ranked according to the number of months used, and the number of feeding visits (table 3.10). The trees used over the longest period were those with several climber species (mainly tertiary sources). These 84% of the 25 trees used in six or more months provided a diversity of plant species, rather than of food categories. 20% of the 25 had strangling figs, and 20% were *Sloetia elongata* (the total is more than 100% because some trees contained more than one). The emphasis of this analysis is in favour of the tertiary food sources, but an analysis of visit frequency increases the emphasis on primary, and even secondary sources. Thus of the 37 trees visited 11 times or more 51% were *Ficus*, and *Sloetia* and vine trees were 11 and 49%, respectively. The high scores for the vines and some fig trees reflect long-term usage, whereas those for *Sarcotheca*, *Antidesma*, *Shorea*, *Eugenia*, and other species of *Ficus* reflect intensive short-term feeding.

Table 3.10. Major siamang food trees at Kuala Lompat ranked by (a) number of months, and (b) number of visits

(a)				(b)			
rank	number of months	tree No.	tree genera	rank	number of visits	tree No.	tree genera
1	12	L096	*Sloetia*, vines	1	28	L126	*Ficus*
2	11	039	*Sloetia*, vines	2	27	074	*Ficus*
		279	vines	3	26	096	*Sloetia*, vines
4	10	077	*Vitex*, vine	4	24	039	*Sloetia*, vines
5	9	060	vines	5	20	041	*Ficus*, vine
		143	vines			077	*Vitex*, vine
		277	vines			277	vines
8	8	045	vine	8	18	322	*Sarcotheca*
		067	*Sloetia*, vine	9	17	233	*Ficus*
		068	vines			279	vines
		099	*Sloetia*, vine	11	16	018	*Ficus*
		123	*Ficus* (2)			047	*Ficus*, vine
		271	vines			049	*Antidesma*
14	7	025	*Ficus* (2)			064	vines
		041	*Ficus*, vine			166	*Ficus*, *Glochidion*
		047	*Ficus*, vine	16	15	045	vines
		064	vines			068	vines
		069	vine			123	*Ficus* (2)

Table 3.10 (continuation)

(a)				(b)			
rank	number of months	tree No.	tree genera	rank	number of visits	tree No.	tree genera
19	6	020	vines			265	*Ficus*
		031	*Ficus*, vine			314	*Ficus*
		062	*Sloetia*	21	14	015	*Ficus* (2)
		073	vine	22	13	001	*Ficus*
		168	*Ficus*			060	vines
		261	vines			271	vines
		444	vines	25	12	017	*Sloetia*
						063	*Shorea*
						143	vines
						284	*Ficus*, vines
						315	*Ficus*, vines
						020	vines
						027	*Eugenia*
						048	*Sloetia*
						168	*Ficus*
						202	vines
						203	*Ficus*
						275	vines
						287	*Ficus*

	Percent of sample		Percent of sample
Trees ranked by			
(a) months		(b) visits	
Ficus	20	*Ficus*	51
Sloetia	20	*Sloetia*	11
Vines (climbing species)	84	Vine	49

Table 3.11. Analysis of food trees at Kuala Lompat in which most feeding time was spent

Rank	Tree No.	Species	Food	Feed time, min	Number of visits	Number of months	Visit duration	
1	L074	*Ficus*	F	1,030	27	5	40	P
2	015	*Ficus* 2 ×	F	986	14	2	70	P
3	203	*Ficus*	F	910	11	3	83	P
4	277	vines	nL	901	20	9	46	T
5	039	*Sloetia*, vine	nL, fl	875	24	11	37	T
6	265	*Ficus*	F	857	15	2	57	P
7	126	*Ficus*	F, nL	823	28	5	30	P

Table 3.11 (continuation)

Rank	Tree No.	Species	Food	Feed time, min	Number of visits	Number of months	Visit duration	
8	096	*Sloetia*, vine	nL, fl	733	26	12	28	T
9	047	*Ficus*, vine	F, nL	664	16	7	42	T
10	027	*Eugenia*	F, nL	632	11	2	57	P
11	063	*Shorea*	st	627	12	5	52	S
12	018	*Ficus*	F	608	16	5	38	P
13	195	*Maranthes*	F	600	12	2	50	P
14	314	*Ficus*	F	580	15	3	39	S
15	001	*Ficus*	F	566	13	3	44	P
16	201	*Ficus* 2 ×	F	558	9	3	62	P
17	049	*Antidesma*	F	557	16	4	35	P
18	123	*Ficus* 2 ×	nL, st	543	15	8	36	T
19	181	vines	nL	538	9	5	60	T
20	275	vines	nL	519	11	3	47	T
21	017	*Sloetia*	nL, fl	517	12	5	43	S
22	064	vines	nL, fl	507	16	7	32	T
23	045	vines	nL	492	15	8	33	T
24	278	*Ficus*	sh	497	10	4	45	P
25	041	*Ficus*, vine	F, nl	487	20	7	24	T
26	311	*Koompassia*	nL, st	474	8	3	59	P
27	143	vines	nL	467	12	9	39	T
28	077	*Vitex*, vine	nL	453	20	10	23	T
29	060	vines	nL	416	13	9	32	T
30	067	*Sloetia*, vine	nL	391	10	8	39	T
31	273	vines	nL, fl	389	10	1	39	T
32	279	vines	nL	383	17	11	23	T
33	315	*Ficus*, vine	F, nL	371	12	2	31	P
34	271	vines	nL	367	13	8	28	T
35	068	vines	nL	342	15	8	23	T
36	016	*Koompassia*	nL, st	326	10	3	33	S
37	228	*Ficus*	sh	320	7	2	47	P
38	118	*Koompassia*	nL, fl	319	8	5	40	S
39	409	*Xylopia*	st	311	4	1	78	P
40	073	vine	nL	302	7	6	43	T

	%		%
Ficus	45	Fruit (F)	38
Sloetia	10	New leaves (nL)	65
Koompassia	8	Flowers (fl)	15
		Stems (st)	13
		Shoots (sh)	5

Food sources: P = primary (40%); S = secondary (13%); T = tertiary (47%).

The interaction of the various factors can be seen more clearly if the 40 most used trees, in terms of time spent feeding, are ranked with scores for visit frequency, visit duration, and number of months visited (table 3.11). Apart from those tertiary food sources used in almost every month, it is the primary food sources which have the highest feeding time scores, with some secondary and the main tertiary sources lower down the order. Thus several trees were used so intensively over a short period, that more time was spent in them than in many of those visited over a much longer period. The preponderance of fruiting trees, in the top half of the order especially, is of interest (see Discussion), although the overall proportions of each food category approximate those for the whole sample (table 3.3).

There was considerable variation between visit frequency and feeding time for these 40 trees, although the mean visit duration of 36.5 min was nearly 10 min longer than the overall score for all food trees. All but four were above the latter (26.6 min) with some figs having a mean visit duration of 60 min.

2. Ulu Sempam

In Ulu Sempam, where figs were even more important as food sources, although many of the same species were present, feeding emphasis shifted from *stupenda*, *sumatrana*, and *virens* to *dubia*, *consociata*, and *heteropleura*. In February 1970 the primary food sources were *F.heteropleura* (S058), *F.pellucido-punctata* (S057), and a species of *Aglaia* (S073), and the new leaf shoots of *Ficus caulocarpa* (S076). In March S057 and S058 were again primary sources, along with the unripe figs in S008 *(F.dubia)*, and the new leaf shoots of *Cynometra inaequifolia* (S087, S090, S095). In April most feeding was on the fruit (grapes) of a *Vitis* sp. (S118), and the fruit of *Bouea oppositifolia* (S107), and the figs *(F.sumatrana)*. Secondary food sources were the small green fruit of a species of *Memecylon* (S099, S110), the leaf stems of *Connarus* sp. (S100), the fruit of a second species of *Aglaia* (S007) and figs of several trees and species (S065, S037 shifting to S114, S117, and finally S008 and S076, the latter having ripened). Tertiary food sources overall were again 'vine trees' (those with climbers) such as S009, 061, 148.

There was considerable variation among the sample of 18 most visited food trees (4 visits or more), which might be expected from the small sample size (3 months, 22 days, cf. 14 months, 148 days). 28% principal food trees had strangling figs, 22% had climbers, 11% were *Cynometra*, and 11% were *Aglaia;* in terms of feeding time the figures were 33, 33, 10, and 5%, respectively (table 3.12). It is noteworthy that in 22 days RS2 visited one tree as often as

Table 3.12. Major siamang food trees ranked by (a) number of visits, (b) feeding time in Ulu Sempam during 22 days

(a) rank	number of visits	tree No.	tree genera	(b) rank	number of min.	tree No.	tree genera
1	28	S058	*Ficus*	1	739	S076	*Ficus*
2	18	076	*Ficus*	2	720	058	*Ficus*
3	17	057	*Ficus*	3	468	057	*Ficus*
4	12	118	*Vitis*	4	269	118	*Vitis*
5	9	090	*Cynometra*	5	241	090	*Cynometra*
6	6	107	*Bouea*	6	223	073	*Aglaia*
		100	*Connarus*	7	194	065	*Ficus*
		073	*Aglaia*	8	188	025	*Ficus*
		008	*Ficus*	9	183	100	*Connarus*
10	5	078	*Eugenia*	10	158	008	*Ficus*
11	4	007	*Aglaia*	11	147	095	*Cynometra*
		025	*Ficus*	12	123	107	*Bouea*
		095	*Cynometra*	13	102	078	*Eugenia*
		096	vine	14	101	153	vine
		108	vine	15	99	151	*Ficus*
		110	*Memecylon*	16	93	140	vine
		140	vine	17	84	094	vine
		162	vine			162	vine
				19	79	115	vine
				20	76	148	vine
				21	75	096	vine

Trees ranked by (a) visits	Percent of sample	(b) feeding time	Percent of sample
Ficus	28	*Ficus*	33
Cynometra	11	*Cynometra*	10
Aglaia	11	*Aglaia*, etc.	5
Vine	22	Vine	22

TS1 visited its most frequented tree over 148 days. This can partly be explained by (1) smaller territory size, (2) less plant diversity, and (3) fewer, shorter feeding visits.

S058 was visited most often, and although occupying the tree for long periods early each morning, feeding was interrupted by resting and grooming bouts, so that feeding visits were relatively short. Unidentified climbers (often tertiary sources) did not figure as prominently as at Kuala Lompat, again because oft the shorter observation period; but their identity was already

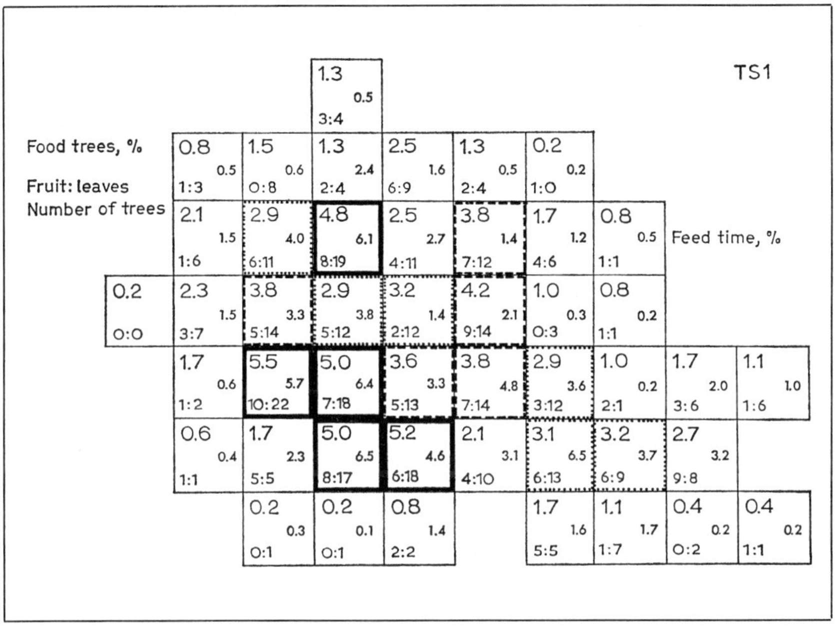

Fig. 3.16. Number of 'fruit' and 'leaf' trees, percent of food trees, and percent of feeding time for TS1 in each hectare quadrat at Kuala Lompat. The quadrats with most food trees are identified by a continuous border (ranks 1 – 5), a dashed line (6 – 10), and a dotted line (11 – 15).

being indicated because of repeated return visits, especially with the limited data from 1969. Although in some trees increased visit frequency was associated with long visit durations in other trees the converse appeared applicable. This could usually be related to tree identity and food category consumed.

Feeding Behaviour

Dispersion of Food Trees

1. Kuala Lompat

There were 15.0 food trees/ha of the 46 quadrats. But they were unevenly distributed with the scores for some squares being more than double the mean, that is, more than 4.4% of the total number of food trees (fig. 3.16). The highest density of food trees was in section B (Chapter 2) at 14.2 trees/ha,

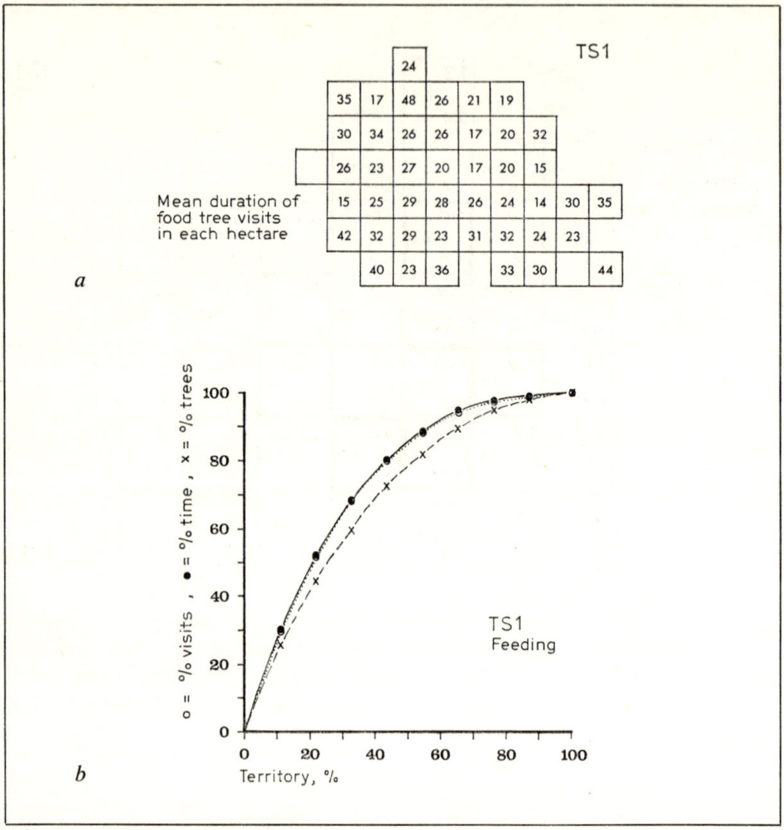

Fig. 3.17. (a) Mean visit duration to food trees in each hectare quadrat by TS1 at Kuala Lompat. (b) Cumulative proportions of territory (ranked hectare quadrats), food trees, feed visits, and feeding time for TS1 at Kuala Lompat.

the lowest in section C at only 7.5 food trees/ha. If it is expected that these food trees were evenly dispersed through the three sections, with 12.8 trees/ha in section A, these differences were highly significant ($\chi^2 = 40.38$). It is not valid to test for the significance of quadrat differences in this way, because of problems in determining expected scores for central and peripheral hectares, and of inter-dependence between quadrats.

40% food trees, and nearly 50% food visits and feeding time were in 20% territory (fig. 3.17). Feeding visits in those squares in which there was most feeding were not especially long; in squares in which little feeding occurred visits were sometimes longer than average, and sometimes of short duration. There was also no special relationship between the proportions of

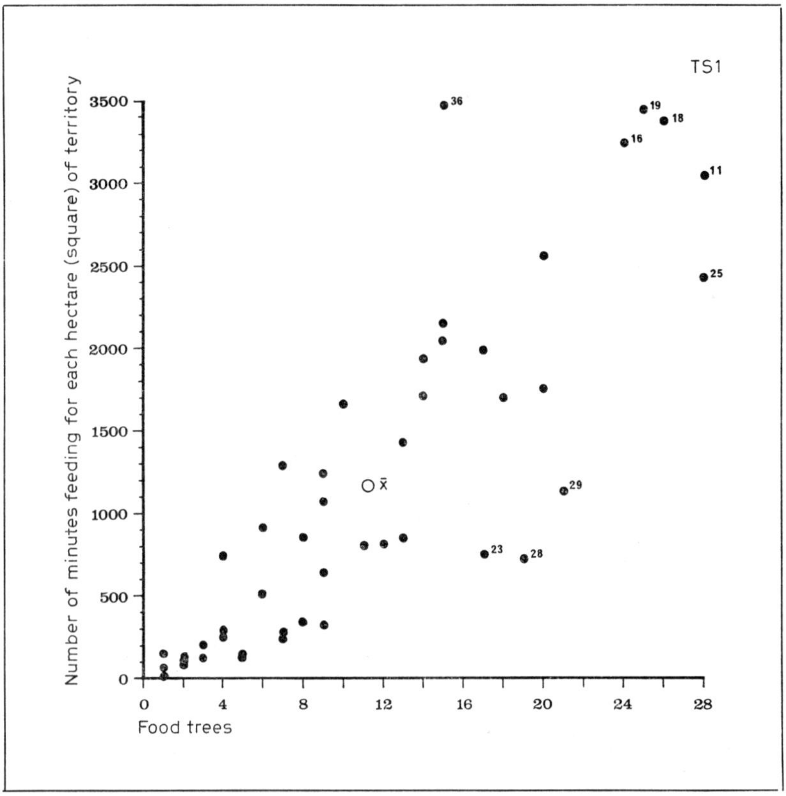

Fig. 3.18. Plots of feeding time against number of feed visits for TS1 in each hectare quadrat of their territory at Kuala Lompat.

different food categories in the most utilised quadrats (fig. 3.16). Usually there were more than two leaf-trees to every fruit-tree, whereas peripherally there was some indication that fruiting trees were more common as food resources.

The number of feeding visits to each quadrat was very highly correlated with feeding time (fig. 3.18, $r_S = 0.96$), whereas the number of food trees was less highly correlated with feeding time (fig. 3.18, $r_S = 0.88$). Although there were many feeding visits in quadrats 23, 28, and 29, they were of short duration; in contrast, feeding visits in quadrat 36 were much longer than expected. This highlights the intensive activity in section B, contrasting with a low level of activity despite reasonable visit frequency in the north-east of the territory. The highest correlation was between quadrat feeding visits and daily activity

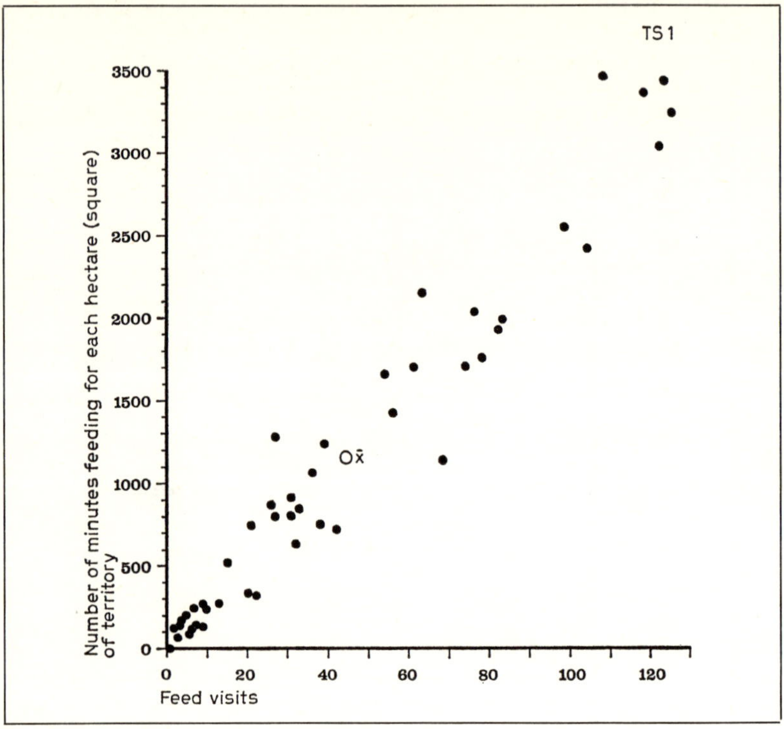

Fig. 3.18 (continued). Plots of feeding time against number of food trees in each hectare quadrat of TS1 territory at Kuala Lompat.

time (r_S = 0.98). This is not surprising as 55% daily activity time was spent feeding, but it would therefore have been extremely interesting if the correlation had been even lower between the distribution of food trees and daily activity time over the territory (r_S = 0.89).

2. Ulu Sempam

In the 22 days over three months nearly 25% feeding time (and feeding visits) occurred in one hectare, which contained two important fig trees (quadrat 6, fig. 3.19) and also accounted for 24% daily activity time. Although day ranges were relatively restricted to the central part of the territory during this period, food trees were again more widely dispersed than activity (fig. 3.20). As at Kuala Lompat the lowest correlation was between food trees and daily activity time (r_S = 0.79). Quadrats 6 and 12 were fed in more intensively than the number of food trees would suggest, quadrats 11 and 17 less so.

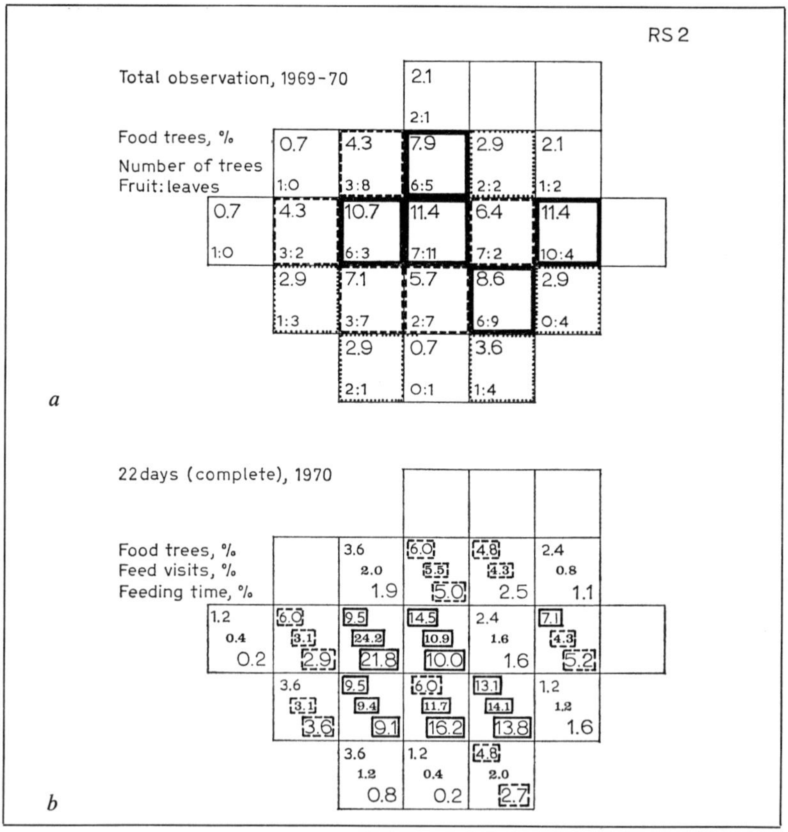

Fig. 3.19. Spatial distribution of food trees in Ulu Sempam: (a) percent of food trees and number of 'fruit' and 'leaf' trees throughout study (many incomplete days of observation; (b) percent of food trees, percent of feed visits, and percent of feeding time over 22 days (complete) in 1970, for each hectare quadrat of RS2 territory.

As at Kuala Lompat (fig. 3.12), more than 50% feed visits and about 65% feeding time were to 20% food trees (fig. 3.21). Similarly, more than 40% food trees, and 55% feed visits and time were spent in 20% territory. The latter refer to the 22 days of detailed observation; only 60% food trees and 80% overall territory area were covered in this period. Because of observation difficulties the distribution of food trees for the whole study is incomplete, especially around the periphery of the territory. It appears that RS2 fed in fewer trees for longer than TS1 and, therefore, as many as 50% of food trees visited in 12 months may have been missed.

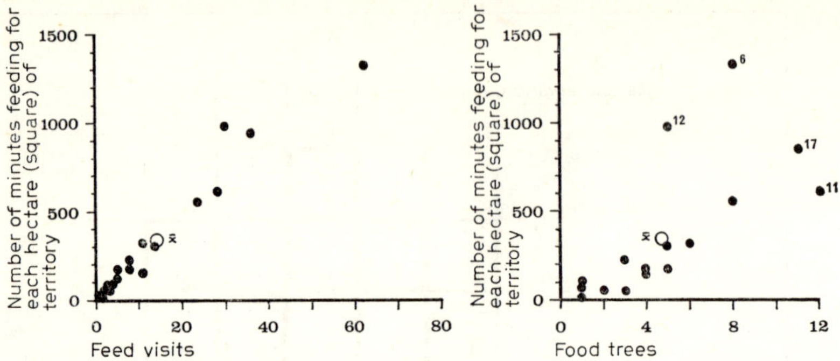

a, b

Fig. 3.20. Plots of feeding time against (a) number of feed visits and (b) number of food trees in each hectare quadrat of RS 2 territory during 22 days in Ulu Sempam.

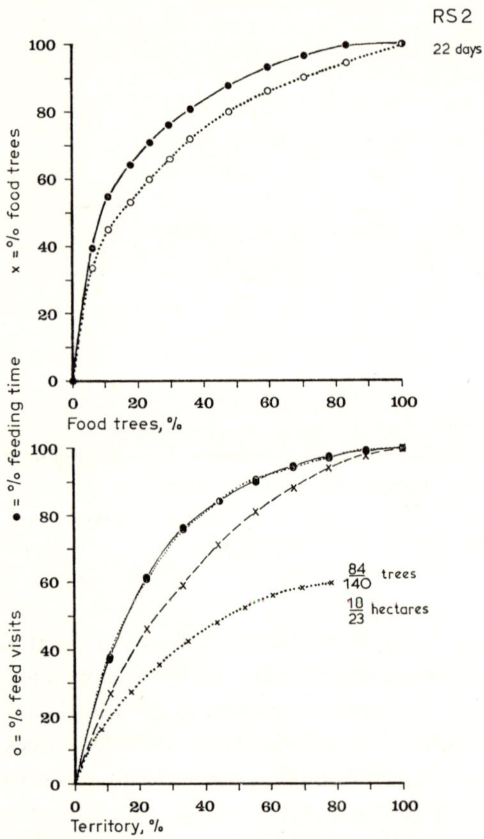

a

b

Fig. 3.21. Cumulative proportions of (a) food trees (ranked), and (b) territory (ranked hectare quadrats), showing feed visits, feeding time, and food trees (for territory only, with correction for data from whole study), for RS 2 over 22 days in Ulu Sempam.

Drinking Behaviour

This was almost only observed at the drier area of Kuala Lompat. Most of the siamang's water intake would appear to be from juicy fruit and succulent young leaves. But on every other day (on average) TS1 visited a 'water tree' for 5 – 20 min. These trees contained one or more holes, caused by rot following branch loss, or hollows at the junction of primary branches. Each animal would sit in turn by the hole, dip his or her hand into the depths of the hole, then raise it rapidly to the mouth (elbow uplifted) to suck the water from the hairs on the back of the hand (plate XIII). It was probably more a matter of water dripping off the hand and forearm, rather than of much sucking and licking.

Only once did a siamang (juvenile) bend its head and drink directly from the hole by mouth. Apart from the possible dangers of such an approach, it seemed that the water was usually too inaccessible for such a method. The other mode of water intake (also seen occasionally in Ulu Sempam) occured after rain, usually in the early morning in a smooth barked tree such as *Koompassia excelsa.* Siamang would sit by a vertical branch and wipe the hand upwards against the flow of water rivulets. The water would then be licked and sucked from the back of the hand. Sometimes it appeared that water was being licked or sucked off leaves. Almost identical drinking behaviour was described for the white-handed gibbon [ELLEFSON, 1967].

The pattern of this behaviour changed markedly during the study; in the intermediate season water trees were visited on only one day in five, but once the rains started (presumably refilling holes) the frequency increased to every other day. Finally, in the dry season, as one might expect, the siamang group was drinking in this way once a day on average. This behaviour occurred earlier in the day then than it had in the wet season (Chapter 4). 40% 'water trees' were *Sloetia elongata*, in which more than 50% drinking bouts occurred; other species of note were *Sarcotheca griffithi* and *Koompassia excelsa*. The latter resembled *Sloetia* in the high frequency of hole formation, whereas the branch structure of the former provided hollows.

'Water trees' were fairly evenly dispersed across the territory (fig. 3.22), especially in terms of activity time; 11% were also food trees. The water holes were usually low at 30 – 50 ft above the ground, with the occasional one up to 100 ft. 76% drinking bouts were in 31% of the 32 water trees; 27.5% bouts were in two trees (L095 and L111 WT). This analysis, although including all trees utilised, concentrated on those drinking bouts involving the major part of the group. It was not uncommon for an individual to have

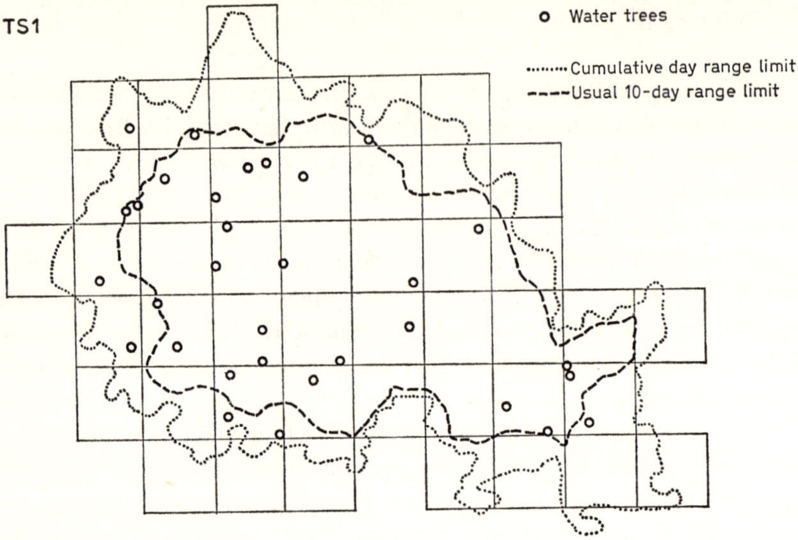

Fig. 3.22. Location of 'water trees' in TS1 territory at Kuala Lompat, indicating cumulative day range limits, and usual limits of ranging over hectare quadrats.

a quick drink as the group passed nearby; in this respect frequencies for this behaviour are somewhat depressed.

Feeding Habits and Posture

When all individuals in the siamang group could be observed feeding continuously for long periods – in primary food sources of open habit – the posture (hang or sit) and method (picking with left or right hand) of each siamang were sampled at 1-min intervals. This was achieved for several trees of different food categories in June, and one fruiting fig in each of September and December 1969.

In the group as a whole there was no significant use of one hand more than the other. But the female was significantly left-handed ($\chi^2 = 5.68$), and the left-handedness of the male and juvenile male were highly significant ($\chi^2 = 21.2$ and 14.8, respectively). These patterns were consistently maintained throughout the study (table 3.13), although there was greater variation in June according to the category of food being consumed. Scores are expressed as the ratio of left- to right-handedness, the greater deviations from unity

Table 3.13. Individual variation in feeding habit and posture at Kuala Lompat

1969, month	Tree	Food	Male	Female	SA male	Juvenile	Infant	Total
Left hand: right hand for picking food								
June	L118	fl	1.45	1.88	0.59	1.08		1.18
	L063	st	1.50	1.03	0.94	0.25		0.98
	var.	nL	1.17	1.18	0.73	1.03		
	L131	F	1.08	1.29	1.25	1.00		
	L074	F	1.13	1.06	0.61	1.14		0.97
			1.21	1.19	0.68	1.06		1.01
	n		823	754	640	359		
Sept.	L203	F	1.57	0.98	0.88	1.81	0.81	1.15
	n		262	227	239	194	172	
Dec.	L265	F	1.30	1.18	0.88	1.40	0.58	1.04
	n		280	292	263	254	217	
Total			1.30	1.14	0.76	1.31	0.67	1.10
n			1,165	1,273	1,142	807	388	
Hang: sit whilst feeding								
June	L118	fl	0.47	0.73	0.35	3.39		0.52
	L063	st	0	0.08	0.57	1.40		
	var.	nL	0.24	1.31	0.96	2.57		0.62
	L131	F	1.29	1.20	0.71	2.40		
	L074	F	1.16	3.20	1.55	4.40		3.04
			0.70	1.58	1.00	3.66		1.47
	n		710	860	716	457		
Sept.	L203	F	0.65	2.04	1.02	1.93	3.31	1.40
	n		278	231	251	199	194	
Dec.	L265	F	16.00	7.63	2.28	3.98	2.55	4.38
	n		306	311	269	284	259	
Total			1.15	2.15	1.19	3.22	2.84	1.89
n			1,294	1,402	1,236	940	453	

Food categories: F = fruit; fl = flowers; nL = new leaves; st = leaf stems.
n = Number per sample.

in either direction being significant. It is possible that this preferred handedness was with respect to suspension within the tree (when hanging) rather than to the picking of food, but behaviour when sitting (table 3.13) does not support this. Siamang changed hands every 2 – 3 min on average (range 1 – 16 min). In all cases the preferred hand was used for nearly a minute longer than the other on average in each session.

There was a greater variation in feeding posture, according to the size of the individual, and the category of food being consumed. When food was widely dispersed throughout the tree (see below), feeding was more active and hanging was more frequent than sitting. Suspension was also necessary because much of the food was located in the terminal branches, upon which it is almost impossible for an animal the size of siamang to sit. And yet it was the largest animal – the adult male – which showed the greatest tendency to sit while feeding (table 3.13). But there was a high significance for the hanging posture for all individuals, especially the female and the juvenile. This was partly because the bulk of the sample was taken from fruiting trees.

In trees in flower or leaf flush (June 1969) sitting was the more usual posture. This confirms the tendency observed in such trees, in which the group usually fed later in the day. Food was plentiful about a firmer substrate, and because of the denser foliage it was difficult to collect more data for comparison – other than for the one study individual sampled at 1-min intervals (Chapter 4). Suspensory behaviour was especially marked in the fig, *Ficus virens* (L265) in December, where food was densely clustered in the ferminal branches of this relatively small tree. Hanging was more than four times more frequent than sitting (more than twice the overall score, and nearly tour times the score in the other *Ficus virens*, L203). By contrast in trees of other food categories the siamang tended to sit twice as much as they hung while feeding (table 3.13).

Feeding Rates

As a first step in the determination of food intake, attempts were made to measure the rate of feeding. Wet and dry weights of food samples were also obtained. But it was not possible to obtain a meaningful estimate of leaf consumption. Even under good observation conditions there were two main problems: (1) several leaves would be picked in one grasp, so that an accurate count could not be made, (2) the whole picking was rarely consumed, preference being shown for some specific part (as described above). In only

Table 3.14. Individual variation in rate of feeding by siamang

Month	Tree	Tree genera	Particles consumed/min				
			M	F	SA	J	I
Fruit							
June	L074	*Ficus*	11.2	19.2	16.0	8.6	
September	L203	*Ficus*	13.8	17.4	16.4	12.2	6.2
December	L265	*Ficus*	13.0	12.6	14.0	10.2	
Total			12.6	18.4	15.6	11.0	6.2
n			210	213	172	107	25
February	L027	*Eugenia*	1.0	1.4	1.5	1.2	
n			30	10	55	25	
Leaves							
June	L126	*Ficus*	3.0	3.3	4.5		
n			22	13	9		

M = male; F = female; SA = subadult; J = juvenile; I = infant.

one situation (L126, *Ficus annulata* in June 1969) was an estimate of feeding rate obtained (but not of food weight). This indicated that on average a siamang consumed 3.4 leaves (medium-small)/min (table 3.14).

Similar problems applied to fruit consumption. Although it was easier to count ingestion rate, often only a part of the fruit was consumed (especially if there was a large seed). But the situation for L027 (*Eugenia* sp.) was more subtle in that variable portions of the skin and underlying structure were eaten. The siamang fed in this tree for long periods each day in February 1970. On average they would pick nearly two fruit each minute, drop 3 out of 20 after one bite, and strip off with their teeth the edible portion, which weighed about 1.8 g wet. Thus 1.3 fruit or 2.4 g were eaten each minute.

Feeding rate and weight of figs consumed were rather easier to measure, particularly for the smaller ones, e.g. *Ficus virens*. These were placed in the mouth one by one, some being discarded by touch. With larger figs, e.g. *Ficus stupenda*, variable quantities of the edible part were consumed before being dropped, and at a distance it was not realistic to attempt an estimate. In three trees in three months – L074 in June, L203 in September, and L265 in December – the overall rate of consumption was 14 fruit or 2.10 g each minute for each individual (only 12 /min in L265). The female and sub-adult male almost invariably ate faster than the male and juvenile and the infant was slowest of all (table 3.14).

Table 3.15. Group food consumption

Tree	Food	Number of visits	Feed time	Fruit/ min	Total number of fruit/leaves × 5	Weight of 1,000 fruit, kg	Total wet weight food, kg
L074	fruit	8	669	14.4	48,150	0.260	12.5
L203	fruit	6	744	14.4	55,850	0.150[1]	8.4
L265	fruit	10	721	12.2	43,950	0.097	4.3
L027	fruit	10	627	1.3	4,080		7.2
L126	leaves	7	307	3.4	1,050		

1 Estimate from L074 and L265 scores.

Considering the six bouts in L203, and the 5 bouts in L265, when all animals were timed (the infant was too elusive in L265), the variation in feeding rate between individuals was highly significant (Kruskal-Wallis Analysis of Variance, $H = 18.3$ and 13.1, respectively). In L203 the infant ate significantly slower than all others, and the female ate significantly faster than the male and juvenile (Mann-Whitney U test). In L265 the female ate unusually slowly, and only the male and subadult ate significantly faster than the juvenile. It was confirmed from day to day that individuals were consistent in their performance, especially in L203 (Kendall coefficient of concordance, $W = 0.80$ (L203) and 0.65 (L265), $p = 0.001$- and 0.01-, respectively). Within 10 days a siamang group can consume 4 – 12 kg of figs (table 3.15).

It was also observed that siamang fed at different rates during a feeding visit. Initially feeding was rapid, with a subsequent steady decline, although in some cases there was a final spell of feeding at an increased rate [Chivers, 1972b]. This whole pattern was well displayed in L074 by the three older siamang on three different days. Conditions did not permit a sample of 5 min to be taken at regular intervals. In L265 the decline in feeding rate was usually maintained to the end. In L203 feeding bouts were longer, and changes in feeding rate, for the female in particular, were less distinct.

Movements within Food Tree

When favourable conditions prevailed throughout a feeding visit, the sequence of movements within the food tree was determined by plotting the position of each individual at 10-min intervals. Such data were obtained on 36 days, between August 1969 and May 1970 inclusive, for 40 feeding visits to five fruiting trees (four figs) and one fig with new leaf shoots (Chivers

Table 3.16. Movements within food tree by siamang at Kuala Lompat

Date	Tree	Start, time	Male n	Male m	Male t	Female n	Female m	Female t	Subadult n	Subadult m	Subadult t	Juvenile n	Juvenile m	Juvenile t	Infant n	Infant m	Infant t
11–5–70	L074	0616	7	5	4	6	6	5	10	7	2	9	5	3	3	6	8
12–5–70		1013	8	5	3	7	4	4	2	3	2	3	9	7	6	5	4
13–5–70		0655	3	6	3	5	6	3	1	4	2	5	5	2	4	5	3

Totals	Male	Female	Subadult	Juvenile	Infant
n (number of movements)	18	18	13	17	13
m (number of metres moved)	5.2	5.4	6.4	5.9	5.2
t (mean time between moves, min)	3.5	3.8	3.2	3.4	4.5
Overall: 78 moves of 5.6 m every 3.5 min (31% are major moves)					
Major moves, %	28	29	38	29	31
Minor moves, %	72	71	62	71	69

[1972b], Appendix VI). They demonstrated that on consecutive visits, and even on return visits after six months or so, individual siamang tend to start feeding in the same part of a specific tree and follow a similar pattern of movement about it. The significance of this will be discussed below.

Towards the end of the study an alternative method was applied – to measure the timing of these movements. This was only achieved for one feeding visit to a fig, S076, in Ulu Sempam, and for three visits to a fig, L074 at Kuala Lompat in April and May 1970, respectively. In just under an hour RS2 individuals each changed primary branches once – a *major movement* – making *minor movements* three to five times at an average interval of 12, 9, and 14 min for male, female, and subadult male, respectively. At Kuala Lompat TS1 individuals moved about 5.5 m every 3.5 min on average. 31% movements involved a major shift in position from one primary branch to another of 10 m or more. The subadult male moved more often than the others, and farther; the infant moved least (table 3.16).

Discussion

It has been shown in this chapter that the dispersion of food trees was highly correlated with the ranging patterns of the siamang family group. This could be because the general nature of their diet enables siamang to find food wherever they happen to be in their territory. But ranging patterns (Chapter 2) and other evidence (see below, and Chapters 4 and 5) suggest that siamang are more selective in their diet. It will also be shown that siamang daily behaviour is orientated about feeding in the way that gibbon daily behaviour is centred on territorial defence [ELLEFSON, 1967].

The siamang also differs from the gibbon in the nature of its eclectic feeding, having about two staple, and three to four supplementary foods at any one time, rather than a diversity of fruit. In this respect they closely resemble leaf monkeys, *Presbytis* spp., both in Malaya [HUNT, personal commun.] and in Ceylon [RIPLEY, 1970], although the siamang spent nearly twice as much time feeding. But a considerable proportion of food was eliminated, and the poorly digested state of leaf material in the faeces (conspicuous in some months) could be indicative of the siamang's inability to exploit a folivorous diet very efficiently. This would help to explain the very high scores for daily feeding time, and the increase in ranging and feeding towards the end of the 1970 dry season, when a greater diversity of foods, especially fruit, was available.

The question of preference has so far been avoided. It is not yet clear whether the siamangs were eating large quantities of young leaves through preference or necessity. Certainly it provides an effective ecological separation from the gibbon, and correlates well with other aspects of behaviour, such as territory size and maintenance. But there remain the physiological problems, and it must be remembered that 1969 followed a year of unusually heavy fruiting (see below). It remains to be seen how scarce food was for siamang at that time.

It appeared that most of the principal food sources yielded preferred foods, and that the 80% of trees visited only five times or less were supplementing the diet when preferred foods were scarce. There was nothing to suggest that most of the latter produced choice foods for a limited period. It has been shown that those principal foods available for only a short period were consumed for as much time over a year as other foods available for much longer. But the evidence indicates that fruits were increasingly consumed when available, as in Ulu Sempam and at Kuala Lompat in 1970. If they represent a generally preferred food there is no reason why siamang could not be as frugivorous as the gibbon is reported to be [CARPENTER, 1940; ELLEFSON, 1967]. For they dominate all other animals within the food tree, but do not consume a very high proportion of the fruit available.

If fruit were a preferred food this would explain their greater frequency in proportion to 'leaf trees' near the limits of ranging. But while fruits (including figs) compose about one third of siamang diet, young leaves, including the succulent young shoots of fig trees, were more prominent in the diet in both areas. This may be a reliable indicator of the preference levels for each category. It has still to be shown whether gibbons in central Malaya can or do subsist as effectively as siamang on young leaves, particularly if fruit are scarce.

It is not clear whether the relative neglect of the north-east part of their territory by TS1 was because of food surplus elsewhere, or behavioural affinities to the most used hectares. Similarly, the very restricted ranging in December (in terms of time), and in August and November (in terms of space) indicates conflicting explanations. Either preferred foods were very abundant locally, or they were plentiful over a wide area. The former is the most likely explanation for August and November, when fruits were locally abundant. But in December the staple food was leaves and flowers of leguminous climbers, shifting to figs, with nearly 25% time spent in one hectare. It was not possible to determine whether food was of a lower quality at this time, and whether it was dispersed throughout the area. The pattern of occasional forays in other parts of the territory suggested that it was not, and that preferred foods were scarce. It has been suggested (Chapter 2) that the more restricted ranging of a siamang group might be a factor of its relatively unspecialised diet. Thus their feeding activity is concentrated in a small area of forest and their ranging is limited, so that contacts with adjacent groups are rare.

The number and duration of feeding visits reflected the abundance and dispersion of food. In certain seasons, when food was less plentiful and/or more dispersed, greater movement was necessary to obtain a sufficiency. No objective assessment of the phenology of the forest flora as a whole is currently available to support this, but the behaviour of siamangs in their small territory can possibly be used as indicators of these factors, i.e. that the siamang's knowledge of its territory is so complete that food resources are fully exploited.

The relative constancy of feeding time throughout the study appears to reflect a basal level of nutritional requirements, which was apparently exceeded in the 1970 dry season by increased feeding time, and so forth. Although ranging increased, the greater diversity of foods, especially fruit, did not suggest a scarcity of preferred foods. But repeated short visits to certain fruiting trees (e.g. *Bouea*, *Sarcotheca*) did not indicate a very efficient utilisation. That only basal levels of metabolism were achieved during 1969 is indicated by the constant low level of activity in all behaviours, followed by a marked increase in 1970 (see later chapters). With regard to feeding, this increase is unlikely to have been a response to a decreased quality or quantity of food.

The available evidence strongly suggests the converse – that the environment became considerably enriched in terms of energy-producing foods. But the question must be left open until it can be demonstrated that increased day range does not invariably reflect decreased food supply. It was surprising that visits to the variety of fruiting trees near the end of the study, especially at Kuala Lompat, were relatively short; this was not usual for preferred foods, and could reflect an energy surplus in this 'time of plenty'. Food scarcity was not indicated at any time, however, because of the less than maximal use of food trees; but it is possible that eclectic and territorial habits dominate, even under hardship, so that the group continues to move about its territory.

The most accurate indication of food intake, among the various measures of feeding used, was feeding time. This was feasible because of the continuous nature of feeding. With data on feeding rate, and the wet and dry weights of food consumed, it is possible to calculate with some precision the volume of food ingested. Nutritional analysis will reveal the value of each food to the siamang, in the light of studies of digestive physiology. It is expected that most siamang foods will share the same (in some cases complementary) characteristics, that will only be reflected partially by plant taxonomy (where different criteria are important). This man-made classification might well have less meaning than food categories in an analysis of siamang diet. The essential components of siamang food, and their required proportions, have still to be determined. But a start has been made that is likely to be concluded.

Apart from fig trees, visits to fruiting trees tended to be relatively short. Fruit consumption (usually in the morning) was almost invariably followed by leaf-eating. Whether or not this has any dietary significance, other than a transition from active to more passive feeding as the day passes, has still to be determined. Figs, and most leaves, had little taste to man, but most fruits were very bitter. This was partly because they were usually consumed when unripe, which appeared to be preferred because feeding decreased when fruit, including figs, were fully ripe. But it is not yet clear whether this was a factor of dietary significance, or of boredom or satiation after successive days with long feeding bouts on a particular food.

Fig trees are an important source of food for animals throughout the tropical world, especially for primates [Reynolds and Reynolds, 1965; Sugiyama, 1968; Hladik, 1968; Ripley, 1970]. And yet they do not seem to dominate the diet of any primate species as they do for the siamang. Although a siamang group can consume 5 – 10 kg of a fig crop (within ten days), other primates, squirrels, green pigeons, and hornbills – in addition to nocturnal mammals – gain considerable nourishment from the same crop. A portion of fruit fall provides an important food source for ground-living mammals.

Thus, when the asynchrony, high density, and short fruiting cycles of the many species in this genus are considered, their contribution to the nourishment of the tropical rain forest fauna identifies them as a major source of food for numerous species. They were the only trees in which leaf monkeys (usually dusky) were seen feeding with siamang, and in which gibbons tried (unsuccessfully) to feed before the siamang had departed. It has already been mentioned (Chapter 2) that animal activity probably signalled to siamang the location of such food sources, that had not been discovered in the course of ranging about the territory.

ELLEFSON [1967] has stressed the importance of suspensory behaviour to gibbons feeding in the terminal branch niche. He further suggested that brachiation might be a feeding rather than a locomotor adaptation. Although gibbons move rapidly and efficiently over short distances, they are apparently less capable of sustained travel – hence the small territory, and relatively short day ranges. The same arguments are applicable to the siamang, although problems are accentuated by their greater body size. ELLEFSON suggested that larger animals (including siamang as well as adult gibbons) could still successfully exploit the terminal branches because of their longer reach and greater strength. He was probably correct in assuming that the siamang represented the maximum size capable of this behaviour. The mechanical advantages of suspensory behaviour in the terminal branches are (1) an expanded sphere of activity and support, (2) reduced branch deformation, partly resulting in (3) reduced energy expenditure for body adjustments and rotations [GRAND, 1972]. His observations on *Macaca sinica*, with those of RIPLEY [1970] on *Presbytis* spp., confirm the view that while the gibbon 'configuration' permits greater utilisation of the terminal branch niche, they do not have exclusive use of it.

The larger and older siamang tend to sit more than hang while feeding; it is not yet clear whether this is of necessity or preference. Certainly the adults appeared to occupy the most favourable parts of the food trees (in terms of food) from the start of a feeding visit. The decreased rate of feeding in L265 in December, when food was concentrated in the small terminal branches and hanging increased markedly, could indicate that the siamang is not as efficient in feeding in such a posture as the gibbon.

The faster feeding rate of the female and subadult male assumes even greater significance when it is discovered (Chapter 4) that these individuals also spend considerably longer feeding each day than the male and juvenile. That it occurs in both areas suggests slow and reduced feeding is not just a feature of age extremes (young and old). It is usual for a male primate to be larger and have a larger appetite and food requirements than other individuals within the group. But this difference from the gibbon, in particular, can be partly explained by the lower level of activity by the adult male in territorial defence (Chapter 2), and in group leadership (Chapter 5).

The habit of picking fruit one by one for transfer to the mouth appeared to be wasteful of energy, compared with the alternative, only occasionally observed, of stripping small fruit from the twigs by mouth. But it became clear that before picking fruit each one was being palpated for degree of ripeness with some being rejected. By testing in this way, food resources in fact were being conserved for future supply, i.e. the next few days; otherwise they would have been spat out and wasted. While feeding there was a low tolerance to the proximity of others. Although actual distances were greater during travel (Chapter 5), temporarily separation was lower since each individual would pass through the same point within minutes. Individual space was especially important while feeding in the way described for leaf monkeys [RIPLEY, 1970], except that the problems of movements about the tree, especially between primary branches, were reduced in the smaller group.

Little is known of the phenology of the Malayan tropical rain forest. MEDWAY [1972b] demonstrated an annual cycle in Ulu Gombak, after 8 years observation with MCCLURE [1966]. 40% tree sample fruited each year mostly between May and November inclusive, following flowering between February and July. But each year it was a different assemblage of species which contributed to the similar overall pattern of flowering and fruiting. Peaks of leaf flush were at the end of the dry season (March to May during the secondary wet season), and less marked from October to December (the main wet season). Thus the plant community exhibited more marked annual cycles than many of the component species, and breeding cycles in some forest birds and mammals could be shown to be adapted to these.

Higher primates, especially the siamang, however, do not appear to be so restricted in their behaviour, although some form of annual cycles are present. This is because they exploit the plant community as a whole rather than the component species, to obtain a diet containing consistent proportions of each food category throughout the year. MEDWAY [1970] recognised three types of periodicity in floral cycles: (1) annually recurrent cycles; (2) non-annual, presumably endogenous cycles, shorter or longer than one year, and (3) sporadic.

It was the 13% of trees with endogenous cycles (usually of less than one year), which were most important for siamang, although those with annual cycles also contributed appreciably to siamang diet. Of special interest from the studies of Ulu Gombak were the records of great increases in fruiting in 1963 and 1968, when 70 – 90% tree sample were productive [MEDWAY, 1972b]. These were years when the dry season was most pronounced, and when those trees with sporadic cycles coincided in their flowering and fruiting. A similar drought was recorded near Temerloh in 1963 [POORE, 1968]. When coinciding with trees exhibiting the other periodicities ripe fruits become extremely abundant from September to November inclusive. Although too irregular to provide a significant long-term dietary contribution, trees fruiting sporadically can still be important to an animal such as the siamang.

Dialium platysepalum fruited in all three study areas at the end of 1968, and in Ulu Gombak on 4-day observation periods in three consecutive months siamang were seen feeding in trees of this species each day. The same behaviour was observed in Ulu Sempam at the beginning of 1969; but these trees did not fruit again during the study. The dipterocarps – dominant tree family in the forest – belong to the same category, and MEDWAY [1972b] points out that there are few examples of exploitation for food, although they are very important to many animals as structural elements in the forest canopy.

The gibbon may resemble the orang-utan [MACKINNON, 1971] in that it prefers fruit in the fruiting season, but at other times shifts its dietary emphasis elsewhere. In contrast, the siamang was more consistent in its diet throughout the year (as shown by all parameters, especially food categories given in tables 3.3, 3.4 and 3.5), because different plant species, in different seasonal cycles, successively contributed the required proportions of each dietic category (tables 3.7 and 3.8). They fully exploited all types of plant periodicity to show consistency in feeding behaviour. But this was interrupted by major irregular fluctuations (as in 1963 and 1968). These years of great food abundance appear to have a profound effect on behaviour within and between groups of all primate species in central Malaysia (Chapters 5 and 8) [CHIVERS, 1971a].

Thus it appears that the great diversity of plants, and floral cycles [WYATT-SMITH, 1953; POORE, 1968], are a major factor allowing the coexistence of many animal species within the tropical rain forest. The siamang appears to be typified by a stability of dietary composition, in which young leaves play a major part. This involves exploiting trees exhibiting a variety of cycles. Thus, while siamang superficially appear to show plasticity in feeding behaviour, this may only apply to tree species, there being a certain rigidity to the proportion of each food category consumed. The diverse habits of fig trees are very favourable to the restricted spatial patterns of siamang behaviour, with the dispersion and floral cycles of these trees having a major effect on these behaviours.

Summary

In the siamang territory at Kuala Lompat there were 290 trees/ha – a somewhat higher density than in the adjacent riverine forest. 8.5% trees were emergents, 28.6 were in the main canopy, and 62.8% in the low level, but by the river there were fewer high and medium trees in particular, with only 16 high, 50 medium, and 155 low trees/ha. In 148 days over 14 months the siamang only fed in 5.5% trees, they spent 27% day time in the high, 63% in middle, and 10% in the low levels of the forest.

TS1 visited about 13 trees each day and spent nearly 6 h feeding on average; 32% time was spent in the 26% trees in fruit, whereas 58% time was spent in the 67% trees with young leaves (also some stems, shoots, and mature leaves) – small quantities of flowers, buds, and insects (termites and caterpillars) were eaten. 24% feeding time was spent in fig trees (13% sample) eating fruit or leaf shoots. The duration of visits to food trees was variable, being shorter for flowers, buds, insects, and some fruit, and much longer than the average of 26.6 min to other fruiting trees, especially figs, and for some young leaves. There was a marked increase in all parameters (except for a decrease in visit duration) during the dry season, from a more regular wet season pattern. RS2 showed very similar behaviour, except for decreases in daily feeding time and visit duration, and an increase in fruit consumption.

While certain tree species dominated the forest canopy, others dominated the siamang diet; nearly half the food trees were of these principal species, which accounted for about two thirds of feeding time. There was a close similarity between the two study areas with a greater diversity at Kuala Lompat. About 80% food trees were visited five times or less during the study, while 7% accounted for 30% feed visits, and 5% were visited in six or more months. These principal food trees were classified as primary, secondary, or tertiary according to their significance (in terms of pattern of use) to siamang diet. They represented both those trees used intensively for a short period, and about which day ranges were orientated (primary, mainly figs), and those used lightly over a much longer period with a less profound effect on behaviour (tertiary food sources).

50% food trees, and 55% feed visits and feeding time were in 25% territory. Activity was concentrated in the south-west of the territory at Kuala Lompat, with somewhat less time spent feeding in the north-east than might be expected from the density of food trees. Such patterns were less distinctive in the smaller territory in Ulu Sempam, although in a short period the central part of the territory was used more intensively (65% feeding time in 25% territory).

Over the year a siamang family group in the lowlands drank from tree holes every other day (once a day in the dry season). 76% bouts occurred in 31% of 32 trees (usually *Sloetia elongata*), with 27.5% in two trees. Water was also actively obtained from leaf and branch surfaces after rain.

Some individuals were left-handed, others were consistently right-handed; there was a change of hands every 2 – 3 min, with the preferred hand sustained for picking up to 1 min longer on average. Most hung rather than sat while feeding, especially in fruiting trees where the food was more dispersed. In fruiting trees hanging was 1.5 – 4.5 times more frequent, whereas in trees of other categories sitting was twice as common as hanging. The subadult male and female in both study groups consistently ate faster and for longer each day than the adult male and young. In a feeding bout rate decreased after a fast start, sometimes increasing again finally.

A siamang group could consume 4 – 13 kg of small figs within 10 days. 31% movements within food trees were between primary branches, with an individual moving 5.6 m every 3.5 min on average. The adults appeared to occupy the most favourable positions (in terms of food abundance and necessary posture) from the start of a feeding bout, with all animals tending to frequent a specific part of the tree on consecutive visits.

Thus siamang are essentially folivorous, with a preference for fig trees and some other fruiting species. There is a marked constancy in proportions of each food category throughout the year. Although the bulk of their food is obtained by suspensory behaviour in a limited part of the territory, they feed in all parts, as indicated by extensive ranging behaviour with which feeding is highly correlated.

Chapter 4: Daily Activity Patterns

Introduction

Siamang were active for over 10 h each day on average – nearly 90% of daylight hours. During this time the family group fed for nearly 6 h, travelled between food sources for about 1 h overall, and rested in between for about 3 h on average. The daily patterns of these 'maintenance' activities [van Lawick-Goodall, 1968] will be investigated within and between months, between seasons, and under different weather conditions. There will be a description of the daily patterns of use of different levels of the forest (see Chapter 3), and of drinking behaviour, following an analysis of the length of, and variation in, daily activity time in relation to day length.

Daily patterns of inter-group conflicts and of ranging have been described in Chapter 2; those of calling will be described in Chapter 6. Other social activities occur briefly during bouts of maintenance activities – grooming, play, aggression, and sex – and daily patterns will be included in a full description of each behaviour in Chapter 5. In addition, there will be an analysis of spatial relations during the day, which will involve the social significance of the maintenance activities described below. Resting behaviour in particular had both maintenance and social components that were sometimes difficult to separate.

The behaviour of the lowland study group, TS1, is emphasised here, but, where possible, comparisons are made with the highland study group, RS2. Observations were made from dawn to dusk on at least ten consecutive days in each month with the behaviour of the group recorded (1) continuously with reference to location and activity, (2) by a 5- or 10-min sample of the activity of each group member, and (3) by a 1-min sample of the behaviour of one individual. That is, each individual was the subject of continuous observation for at least two consecutive days in each observation period.

The daily pattern of each maintenance activity can then be analysed in terms of (1) number of minutes in each hour (group scores were obtained according to the first to change activity), and (2) number of bouts starting in each hour with reference to the group rather than the individual. This form of analysis was facilitated by the regular habits of the group pursuing a given activity in concert for a considerable period once initiated (see below and Chapter 5).

Each maintenance activity was characterised in part by particular patterns of muscular coordination producing one or more of the following basic postures (sit, lie, hang, stand), locomotion (brachiate, walk bipedally, climb up or down, bridge, jump), or expressions (open and/or close mouth, head or hand movements) [BALDWIN and TELEKI, in press]. The diagnostic character of such patterns for different species requires further quantitative investigation.

Daily Activity Time and Day Length

Two measures of the start of daily activity were recorded: (1) the time of first movement within the siamang group from the sleeping positions, which was clearly marked by the communal urination and defaecation that usually followed within minutes, and (2) the time that the group started to leave the night sleeping tree. Each measure was useful for indicating different facets of siamang behaviour. Similarly, two measures were used to mark the finish of daily activity: (1) the time the group started to enter the night sleeping tree, and (2) the time the first individual reached its sleeping position within that tree. Each had one or more specific positions within each sleeping tree that they invariably used – infant with female, juvenile with male, and sub-adult alone.

Following the first movement, the group would either start leaving the tree, or rest some more. But the different individual postures and spacing of the group clearly separated the latter from night activity (or inactivity!). Thus the time from first movement to last was the most useful measure of daily activity time (or alert period), although from some viewpoints, such as ranging behaviour, it was more relevant to consider the time between leaving and entering night sleeping trees. Problems arose when the group fed in a night tree before sleeping, although it was quite clear when they moved from one part of the tree to another with change of activity.

Observers were always in position before the sky began to lighten. The first movement usually occurred as day dawned – at 0618 h or 21 min before sunrise on average over 14 months at Kuala Lompat (scores ranged between 0555 and 0650 h). The group started to leave the night tree a few minutes after sunrise at 0642 h on average (0600 – 0820 h). There was greater variation in the latter parameter, but this was reduced for both when plotted in relation to sunrise, rather than by clock time (fig. 4.1). The siamang group entered the night sleeping tree between 1410 and 1830 h (monthly means 1610 – 1735h), which was about 120 min before sunset on average; sleeping positions were usually reached about 7 min later.

Sunrise in central Malaya during this study varied between 0627 and 0657 h – a difference of 30 min (Survey Department, Kuala Lumpur). Similarly, sunset varied between 1827 and 1857 h. But although they varied in phase bimodally between the same limits, the changes were of different amplitude (fig. 4.2). Thus day length (sunrise to sunset) followed a sine curve between 717 and 738 min (about 12 h) – an annual difference of only 21 min. The time of first movement was highly correlated with the time of sunrise (Spearman rank correlation on monthly means, $r_S = 0.75$, $p = 0.01 - 0.001$), so were the measures of cessation of activity ($r_S = 0.64$, $p = 0.02 - 0.01$), but the time of leaving the night tree was not ($r_S = 0.33$). Neither were times of entering night tree and reaching night position significantly correlated with sunset ($r_S = 0.39$ and 0.43, respectively). The most constant measure was the time of first movement.

The correlation of the finish of daily activity with sunrise suggested a constancy of daily activity time that was not confirmed by the lack of correlation between first movement and reaching night position ($r_S = 0.33$), and especially between leaving and entering the night tree ($r_S = -0.03$). This can be explained by the change in behaviour in the 1970 dry season, when siamang daily activity time, or alert period (from first to last movement) increased markedly with activity starting earlier and ending later (fig. 4.2). The constancy of the length of the alert period until December was demonstrated more clearly as a proportion of day length, with the group active for about 10.5 h daily (551 – 753 min) or 86% (75 – 103%) day length (sunrise to sunset, rather than dawn to dusk). For the last few months group activity was prolonged up to 11.25 h on average or 93% day length (fig. 4.2). Although correlated with day length throughout, the increase in activity time was highly significant (Kruskal-Wallis analysis of variance, $H = 33.8$, $p = 0.001 -$).

The length of the alert period was highly correlated with the time of entering the night tree ($r_S = 0.95$); the longer they were active the later they settled for the night. There was no correlation between the alert period and day length ($r_S = -0.07$), although both increased simultaneously at the end of the study.

At Ulu Sempam in 1970 RS2 was active for 84% of day length or just over 10 h; from about 12 min before sunrise until about 126 min before sunset (0637 – 1648 h on average, table 4.1). If there was an increase in activity in Ulu Sempam at this time, it was not of the same order as at Kuala Lompat during the same period.

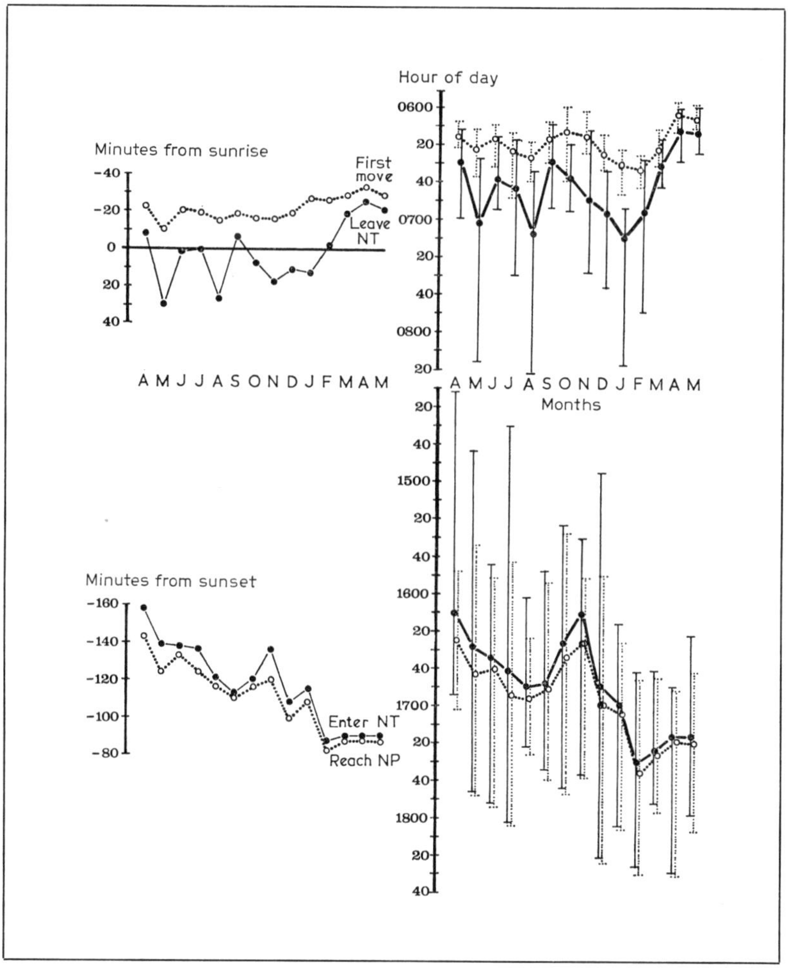

Fig. 4.1. Mean times (with range of observations) in each month for starting (first movement and leaving night tree) and finishing (enter night tree and reach night sleeping position) daily activities by sun time, and local time at Kuala Lompat.

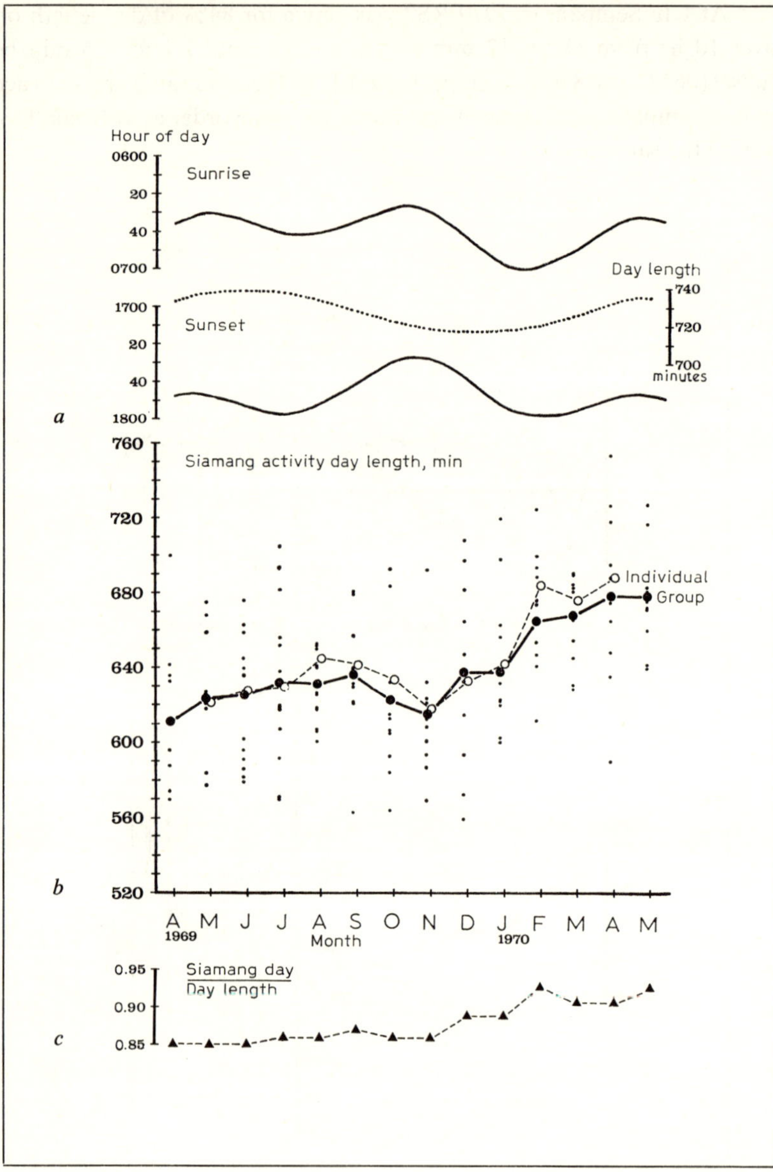

Fig. 4.2. Mean times of (a) sunrise, sunset, and day length, (b) the duration of siamang daily activities – alert period (for individuals and group, with all observations for the latter), and (c) the proportion of daylight hours in which the siamang were alert in each month at Kuala Lompat.

Table 4.1. Day length and daily activity time (alert period) in Ulu Sempam

	Ulu Sempam 1970							Kuala Lompat 1970
	February		March		April		total	
	$\bar{x}$	R	$\bar{x}$	R	$\bar{x}$	R	$\bar{x}$	$\bar{x}$
Number of days	5		7		6		18	49
Sunrise	0657		0651		0639		0649	0645
Sunset	1858		1856		1850		1855	1851
Day length	721		725		731		726	726
First movement	0641	0633–0657	0632	0626–0637	0641 (0628)[1]	0615–0647	0637 (35)	0617
Minutes from sunrise	–16		–20		+1 (–9)		–12 (–15)	–29
Reach NP	1658	1617–1732	1702	1545–1742	1623	1552–1720	1648	1721
Minutes from sunset	–120		–114		–147		–126	–90
Daily activity time	618	560–657	631	557–671	581 (545)[1]	514–648	610	666 600–753
DAT: DL	0.86		0.87		0.80[1]		0.84	0.93

1 On one day the group was located later than usual sleeping in night positions because rain was falling. Movement had probably occurred earlier so calculations were made with and without (in parentheses) the record for this day.

Differential Use of Forest Levels

Having analysed in detail the ranging patterns of siamang throughout their territory in the horizontal plane (Chapter 2), reference must be made to the vertical component of their movements. The emergent trees of the high level mainly provided positions for sleeping, resting, and calling. Most arboreal pathways and food were found in the main canopy layer at 80 – 120 ft (Chapter 3). The small branches of low trees did not provide sufficient support for prolonged travel, but they did contain a certain amount of food. Siamang were never observed less than 30 ft above the ground.

The level of the siamang group was recorded at 10-min intervals throughout the day from 0600 to 1800 h on 75 days in October and December 1969, and January to May 1970 inclusive at Kuala Lompat, and on 16 days in Ulu Sempam in February, March, and April 1970.

TS1 spent 25% of the day high, 63% medium, and 10% low, whereas in Ulu Sempam (where many high trees have been felled) RS2 were observed high 36%, medium 57%, and low 17% of the day time. These differences can largely be explained by the tendency to sleep high, and by RS2's shorter alert period; also, the felling of high trees had encouraged growth in the lower storeys. Scores stayed remarkably constant from month to month at Kuala Lompat (fig. 4.3), and the only marked discrepancy in Ulu Sempam was in February, when RS2 were spending most of their feeding time in two high fig trees, so that they were high for 58% of the day (27 and 25% in other months) and at medium height for only 40% of the day (65% in March and April).

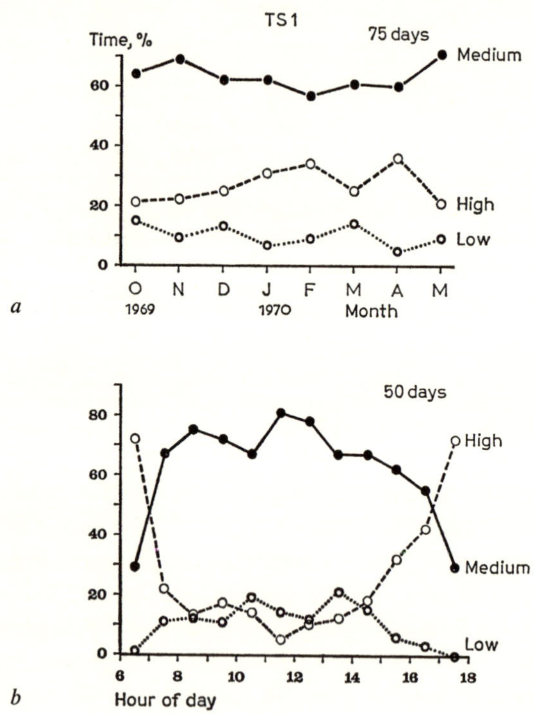

Fig. 4.3. (a) Monthly patterns of use of different forest levels by TS1 on 75 days from October 1969 to May 1970 inclusive and, (b) daily pattern of use of different forest levels by TS1 on 50 days in October and December 1969, and January, March, and April 1970.

Thus the siamang descend from their sleeping positions in the emergents to spend most of the day in the main canopy, descending to the low level before or after noon. There was a highly significant concordance for this daily pattern of each level between the five months at Kuala Lompat (fig. 4.4; Kendall coefficients of concordance, W = 0.59, 0.66, 0.55; χ^2 = 32.4, 36.4, 30.2 for high, medium, and low, respectively). TS1 spent more time in the upper parts of high trees in February and April, and in low trees in October, December, and March – but these fluctuations were small. Variations can be explained by the distribution of preferred foods (Chapter 3).

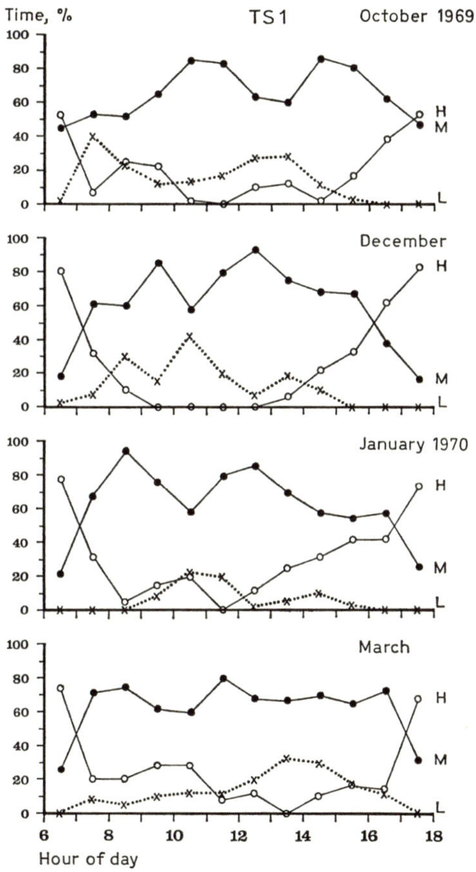

Fig. 4.4. Daily pattern of use of the different levels of the forest – high, medium, and low – by TS1 in October, December, January, and March (for 10 days and 720 observations/month).

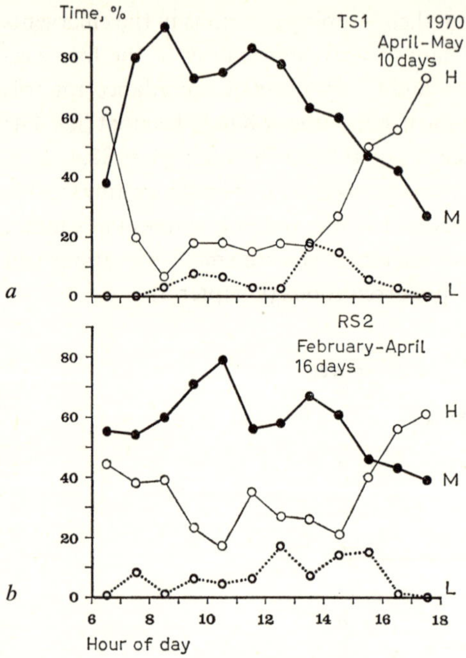

Fig. 4.5. Daily pattern of use of different levels of the forest by (a) TS1 in April and May 1970, and (b) by RS2 in February, March, and April 1970.

RS2 showed similar patterns to TS1 (fig. 4.5), but that for the high level was significantly different from TS1 (April to May). RS2 spent more time in fewer high trees, especially in the middle of the day.

Because of the limitations for analysis of these data in the form of time, they were analysed in terms of the number of entries into each level, and the total number of level changes, each day [Chivers, 1972b]. But these measures of rate of movement in a vertical direction introduce new dimensions that produce differences in daily pattern from the time spent in each level, and from movement horizontally about the forest. The increase in mean number of level changes per day, from 6.1 to 14.9, was correlated with an increase in day range (Chapter 2), but the daily pattern of these changes (10.7 per day at Kuala Lompat on average) was not. The peak of horizontal movement was in the first quarter of the day; subsequent movement occurred more in the vertical plane.

Activity Bouts

It was feasible to partition siamang daily activity time (the period of continual alertness between the first movement and reaching the night position) into bouts of maintenance activities. This was because of the high degree of behavioural synchrony between members of a siamang family group (see also Chapter 5), and because once all individuals were participating in the same activity it was likely to be sustained for a minimum of 10 min. Bouts were recorded, as a group measure, from the time the first animal changed activity, as long as most of the group joined in. No activity change was recorded if the others did not follow the first and sustain the activity for at least 10 min. It will be shown below that there were few problems in identifying such bouts of activity, and that, being more readily used in statistical analysis, this method based on the concept of 'bouts' is particularly useful for investigating daily patterns of behaviour, and for searching for any periodicities therein.

Behavioural Synchrony in Group Activities

Siamang individuals in the same family group started daily activities (first movement) within 14 min of each other on average, and ceased (reach night position) within 19 min; the former was less variable during the study (fig. 4.6). These aspects of behaviour were most closely synchronised at the height of the wet season (November 1969), and in the latter part of the dry season (March to May 1970). At these times mean intervals for starting and ceasing activities fell below 10 and 15 min, respectively. Leaving and entering the night sleeping tree were closely synchronised – this counted as travel (see below).

The maximum mean monthly interval between first and last to start feeding and travel bouts was about 12 min (fig. 4.6). These scores decreased from 9.5 min for feeding, and 11.2 min for travel, in the intermediate season, to 5.8 and 4.8 min, respectively, in the dry season. Thus an increased rate of activity (see Chapter 2, ranging patterns) resulted in increased synchrony within the group. Although scores ranged consistently up to 40 min for each activity (and occasionally even higher) the depressed means indicate the concentration of observations at the lower end of the scale (fig. 4.6).

This was confirmed by plotting the distribution of observations in three categories (fig. 4.7). In more than half feed and travel bouts all siamang were

a

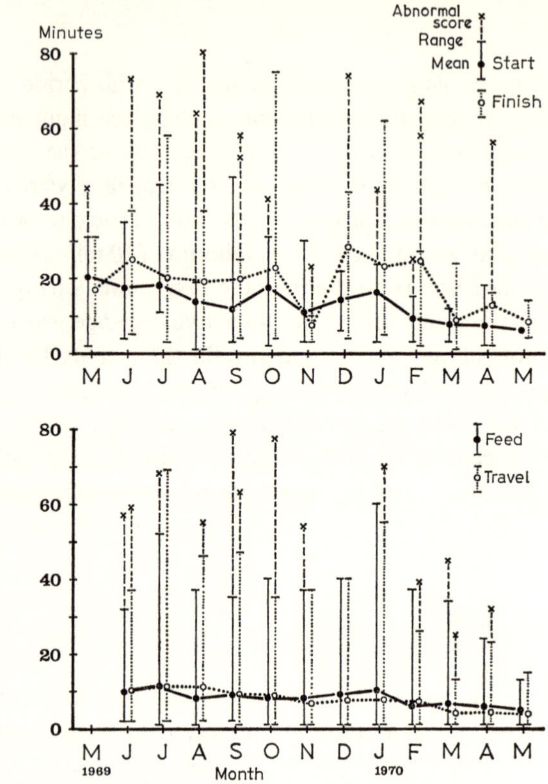

b

Fig. 4.6. The mean time (min) and range of observations (indicating unusually high scores) between the first and last group member (a) start and finishing daily activities, and (b) starting bouts of feeding and travel in each month at Kuala Lompat.

engaged in the new activity within 5 min of its initiation; about one quarter of feed and travel bouts were fully engaged between 6 and 10 min, and only about one fifth were longer than 10 min. In the dry season the increased synchronisation of individual behaviour was reflected by two thirds of bouts being fully underway within 5 min; this was highly significant (k-sample χ^2 test). The close similarity between the two activities, and the two study areas provides further confirmation of the strong tendency for members of a siamang group to act in concert. In 1970 RS2 were all feeding within 6 min, and travelling within 4 min on average; 66% of feeding bouts, and 80% of travel bouts were fully underway within 5 min.

47% feeding bouts, and 79% travel bouts from 148 days of observation of TS1 were included in this analysis; but only 24% feeding and 75% travel

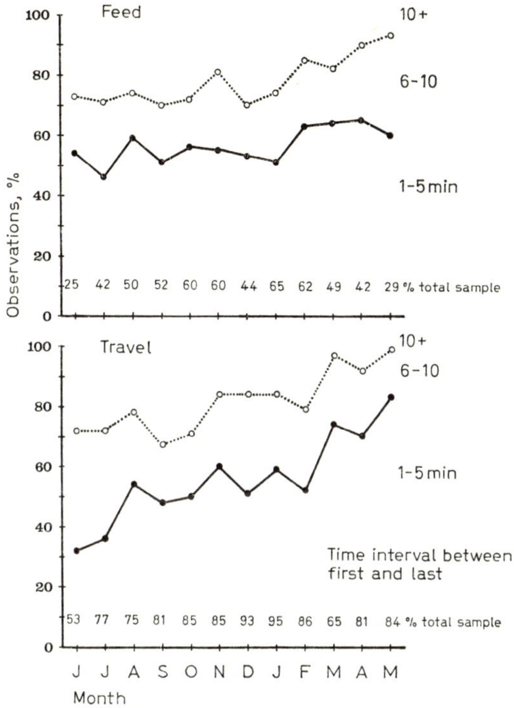

Fig. 4.7. The proportion of observations of the start of feeding and travel bouts in each month at Kuala Lompat, in which all group members were within 1–5, 6–10, and 10+ min of each other, indicating the proportion of bouts sampled.

bouts by RS2 in 1970. These can be regarded as representative samples, because although observations were omitted which reflected very poor synchronisation, others were not recorded because the whole group started a new activity almost simultaneously. The results for the small sample of feeding by RS2 probably cannot be more widely applied, however, because either the subadult male was absent from the immediate vicinity of the group, or two might feed while the other two rested and groomed in the same tree.

No data were analysed on the start of resting bouts; it seemed that this was not so closely synchronised. But this was partly because it did not involve the same degree of change in spatial relations that occurred when feed and travel bouts started. Thus individuals of a siamang group were closely together in terms of either time or space, or both. This will be elaborated during a description of social organisation (Chapter 5).

Daily Pattern of Activity Bouts

During the study at Kuala Lompat the total number of daily activity bouts varied between 4 and 18; these scores were more variable than those for the specific maintenance activites – rest, feed, and travel (fig. 4.8). On average daily activity time was divided into 11.7 bouts, of which 3.5 (30%) were rest, 5.1 (44%) were feed[1], and 3.1 (26%) were travel. There was little increase within and between months, except for an increase in the daily number of feed and travel bouts towards the end of the 1970 dry season.

But there was greater fluctuation between months in the duration of bouts of each activity (except travel), which counteracted in part the increase in bout number. Rest, feed, and travel bouts lasted for 66, 72, and 34 min, respectively, on average (fig. 4.8). The increase in feeding bout duration in September 1969 represented long feeding bouts in the fig, *Ficus virens* L203 (similarly, in December for L265). The very long rest bouts in February 1970 reflected a change in activity pattern whereby the siamang rested for an unusually long time in the middle of the day; this was the hottest and sunniest month of the study (see below and Chapter 1).

The frequency distribution of the starting times of bouts of each activity for each hour of the day shows no very distinctive patterns (fig. 4.8). With the maximum score for feeding soon after dawn, the frequency for subsequent hours fluctuated (with a second peak between 0900 and 1100 h) but generally decreased; in contrast, the incidence of resting increased to a maximum in mid-afternoon. There were two peaks of travel – in early and late morning – with a decreased frequency of bouts in the afternoon. But the lumping of data from throughout the year in terms of clock time might well obscure any activity rhythms, even though sunrise varied by only 30 min.

Periodicity of Maintenance Activities

The fluctuations from hour to hour in the distribution of the starting times of activity bouts indicate the possible presence of rhythms in the pattern of a relatively constant number of bouts of each maintenance activity from day to day. This will be investigated in terms of (a) alternative daily

1 Feeding bouts must not be confused with feeding visits (Chapter 3). One feeding bout usually contained more than one visit, because the group moved from one food source to another well within 10 min, i.e. without a noticeable change in activity.

a
b
c

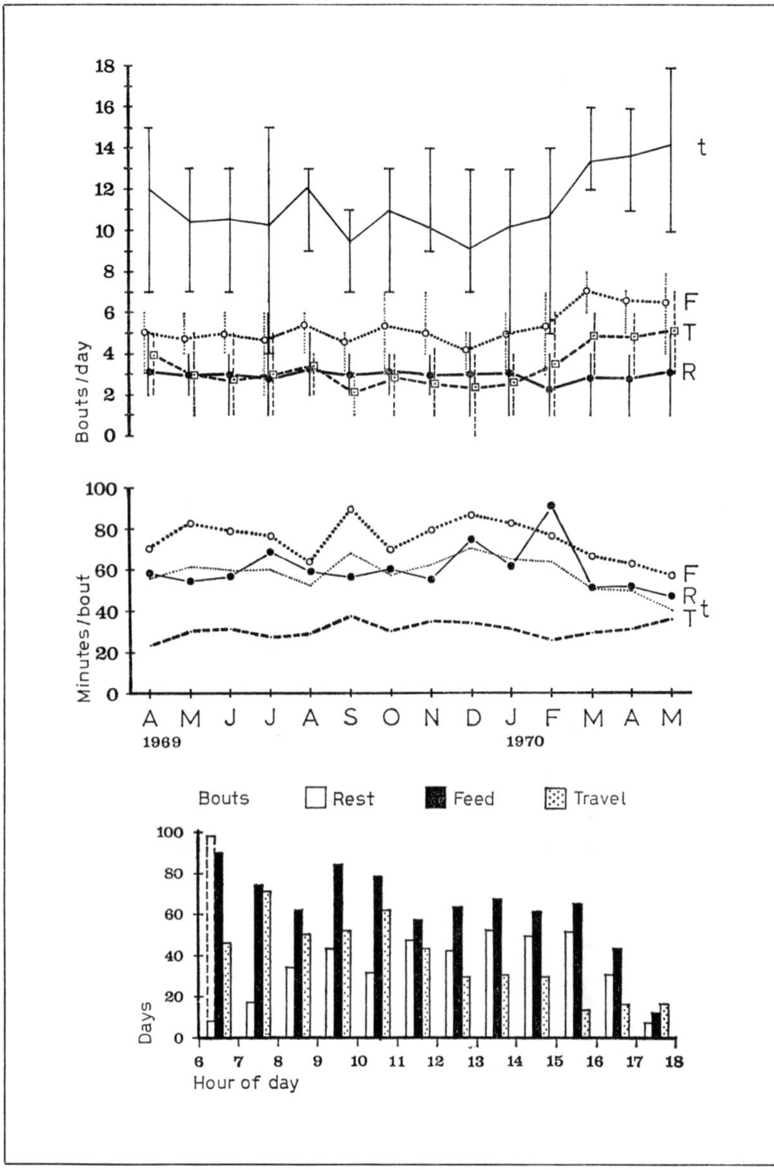

Fig. 4.8. (a) The mean number of activity bouts/day – overall, rest, feed, and travel; (b) the mean duration of bouts in each month at Kuala Lompat, and (c) the frequency of incidence of bouts of each maintenance activity in each hour of the day during the study.

scales, and (b) the time interval between bouts of each activity for data from 147 days observation of TS1 at Kuala Lompat.

Apart from clock time, three other time scales was chosen. The frequency of starting times of each maintenance activity bout were counted for each 50-min interval through the day from (1) sunrise, (2) the time of leaving the night sleeping tree, and (3) the time of the first feed (fig. 4.9). It has already been shown that the time of onset of daily activities – the start of the alert

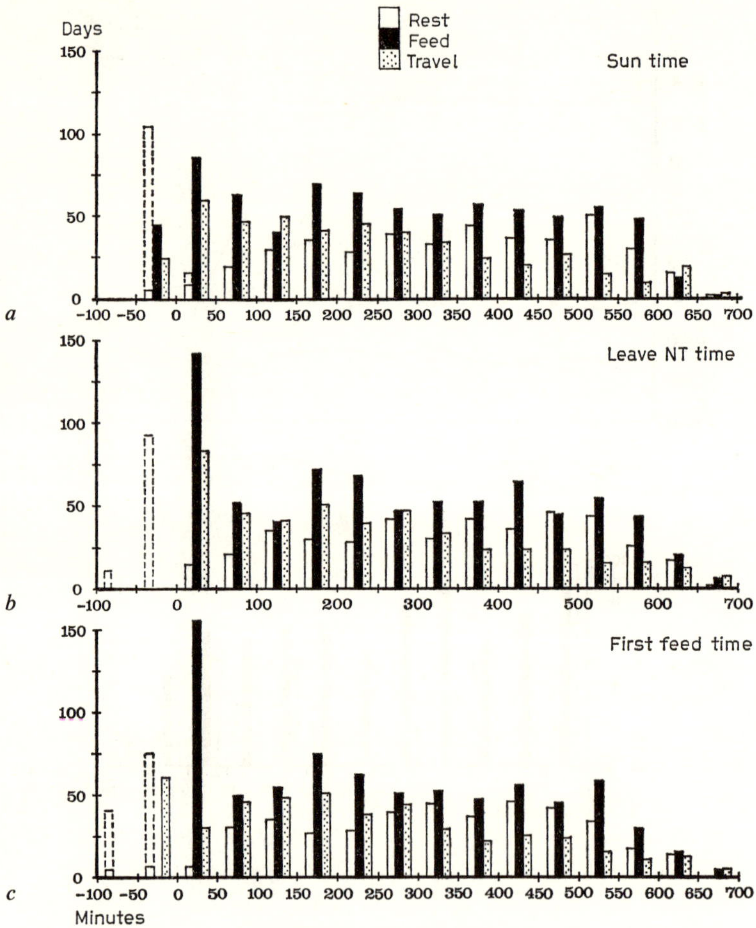

Fig. 4.9. The number of days during the study at Kuala Lompat on which bouts of rest, feeding, and travel started in each 50-min interval from zero. Zero equals (a) sunrise, (b) departure of group from night sleeping tree, and (c) the start of the first group feed (dashed bar = resting in night tree).

period – was closely correlated with dawn, and therefore with sunrise. But the pattern of daily activities was no more distinctive relative to sunrise, than to clock time.

The timing of activity bouts showed a greater periodicity with respect to the time of leaving the night tree and starting the day range (fig. 4.9). The siamang group invariably started feeding within 50 min of leaving the night tree; the second peak of onset of feeding was around 200 min later, with a final less distinctive peak at 400 or 500 min (apparently varying from day to day). The incidence of travel bouts was also greatest after leaving the night tree. Following that there was a steady decline throughout the day, except for a secondary peak at 200 or 300 min.

There was no real indication of peaks of resting, which steadily increased as the day passed. Because the onset of feeding was closely correlated with leaving the night tree, there were similar daily patterns for the incidence of bouts of each activity. But it can be seen that the first feed of the day was usually preceded by travel. The daily patterns of feed and travel were highly correlated, but those of rest and feed, and rest and travel were not, in terms of frequency of bout incidence (Spearman rank correlation coefficient, [Chivers, 1972b]).

Data were analysed for each season in case, even with the three corrections above, summation of data from the whole study was obscuring short-term rhythms, but although this was partly true, there were no striking results Chivers, [1972b].

Although it has been shown that bouts of activity, particularly of feed and travel, were not initiated at random throughout the day, starts being grouped significantly at certain intervals, no evidence of marked periodicity has yet been produced. One problem was that one long bout on one day might be equivalent to two bouts on the following day. This source of day-to-day variation could be one of the factors obscuring marked peaks of activity. In an attempt to investigate this methodological problem, the maintenance activities were investigated in terms of the interval between bouts, and the duration of these rest, feed, and travel bouts.

More than 60% feeding bouts commenced within 50–150 min of each other – over 80% between 50 and 200 min. There was a similar pattern for travel bouts, but only about 50% of rest bouts followed each other within 50–200 min – 80% between 50 and 250 min. Thus the interval between rest bouts was longer and more variable; while the peak for rest bout intervals was 100–150 min, those for feed and travel were clearly between 50 and 100 min. This suggests some rhythm of activity change after this order of time

Table 4.2. Mean time interval (min) between the start of activity bouts

Activity bout		Season							
		intermediate		wet		dry		total	
		$\bar{x}$	R	$\bar{x}$	R	$\bar{x}$	R	$\bar{x}$	R
Feed	1st–2nd	129	40–276	143	49–305	107	42–232	128	40–305
	2nd–3rd	129	53–469	154	29–377	97	44–374	129	29–469
	3rd–4th	138	36–297	147	30–314	103	32–328	131	30–328
	Penult.	140	27–469	136	58–314	106	41–374	130	27–469
Total		134	27–469	145	29–377	103	32–374	129	
Mean bout duration		77		80		62		72	
Mean interval between bouts		57		65		41		57	
Travel	1st–2nd	134	30–426	179	36–476	112	35–455	140	30–476
	2nd–3rd	135	35–326	126	38–354	136	35–494	135	35–494
	3rd–4th	153	41–424			112	39–252	132	39–424
Total		138	30–426	160	36–476	121	35–494	136	
Mean bout duration		28		33		40		34	
Mean interval between bouts		110		127		81		102	
Rest	1st–2nd	178	50–412	175	43–391	202	33–541	183	33–541
	2nd–3rd	160	41–357	142	30–326	151	62–274	152	30–357
Total		171	41–412	162	30–391	182	33–541	171	
Mean bout duration		60		61		59		66	
Mean interval between bouts		111		101		123		105	

interval. The mean interval between the start of rest bouts was 171 min; thus with a mean bout duration of 66 min, the average interval between bouts was 105 min. Similarly, bouts of feed and travel recurred on average 57 and 102 min, respectively, after the cessation of the previous bout (table 4.2).

There was no seasonal variation in the mean duration of rest bouts, but they were more widely spaced in the dry season. Feeding bouts were shorter and more frequent in the dry season, when travel bouts were longer and more frequent (table 4.2). Plots of the monthly means for (1) time interval in minutes between the start of bouts, (2) time interval between bouts (from finish to start), and (3) bout duration for each maintenance activity (fig. 4.10) confirm the greater constancy of bout duration relative to the spacing of bouts

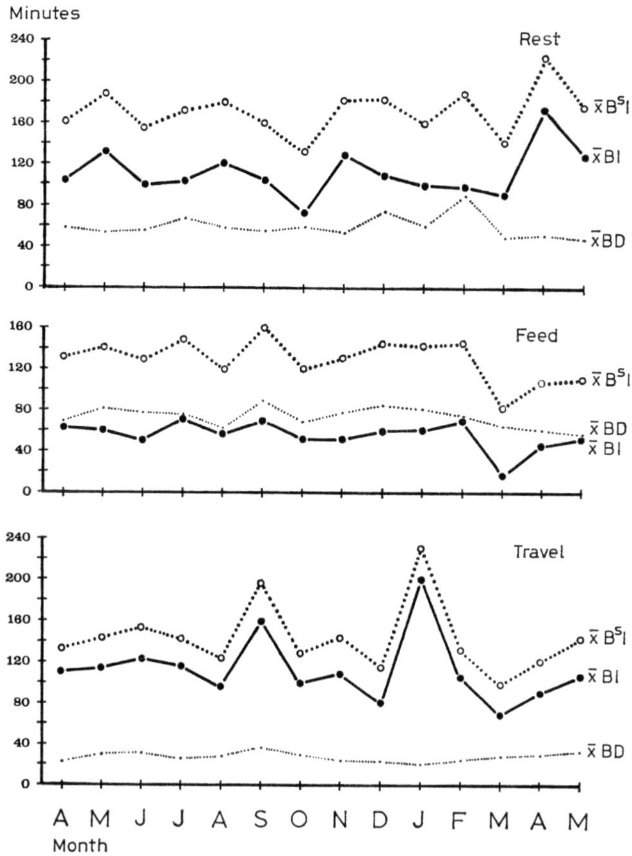

Fig. 4.10. The mean time (min) in each month of the study at Kuala Lompat for the duration of bouts (BD), the interval between the starting times of bouts (BsI), and the interval between bouts (BI) for each maintenance activity – rest, feed, and travel.

through the day. Thus resting bouts were more frequent in October and March, and less frequent in April 1970; feeding bouts were much more frequent in March; and travel bouts were markedly less frequent in September and January, with increased frequency in March, as indicated by the shorter interval between bouts (fig. 4.10) and small variation in number of bouts each day (fig. 4.8).

Bouts of each maintenance activity were categorised as 1st, 2nd, 3rd, and so forth. There were sufficient observations of feeding and travel only to investigate the interval between 3rd and 4th, and penultimate and ultimate bouts of feeding for each day were also investigated (in some cases the same

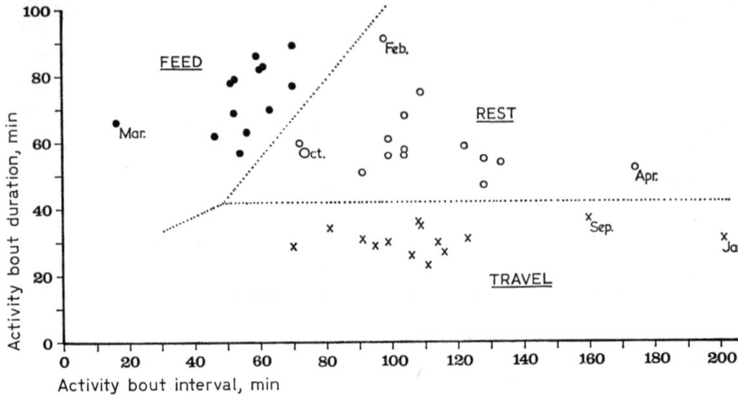

Fig. 4.11. Bivariate plots of the mean intervals between bouts, and mean bout duration of bouts of each maintenance activity in each month at Kuala Lompat.

observations fell in both of the latter categories). The interval between 1st and 2nd tended to be longer than that between 2nd and 3rd rest bouts, but not significantly (Kolmogorov-Smirnov two-sample test); the converse was true for travel, with the small sample of 3rd and 4th intervals even longer (table 4.2). The interval between feeding bouts was even less variable with regard to time of day (or category). The interval between 2nd and 3rd bouts appeared to be more uniform, with 1st and 2nd bout intervals more widely distributed. But there was surprisingly little variation irrespective of season with regard to the interval between bouts of each activity.

Plots of the monthly means for the duration and interval between onset of each maintenance activity (fig. 4.11) confirm that feeding bouts occur more frequently than rest bouts, but do not last much longer. In contrast, travel bouts occur as often as rest bouts, but are of a shorter duration. Exceptional scores occurred in March for feeding; in October, February, and April for resting; and in September and January for travel. But while the duration of travel bouts appears to be independent of the interval between bouts, the duration of feeding bouts increased with longer intervals between bouts.

Drinking Behaviour

Eighty bouts of group drinking at water holes were observed in the 147 days of observation of TS1 at Kuala Lompat. Unlike the other maintenance activities, including feeding of which it is a part, individuals participated sequentially for only a few minutes each. The frequency observed was once every 4 – 5 days in the first half of the study, with a marked increase with the advent of the rains in November (following a comparatively dry October). At the height of the dry season, in February, the rate had risen to 1.6 bouts/day (fig. 4.12). In the intermediate season bouts were mainly concentrated in the early afternoon; in the wet season they were even more concentrated in the middle of the day (79% bouts from 1100 to 1400 h). Drinking occurred earlier in the day in the dry season, with nearly 40% bouts before noon, but this was when the spread was most even throughout the day (fig. 4.12). Overall, 56% bouts were between 1100 and 1500 h; this coincides with the hottest time of the day, and with the peak of resting (see below).

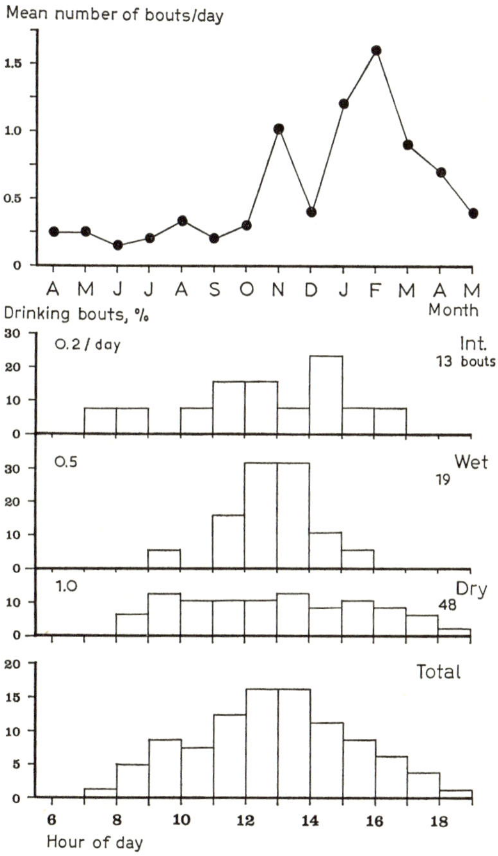

Fig. 4.12. The daily frequency of drinking bouts in each month, and the pattern of drinking (hourly) during the day – overall, and in the intermediate, wet, and dry seasons at Kuala Lompat.

Activity Time

While a study of the frequency and timing of activity bouts provides many advantages for analysis, the more precise measure of time spent in each activity probably provides greater insight into the significance of the behaviour (e.g. feeding time in a study of diet, Chapter 3). Accurate results can only be achieved by dawn-to-dusk observation. Group scores were obtained by recording throughout the day the time that the first member of the group changed its activity. Summation provided the total number of minutes spent in each maintenance activity. A more precise measure was obtained from the 1-min sample of a specific individual throughout the day. Having watched each independent member of the group for two consecutive days, a score representative of group activity was obtained for each observation period.

This section will investigate the daily pattern of maintenance activities in the two study areas resulting from these methods; they are compared with a 5- and 10-min sample of all group members throughout the day (CHIVERS [1972b], Appendix IX). Further reference will be made to these different methods here, followed by a comparison of the behaviour of individual group members, and an analysis of consecutive days of high and low activity.

Kuala Lompat (TS1)

1. Group Behaviour

The mean and ranges of daily aggregate durations for each maintenance activity in each month are presented in figure 4.13. Although there was considerable day-to-day variation, the patterns showed much less fluctuation from one month to the next. The group scores (continuous method) were less variable than the individual records (1-min sample). It must be noted that rest scores by the 1-min sample are inflated by about 35 min/day by those observations of 'resting' which occurred during travel.

Over the whole study (continuous method) there was no significant variation in feeding (Kruskal-Wallis analysis of variance, H = 15.1 and 15.8, respectively), but a highly significant increase in travel towards the end of the study (H = 70.3, p = 0.001, see Chapter 2). But while dietary composition showed some variability from month to month, daily feeding time remained relatively constant, with an increase in ranging where necessary to maintain the required food intake. The lowest scores of daily feeding occurred at the

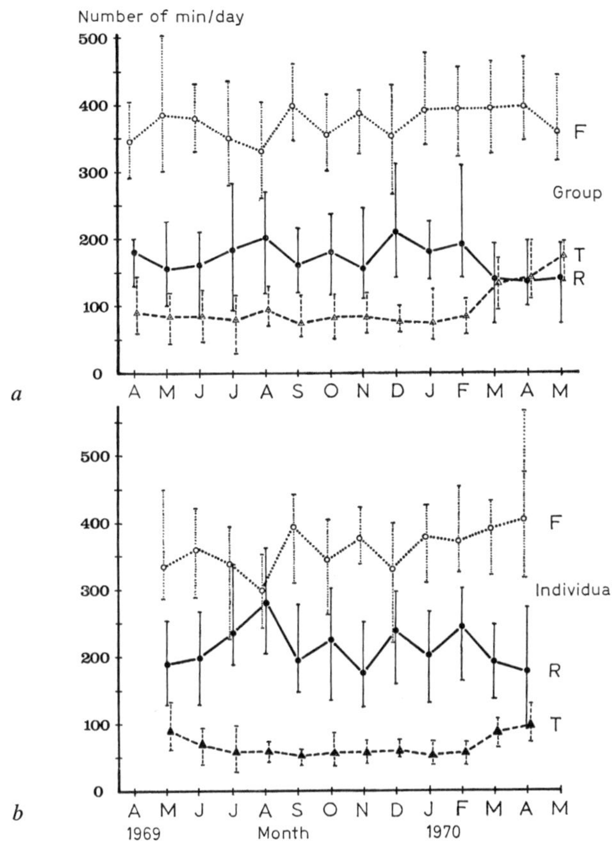

Fig. 4.13. The mean number of minutes spent daily in each maintenance activity by the siamang of Kuala Lompat in each month in terms of (a) continuous record of group activity, and (b) 1-min sample of individual behaviour.

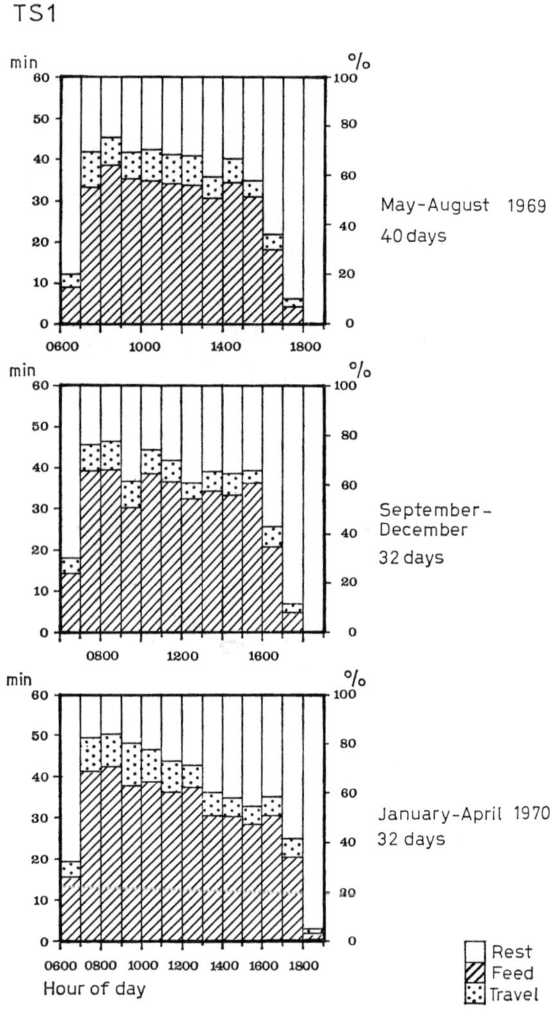

Fig. 4.14. Daily patterns of resting, feeding, and travel in each hour for siamang individuals in the intermediate, wet, and dry seasons at Kuala Lompat.

end of the intermediate season, the highest at the end of the dry season; the latter reflected increased food intake rather than quicker eating.

The daily patterns of each maintenance activity in each month (from cumulative 1-min samples) are described in Appendix D. They confirm that in most months time spent in feeding and travel were greatest early in the day, steadily declining to be replaced by increased resting in the afternoon. These patterns, and monthly variations, will be amplified in a discussion of the effects of weather on behaviour (see below). Both methods demonstrated a highly significant concordance between hourly scores for each month within seasons, and overall – and between daily patterns in each season, except for travel (Kendall coefficient of concordance).

But although daily patterns of resting were highly correlated negatively with feeding in all months, travel was only positively correlated with feeding in September, February, and April (Spearman rank correlation coefficient). Seasonal scores gave a more significant result, and showed that resting was negatively correlated with travel.

The seasonal summaries of the 1-min sample data (fig. 4.14) show the most distinctive pattern of early feed and travel with afternoon resting, and a final feed, in the 1970 dry season. It was much less emphasised in the intermediate season, although the group settled earlier for the night. Between 0700 and 1100 h in the dry season feeding occupied 60 – 70% time, and travel about 15% time, i.e. about 40 and 10 min in each hour, respectively. In the afternoon feeding and travel occupied less than 60% time. Overall feed and travel time diminished from 47 min in the hour at 0800 h, to 35 min in the hour at 1500 h.

Thus, on average the group rested for 170 min (27%), fed for 374 min (58%), and travelled for 99 min (15%) of the 643 min of daily activity time. Scores from the 1-min sample of individual behaviour for rest, feed, and travel averaged at 214 (33%), 362 (56%), and 67 (10%) min, respectively.

2. Individual Behaviour

Whereas 1-min records on individual behaviour provided information for only two days in each observation period, the 5- or 10-min sample of all individuals provided a fuller comparative measure of the behaviour of each group member throughout the study. Mean monthly scores (table 4.3) indicated that the adult male and adult female were active for about 15 min less each day than the younger individuals (except for the infant). But the adult female fed for about 60 min longer each day on average than the adult male (fig. 4.15); it has already been shown that she ate nearly twice as fast (Chapter

Table 4.3. Individual variation in daily amounts of rest, feed and travel, with comparision for each group member between days with 1- and 5-min samples, correcting for resting pauses in travel bouts, and comparing with complete 5-min sample

Individual	Sample, July–April (n = 21) 1-min	5-min	Rest during travel[1]	Rest including travel pauses	Travel excluding rest pauses	Error of 5 min	5-min overall (n= 109)[2]
Rest							
Male	251	220	32±1	252	60	+ 1	203
Female	193	158	31±4	189	60	+ 4	155
SA male	200	160	37±3	197	65	− 3	178
Juvenile	230	188	40±2	228	66	− 2	169
Infant	217	183	32±2	215	83	− 2	180
Feed							
Male	323	320				− 3	332
Female	399	383				−16	379
SA male	398	413				+15	383
Juvenile	355	356				+ 1	376
Infant	300	305				+ 5	328
Travel							
Male	59	92				+ 1	105
Female	64	91				− 4	108
SA male	67	102				− 2	102
Juvenile	67	106				− 1	108
Infant	80	111				+ 3	93
Total							
Male	636	640				+ 4	
Female	639	639				0	
SA male	664	663				− 1	
Juvenile	653	653				0	
Infant	597	598				+ 1	

1 Mean differences between 1- and 5-min rest and travel times to give resting time, inclusive of resting during group progressions, for comparison with 1-min sample.
2 Calculated from mean active day length in small 5-min sample.

3). The subadult male fed for about 20 min longer than the female each day, with the juvenile scores intermediate between male and female.

There was little variation in travel time between individuals, because the group travels around its territory together, even though immature animals were more active in movements. The rise in the infant travel scores to the same level as the others (and higher) by the end of study indicate her increasing independence (she was not scored as travelling when carried by the male). Resting varied inversely with feeding (and travel) so that the male

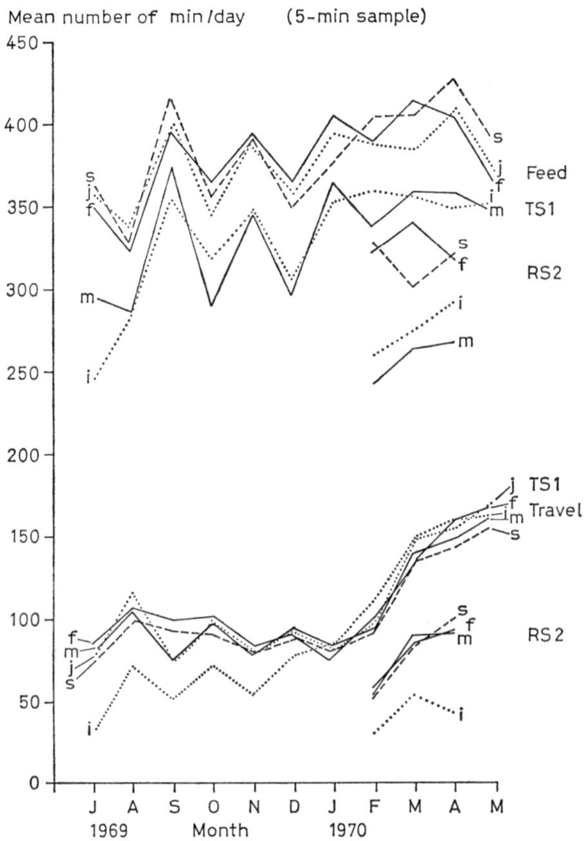

Fig. 4.15. The cumulative time each day spent feeding and travelling by individuals of TS1 and RS2 in each month of detailed study, illustrating differences according to age and sex.

and infant rested most, and the female and subadult male least. Similar patterns at a lower intensity were observed in Ulu Sempam (see below).

Further insight into the performances of individuals was gained by an analysis of the 5-min data for 106 days at Kuala Lompat – a larger sample, so that the distorting effect of some day-to-day fluctuations in the small sample was eliminated. Each individual was ranked for each activity on each day from highest to lowest [CHIVERS, 1972b]. The adult male rested most (ranked 1 or 2 on 91% of days), especially in the dry season. The subadult male fed more than the others in the intermediate and dry seasons (ranked 1 or 2 on 73% of days overall), whereas the female was ranked highest more often in the wet season (ranked 1 or 2 on 71% of days overall).

Rank	Rest					Feed					Travel				
	M	F	SA	J	I	M	F	SA	J	I	M	F	SA	J	I
1 or 2	91	10	8	11	89	8	71	73	58	6	45	52	41	59	40
4 or 5	4	49	55	59	3	82	5	6	10	86	30	21	41	22	52

Maintenance activity	Rank for each individual over 106 days				
	1	2	3	4	5
Rest	I	M	F	SA	J
Feed	SA	F	J	M	I
Travel	J	F	M	SA	I

M = male; F = female; SA = subadult; J = juvenile; I = infant.

Because the group travelled together, there was less inter-individual difference in travel time, and greater variations in ranks (but not times) between seasons. No individual was consistently first or last in the rankings. Although carried by the male, especially in the early months, the infant increased its travel time by movements while the adults were resting. Similarly, the adults were recorded as resting (by 1-min sample, but not by continuous method) during group progressions, while the infant, which travelled more slowly, caught up.

The subadult male was almost invariably active for the longest period of daytime, but while the adult male (in particular) and the female had the shortest daily activity times on more than half the days, they had the longest on more than one third; the converse was true for the juvenile. This illustrates a possible dichotomy of daily activity patterns, which will now be further investigated.

Two measures of days of high and low activity were taken: the proportion of days with groups scores above or below (1) the monthly mean, and (2) the overall mean of the daily group score for feed, travel and alert period by the continuous method. Days of 'high activity' were those with a prolonged alert period, but mainly with an increase in time spent in feeding and travel. For feeding July and August were the most active by (1), with more than two thirds days above mean, whereas September, November, February, March, and April had this unusually high proportion of active days by (2). For travel only May and February had high scores by (1), and March, April and May by (2). Daily activity time (DAT) was consistently unusually high in July (and April and May 1970) by (1), and from February to May 1970 inclusive by (2).

From the overall mean an analysis of consecutive days of high and low activitiy indicated that on average two (1 – 8) days of prolonged DAT (alert period) were followed by two (1 – 6) days of reduced activity. Two to three days (range of 1 – 5 days) of intensive feeding were followed by one to two (1 – 3) days of reduced activity, whereas one (1 – 2) day of prolonged travel was followed by three to four (1 – 7) days of reduced daily movement. Thus, it can be inferred that following a day of increased movement, there were two or three days of prolonged alert periods with much increased feeding; this can be compared with patterns of daily ranging (Chapter 2).

Ulu Sempam (RS2)

In the 1969 wet season RS2 were still not habituated sufficiently for dawn-to-dusk observation. But on 14 days from October to January inclusive continuous observation was maintained for 4 – 10 h to provide a 5-min sample of the behaviour of all four group members (no juvenile in RS2). The daily pattern of maintenance activities constructed from this (fig. 4.16) showed the opposite of the TS1 pattern in most months. It most resembled the November pattern at Kuala Lompat (Appendix D) with the amounts of feeding and travel increasing as the day passed to peaks between 1300 and 1600 h.

On 21 complete days of observation in 1970 activity patterns resembled those of TS1 in 1969, with an increase in feeding and travel, and a decrease in resting, in the morning. But, although the 1970 patterns for rest and travel were not correlated with those for 1969 (Spearman rank correlation coefficient, r_S = 0.48 and 0.49, respectively), that for feeding, perhaps the most important, was still significantly correlated ($r_S = 0.62$, $t = 2.47$, $p = 0.05 – 0.02$). But there were no other correlations between RS2 and TS1 in or between wet and dry seasons.

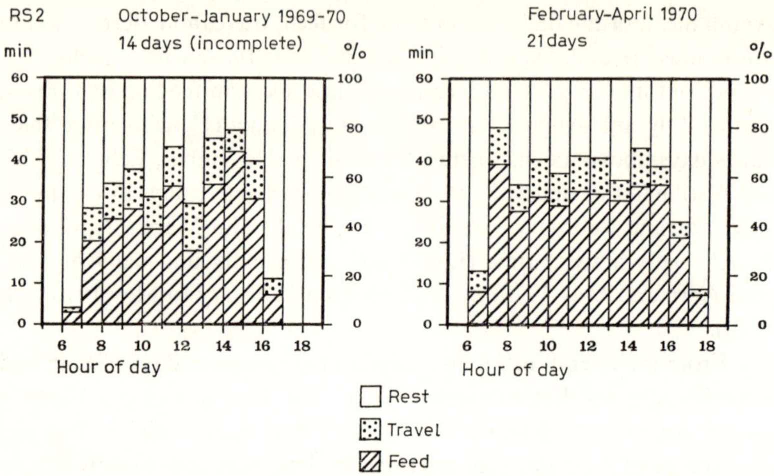

Fig. 4.16. Hourly patterns of resting, feeding, and travel by siamang individuals in the wet and dry seasons in Ulu Sempam – based on 5-min sampling of the behaviour of all group members.

From mean hourly scores for each maintenance activity, it was possible to estimate the daily totals. An unknown proportion of the 314 min of resting must be excluded from the alert period, because they represent sleeping in the night tree. There were 263 min of feeding, and 84 min of travel. The 1970 scores of 204 min resting, 322 min feeding, and 84 min travel in a daily activity time of 610 min, show an increase in feeding time. But although there are similar proportions of the alert period spent in each activity, the actual scores for DAT, feed, and travel were considerably below those of TS1 in the 1970 dry season at Kuala Lompat (table 4.4).

The time spent in each maintenance activity showed little variation in the three months of detailed observation in Ulu Sempam. There were 3.9 rest, 5.6 feed, and 2.8 travel bouts on average each day, of 52, 58, and 30 min mean duration, respectively (table 4.4). Thus bouts tended to be shorter for RS2, except for travel, and there were more rest and fewer travel bouts than for TS1 in the same period. But although feeding bouts were shorter, their frequency was the same. In no month was there a distinctive daily pattern (Appendix D); if anything, there was a tendency for all activities to increase early and late in the alert period.

There were only nine days in Ulu Sempam when a compete 5-min sample of the maintenance activity of each group member was achieved. The most elusive individual was the subadult, who was ranging further from the rest

Table 4.4. Siamang daily behaviour in Ulu Sempam

	Ulu Sempam 1970								Kuala Lompat 1970
	February		March		April		total	%	
	$\bar{x}$	R	$\bar{x}$	R	$\bar{x}$	R			
Number of minutes									
Rest	212	145–257	208	155–277	194	149–296	204	33	157
Feed	331	282–494	318	231–375	321	228–360	322	53	388
Travel	64	44– 87	94	60–123	89	57–120	84	14	120
Total	607	520–690	620	558–671	604	547–647	610		665
Number of bouts									
Rest	4.0	2–5	4.3	4–6	3.3	2–5	3.9	32	2.6
Feed	5.7	5–7	5.9	5–7	5.3	4–6	5.6	46	5.3
Travel	2.5	2–4	2.5	1–4	3.4	1–6	2.8	23	3.9
Total	12.2	11–15	12.7	11–16	12.0	7–17	12.3		11.8
Number of min/bout									
Rest	53.0		48.3		58.7		52.3		60.4
Feed	58.0		53.8		60.5		57.6		73.2
Travel	25.7		37.4		26.1		20.0		30.8
Mean	49.8		48.8		50.3		49.6		56.3
Number of days	6		8		7		21		49

of the group with increasing frequency (Chapter 5). Data were substantially complete for most group members on 18 days. But the daily scores for each individual confirm the findings from Kuala Lompat that the female and subadult consistently spent more time feeding than the male and infant; the converse was true for resting. Scores for travel and DAT were again more variable for each individual, but little separated the individuals in each month.

Weather and Daily Activity Patterns

The daily patterns of weather – sun, cloud, and rain – in each month in each study area are illustrated in Appendix D. Particular consideration is given to the presence or absence of marked peaks of sunshine, or cloud cover, and whether these occurred in the morning or afternoon. Rainfall was mainly confined to the hours of dawn, dusk (and sometimes late afternoon), and darkness. The aim of the section is to determine what effects, if any, weather exerted directly on the daily patterns of siamang behaviour.

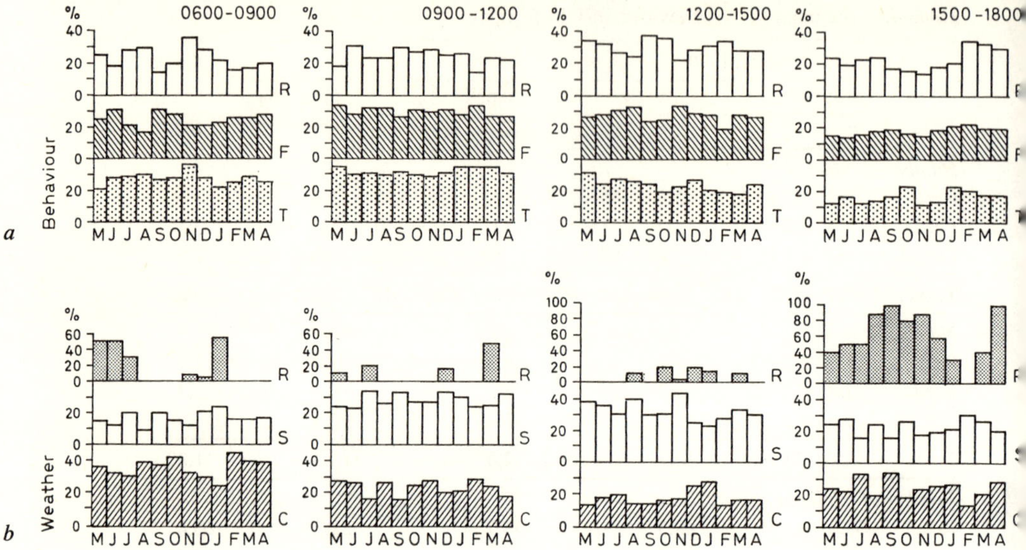

Fig. 4.17. Mean proportions of observations of (a) rest, feed, and travel, and (b) rain, sun, and cloud in each quarter of the day during each month at Kuala Lompat.

The daily patterns of sunshine were significantly similar in the 'intermediate' and 'wet seasons' (Chapter 1), and overall (Kendall coefficient of concordance, p = 0.01 –); the similarity was less marked (p = 0.05 – 0.02) in the 'dry season', because of a differing pattern in January. The concordance of the daily patterns of cloud cover were highly significant overall, and in the 'intermediate' and 'dry seasons', but a differing pattern in December reduced the significance of the 'wet season' concordance. Rainfall was concentrated in the late afternoon, with frequent absence in the other hours. The early afternoon tended to be the sunniest time of the day, and the early morning the cloudiest (fig. 4.17).

Behaviour and Weather

The daily patterns of maintenance activities have already been described (see above). There was less variation between months, and during the day – in terms of 3-hour periods (quarter days) – than for weather (fig. 4.17), whatever the daily totals. Feeding and travel were most frequent in the late morning, declining during the afternoon; the peak of resting was in the early afternoon, which was also the sunniest (and hottest) part of the day (see also

Appendix D). It was more often cloudy early, with rain towards the end of the day. It remains to be seen whether these associations between weather and behaviour patterns are other than fortuitous.

The late morning travel peak was more marked in the dry season, with a decrease in morning resting; the feeding pattern remains almost unchanged. It is likely that this crude analysis of large samples masks any important effects of weather on behaviour patterns, which are, therefore, of a more subtle nature if present. Variations in monthly patterns (Appendix D) indicate that it is at this short-term level that any effects may operate. They might also be detected by investigating specific days of contrasting weather.

The relationships that emerge from a scrutiny of monthly daily activity patterns are of feeding and travel being maximal in the morning except in cloudy and wet months, when morning resting increases to give way to afternoon feed and travel peaks. But comparisons of daily weather and activity patterns in the months of the wet and dry seasons produced few significant results (Spearman rank correlation coefficient). At Kuala Lompat in the wetter months of November, December 1969, and May 1970 feeding patterns were correlated with the daily patterns of sunshine; at other seasons this was not the case.

The daily patterns of rest and feed in Ulu Sempam were not correlated with sun or cloud patterns in any month in 1970; travel was only correlated with sunshine in April. Thus it is difficult to demonstrate any direct effect of weather on behaviour, except under extreme conditions. High temperatures, or wet cloudiness, sustained over several days, exerted a significant effect on resting and feeding patterns respectively to produce characteristic short-term activity patterns. Heavy rain inhibite dall activities other than resting (solitary in a hunched position), but because such rain fell only for short periods, and usually at times when siamang were in, or preparing to enter, night sleeping trees, its effect on behaviour and daily activity patterns was minimal.

It seemed significant that not only were there no correlations between the daily patterns of maintenance activities between RS2 and TS1 in and between the wet and dry seasons, but in almost all months there was no correlation for sunshine and cloud cover patterns (Chivers [1972b], Appendix VII. 10). The exception was March 1970 (for sunshine), and in this month the daily patterns of rest and feed were correlated between the two groups ($r_S = 0.62$ and 0.67, $p = 0.05-0.02$ and $0.01-0.001$, respectively). It therefore appears that the different activity patterns of RS2 and TS1 represented responses to different meteorological conditions, rather than to intrinsic behavioural factors.

Only 22% of 150 days at Kuala Lompat were either persistently hot and sunny or cloudy and wet (75% level – see Chapter 2, effect of weather variation on ranging). Activity pattern data were available in the former of continuous group scores, and 1-min samples of individual behaviour. Although more precise, in the small sample of 15 hot and sunny, and 14 cloudy and wet days, observations were biased towards certain individuals.

	Male	Female	Subadult	Juvenile	Infant
Hot and sunny days	4	2	4	3	2
Cloudy and wet days	7	1	1	2	3

Thus measures on cloudy and wet days in particular were biased towards the low level of feeding and the high resting scores of the male and young, with travel scores less affected; observation on hot and sunny days were more evenly balanced. Although the results from the two observation methods were highly correlated for each observation period, this was not true for this sample. Thus greater reliance must be placed on the results of the continuous group records.

Hourly aggregates of the mean time spent in each activity produced an extended alert period on hot and sunny days, with feeding and travel peaks in the morning, and a tendency to rest at midday. On cloudy and wet days, however, the alert period ended earlier and feeding was increased in the early afternoon (fig. 4.18).

Finally account must be taken of the spread of 20 hot and sunny days across all seasons (including overall increased activity at the end of the dry season), whereas the sample of 17 cloudy and wet days was concentrated towards the end of 1969:

	Apr.	May	June	July	Aug.	Sept.	Oct.	Nov.	Dec.	Jan.	Feb.	Mar.	Apr.
Hot and sunny days	2	1	1	2	1		3			1	3	3	3
Cloudy and wet days		2	3	2			1	4	3			2	

But although the expected pattern was more marked in the dry season, it was present in the intermediate season, with the daily patterns for feeding, but not travel, highly correlated.

Siamang fed more in the mornings on hot and sunny days, and after noon on cloudy and wet days. Resting was increased early and late in the day

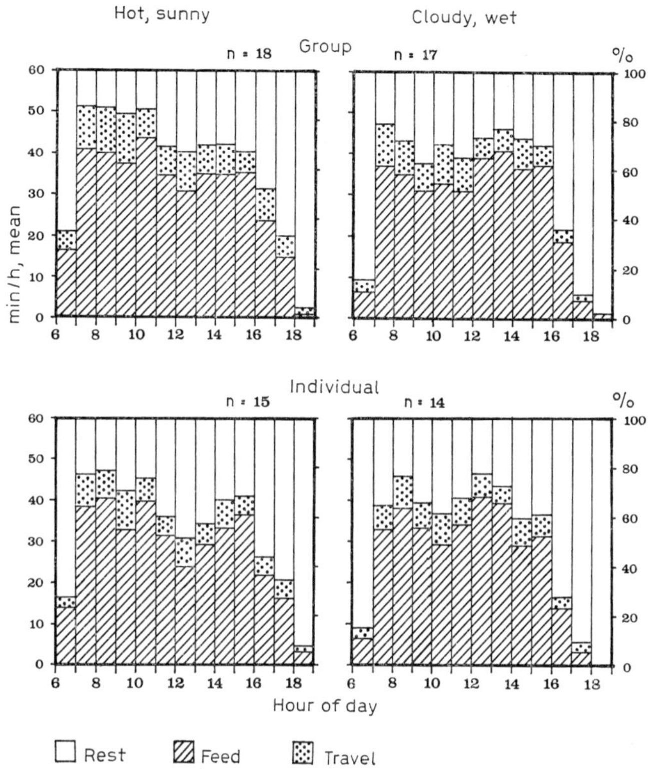

Fig. 4.18. The daily patterns of rest, feed, and travel in each hour on hot and sunny days, and cloudy and wet days at Kuala Lompat, by continuous observation of the group, and 1-min samples of each individual.

when cloudy and wet, with travel depressed in the afternoons. There was little seasonal difference in pattern on hot and sunny days.

Thus on cloudy and wet days the group became alert 5 min later on average, than on hot sunny days, resting about 20 min longer before leaving the night sleeping tree (table 4.5). The group (TS1) entered the next night tree to reach sleeping positions an hour earlier than on hot and sunny days. There was also a 50-min difference in the length of the alert period for hot and sunny days between seasons, but the intermediate season score was still 30 min more than that for cloudy days.

There was little difference in the mean number of activity bouts/day; there were two more on hot and sunny days, of which one was feeding. There was little difference between intermediate (hot and sunny) and the wet season. Resting times were very similar but daily scores for feeding and travel on hot and sunny days showed an extra 30 and 15 min, respectively. But resting and feeding bouts were about 10 min shorter on average, and travel bouts were about 5 min longer than on cloudy and wet days. But the only significant difference that could be detected for any of the parameters was the increase in travel time on hot and sunny days (Mann-Whitney U-test).

Table 4.5. Difference in behaviour on hot, sunny and cloudy, wet days

Behaviour		Hot and sunny days				Cloudy and wet days, total	
		intermediate season	dry season	total $\bar{x}$	total R	$\bar{x}$	R
Time 1st move		0616	0617	0617	0554–0633	0622	0607–0646
Time leave NT		0638	0633	0633	0600–0723	0656	0628–0734
Time enter NT		1646	1734	1710	1523–1829	1617	1455–1802
Time reach NP		1649	1738	1714	1528–1830	1626	1550–1804
Active day length		633	681	657	564– 753	604	551– 704
Siamang active day/day length		0.87	0.93	0.90	0.81– 1.03	0.85	0.75– 0.96
Number of bouts of rest		3.0	2.6	2.8	1–5	2.5	1–4
Number of bouts of feed		4.7	6.2	5.5	3–8	4.5	3–7
Number of bouts of travel		3.1	3.8	3.5	1–6	2.9	1–5
Total		10.8	12.6	11.8		9.9	
Time rest	I	189	220	229		222	
	G	149	182	167		174	
Time feed	I	384	385	361		330	
	G	385	386	385		371	
Time travel	I	66	79	71		63	
	G	93	110	103		75	
Extreme sample							
Time rest	I			204		193	
	G			159		158	
Time feed	I			389		350	
	G			410		377	
Time travel	I			66		56	
	G			88		71	
Mean bout length							
Rest				59.7		69.7	
Feed				70.1		82.5	
Travel				29.4		25.8	

I = individual; G = group.

Siamang were generally rather less active in bad weather, with a shorter daily alert period, and fewer somewhat longer bouts of each maintenance activity. Even so, weather cannot be said to exert directly a profound effect on behaviour. It is clear that other factors, such as the location and dispersion of food trees, are moulding daily patterns of activity to varying cycles that are only secondarily altered by the weather.

Discussion

The daily pattern of activities whereby feeding and movement are concentrated early and late in the day, with long rest periods around noon is characteristic of diurnal animals especially in the tropics. This pattern is found among most primate populations studied so far [see DeVore, 1965, for examples], and is readily interpreted in terms of daily cycles of weather. So too are the exceptions [Chivers, 1969], for this daily pattern tends to be disrupted in cloudier environments. But it is not always clear to what extent it is high temperatures or precipitation, or both, that tend to inhibit all activities other than resting.

The typical daily activity patterns of leaf monkeys [Chivers, 1973, confirmed by Hunt, personal commun.] and apparently of macaques [Chivers, 1973] in the same habitat, indicate that in most months of the year Malaya does not provide a particularly cloudy or wet environment. Although the gibbon shows some tendency towards a midday siesta [Ellefson, 1967] its daily pattern of activities more closely approximates that of the siamang. It is pertinent to enquire why these hylobatids show this distinctive pattern of steady decline in feeding and movement after intensive activity during the morning, and why they settle for the night so much earlier than other primates. Even in adverse weather conditions the pattern, although reversed with peaks in the afternoon, remained equally distinctive. It could be that siamang achieved a similar degree of thermoregulation in the heat of the day by moving lower in the forest canopy (where it was 10°F cooler), rather than by reducing energy expenditure in feeding and travel.

Siamang appeared to have unusually high feeding requirements in terms of time; no primate studied so far has been shown to spend more than half the active day (or more than 6 h cumulative) feeding. It may be that this is dictated by the nutritional values of the food (Chapter 3), and that intrinsic factors amongst the highest primates (hominoids – apes and men) will often overcome environmental conditions to produce such deviations from the 'typical' pattern.

The dispersion and abundance of food influenced daily patterns of activity, as well as ranging behaviour. The onset of feeding occurred earlier when certain food trees were close to the night sleeping trees; if no principal food sources were close, the departure from the night tree was often delayed. Resting bouts started earlier, and were longer, after more active mornings of feeding and travel. A final feed before settling for the night appeared to depend not only on food availability, but on the degree of satiation or fatigue of group members. The marked constancy of feeding time and patterns throughout the study, relative to resting and feeding, emphasised the fundamental significance of this behaviour.

ELLEFSON [1967] indicated the activity of 'foraging' to be an important element of gibbon daily behaviour. It was in this way that group members moved in loose association between principal food sources, consuming small quantities of a variety of foods as they passed. White-handed gibbons on average spent 2.5 h feeding and 5.5 h foraging daily (80% DAT); less than half foraging time was spent actually plucking or ingesting food, the remainder being spent moving, resting and looking. This behaviour was not applicable to siamang as a maintenance activity. Although siamang group progressions involved food searching and occasional brief sampling of certain foods, in general the group moved together from one food source to the next. It is shown that they sit and watch each other and the forest for a substantial part of travel bouts but picking food was not a noticeable component.

Dawn-to-dusk observation of primate groups is essential, not only for plotting ranging, but also for constructing daily patterns for each activity. Because of the many environmental variables, which produce considerable differences in behaviour from one day to the next, the incomplete sampling of behaviour will always be unsatisfactory, especially for elucidating the characteristic patterns of each age and sex class. The continuous observation of one individual for two or more consecutive days assumes special importance in a study of social organisation (Chapter 5).

The continuous observation of one individual from dawn to dusk, with behaviour sampled at 1-min intervals at least, provided a precise description of the daily behaviour of that individual. But unless observations were sustained for several days, the problem of day-to-day fluctuations still remained. To obtain this detailed picture of group behaviour at a given time, it was not possible to spend more than two days on each individual of the family group. Thus in some monthly observation periods it appeared that the male spent as much time feeding as the female, simply because the two days of male observation coincided with a time of high group activity, followed by the almost inevitable low level while the female was observed. This was one of the more easily detectable variables which was, in fact, overcome by averaging over the whole study.

To make a full assessment of group behaviour in the short term, the sampling of the behaviour of all group members at 5- or 10-min intervals throughout the day provided a most effective compromise between the 1-min individual sample and less precise methods. Attention could then be focused on social interactions as and when they occurred. This method has the greatest potential for small social groups; in larger aggregations it is necessary to develop observations on specific individuals, or failing that, on specific age and sex classes – with less emphasis on interpreting short-term behaviour patterns. The sampling of behaviour at 10-min intervals proved adequate for investigating spatial relations (Chapter 5), and movements between different levels of the forest canopy.

Attention has been drawn to the analytic advantages of investigating behaviour patterns in terms of bouts, but the significance of each behaviour studied, especially of its daily pattern, was more clearly appreciated when measured as time distributed over the day. Because bout duration stayed relatively constant for each activity, interest centred on the timing of, and interval between bouts.

There were problems in defining the limits of 'activity' each day. These were resolved by referring to the duration of DAT, which was the time interval between first movement and reaching the next night sleeping position, as the alert period. This avoids confusion with specific social and maintenance activities. But although siamang rapidly became 'alert' as day dawned, it was not clear at what point their continual vigilance – even while resting – diminished in the evening. It can clearly not be strictly equated with reaching the night position on most days, because of the grooming that sometimes occurs subsequently.

There was evidence of shifts in position during the night, especially when the moon was full. Because of this, which could have brought the male and female together, and the absence of sexual behaviour in the first 10 months of observation, some observations were made in the hours of darkness. Nothing was observed other than relatively minor adjustments in position up to 2 h after dusk. On most nights the juvenile was sleeping about 1 m from the male at dusk (plate XVI), but they were always closely huddled together at dawn (the coldest hour of the day). Thus alertness declined more gradually at the end of the day, than its rapid increase at dawn. It was this difference in alertness that distinguished daytime resting from night tree sleeping. But there is still a need to investigate nocturnal behaviour more closely.

Plate XIV. Locomotion: subadult and juvenile males (TS1) walk bipedally, and adult male (GS16B) brachiates (upper left to right). Communication: adult female (TS1) silently watching siamang above (lower left), and subadult male (RS2) calling noisily – note half-inflated throat sac, and protruded rounded lips (lower right).

Plate XV. Group travel. Adult female (RS2) brachiates through fig tree followed closely by adult male, with infant male behind, walking bipedally – note change in locomotion as they climb up to the branch above.

Plate XVI. Day and night. Adult male (RS2) rests in fig tree, while adult female feeds below, and the infant moves the 10 m from female to male (upper left – illustrating daytime spacing). Adult male (TS1) grooms juvenile male in sleeping tree, *Koompassia excelsa;* the adult female and the infant are at the top centre, and the subadult male out of view to the left, on the west side of the tree (upper right). The adult male and juvenile eventually separate to sleep (lower right – note sleeping postures and position on branches). During the day the male carries the infant; here he walks bipedally along a primary branch of *Dipterocarpus grandiflorus* (corrugated leaves) – note arms extended for balance and flexed legs (lower left).

Returning finally to weather, it was shown that in most months the feeding peak was in mid-morning, and most resting in the afternoon. This was particularly striking in the dry months of September, February, and April. In wet months the peaks were reversed, as in November, or bimodal, as in August, December, and in Ulu Sempam, Thus, as confirmed by observations from days of contrasting weather, overcast conditions and light rain tended to suppress activity in the early part of the day, until such times as intrinsic factors superceded extrinsic ones. The difficulty in demonstrating precise correlations indicates the extent to which siamang – compared with howling monkeys, for example [Chivers, 1969] – are able to organise their activities into a characteristic pattern relatively independent of such environmental variables.

Summary

Siamang activity started as day dawned – about 20 min before sunrise. The time of leaving the night sleeping tree and commencing the day range was more variable, and not correlated with sunrise, although on average it was just after sunrise. Entering the next night tree about 2 h before sunset (and reaching the night position within 10 min) occured on average a fixed period after sunrise, rather than before sunset, and was also variable. TS1 were active for 10.5 h, RS2 for 10 h, which was 88 and 84%, respectively, of daylight hours (sunrise to sunset). There was a significant increase in daily activity time in the dry season at Kuala Lompat to 92% of daylight hours – just over 11 h.

Both groups were usually in the emergent trees early and late in the day, descending to the low level before or after noon, and otherwise spending most time in the main or middle level of the canopy; RS2 spent more time in the fewer high trees, especially in the middle of the day. A somewhat different pattern emerged on analysing the frequency of entry into each level, rather than the level at 10-min intervals. On average there were 10.7 changes in level rather evenly spread through the day. RS2 did not start high as often as TS1, but both groups entered the middle level twice as often as the other two. But peaks of entry to each level tended to coincide around 0900 and 1100 h at Kuala Lompat, and 0700, 1100, and 1400 h in Ulu Sempam. There was greater difference between months for TS1 than between areas. Thus the daily pattern of movement in a vertical direction appeared to differ from that in the horizontal plane as measured by travel time and distance moved (see below).

It was feasible to analyse daily patterns of maintenance activities for the siamang group in terms of bouts, because in 80% observations all group members were engaged in the same activity within 10 min of it having been initiated (nearly 60% within 5 min); this synchrony increased in the dry season. Both groups had about 12 bouts each day (of which half were feeding), each lasting for about 50 min. There were fewer rest bouts, more shorter feeding bouts, and more longer travel bouts in the dry season.

But the frequency of bouts was more variable than their duration. The three to four rest bouts and five to six feed bouts tended to be around 0600, 0900, 1300 and 1500 h, and the three travel bouts at 0700, 1000 and 1400 h. But the daily patterns of feeding and travel durations in each hour were correlated positively, and of feeding and resting negatively. This was more clearly demonstrated in terms of time, rather than bouts. On average feeding was recommenced after 57 min, and rest and travel after just over 100 min, from the end of the previous bout. Such periodicity as was initially present became increasingly indistinct as the day passed; it therefore seemed that no powerful rhythms were inherent in siamang behaviour.

The daily patterns of maintenance activities were more clearly appreciated by an analysis of the aggregate duration of each, derived in a variety of ways. The peaks of feeding and travel occurred early in the day (70 – 85% of minutes in each hour between 0700 and 1100 h declining during the afternoon to less than 60%) to be replaced by a resting peak. But although there was seasonal variation in resting and travel time (the latter was doubled to 120 min, 18% in the dry season), in both study areas the siamang group consistently fed for just under 60% daily activity time. One day of increased ranging was often followed by two to three days with an increased alert period and more intensive feeding. Thus, although there was considerable day-to-day variation, these fluctuations were effectively counteracted within a 10-day period, proportionately if not absolutely.

In both groups the male and infant spent the most time resting and the least time feeding (by about 60 min on average); the subadult male, and female fed most, with the juvenile scores intermediate (if present). Individual travel performance was more variable within much smaller limits, because of the way the group kept close together.

It was difficult to correlate behaviour patterns with weather, but consistent similarities between wet season months and cloudy, wet days, and between dry season months and hot, sunny days led to the following conclusions. Siamang were less active in bad weather, with a shorter day (by 50 min on average), and fewer longer bouts of each maintenance activity, with the peak of feeding in the afternoon and increased resting in the morning. There were nearly 30 min less feeding and travel each day on average; rest and feed bouts were 10 min longer, and travel bouts 5 min shorter. The different activity patterns of TS1 and RS2 apparently represented responses to different weather conditions, because in the one month when sunshine patterns were correlated, so were the daily patterns of feeding and resting. RS2 generally exhibited behaviour approaching the bad weather activity patterns.

But behaviour patterns were not profoundly influenced by weather. While having an effect under more extreme conditions, the daily pattern of intense morning activity – in which feeding predominated – appeared to be moulded mainly by intrinsic and other extrinsic factors.

Chapter 5: Social Organisation

Introduction

Few primate species, other than man, are monogamous. The main exceptions are species of Indriidae [PETTER, 1965], *Callicebus moloch* [MASON 1968], and hylobatids [CARPENTER, 1940; ELLEFSON, 1967; KAWABE, 1970]. MASON describes the characteristics shared by these species as no clear dominance, marked social tolerance, and the continual maintenance of social proximity. These stable social groups occupy an almost exclusive area of habitat, and have rich vocal repertoires that play an important part in the maintenance of these territories. It appears that many of these traits are part of a single pattern of adaptation. Because of the considerable phyletic separation of the species involved, this must represent evolutionary convergence resulting from similar selective pressures or environmental opportunities [MASON, 1968].

The territorial behaviour of siamang was referred to in Chapter 2, and the pattern of loud calls between groups will be described in Chapter 6. The organisation of the family group will be investigated here, with the emphasis on spatial relations and the frequency with which each individual initiates various social activities (and the sequence in which the others join in). This will illustrate the degree and nature of cohesion of the siamang family group, in terms of synchronous activity and the distance between individuals. There will be descriptions of grooming, play, sexual, and aggressive behaviour. Although mainly concerned with intra-group behaviour, reference is made to relations between groups, mainly in connection with group formation; otherwise this aspect of siamang behaviour will be described in Chapter 6.

Spatial Relations

It is likely that the pattern of spacing between individuals of the family group – throughout the day and during different activities – will reflect important aspects of social organisation [KUMMER, 1968]. It is of great interest to know how the kind and frequency of social interaction varies with the distance maintained between individuals, so that the significance of the spacing data can be better appreciated. During four months in the wet and dry seasons at Kuala Lompat, as an adjunct to the detailed observations of the behaviour of each group member for two consecutive days, the distance of each other individual was estimated at 10-min intervals [CHIVERS, 1971 b]. The number in the same tree as the subject, and the activity of each individual was also recorded. In Ulu Sempam it was only possible to collect these data for several days in each month in relation to the adult male; similar data were collected for TS1 in April 1970 to assist comparison.

Observations were made from 0600 to 1750 h inclusive in October, December, January, March (and April) at Kuala Lompat. This represents 40 days and 72 observations each day; in only 16% of observations were TS1 in their night sleeping positions. The information collected was analysed in three main ways – in terms of group cohesion (synchrony of activity), group spread (the daily pattern of distance between all individuals), and inter-individual distance (according to age and sex), from dawn to dusk and during each maintenance activity. Temporal changes in these aspects of social relations were investigated. Thus the following were ascertained: (1) the proportion of observations in which the whole group and all but one were engaged in the same activity, and how this synchrony varied with time of day; (2) the proportion of time the whole group and all but one were observed in the same tree, and how this close proximity varied with time of day; (3) the mean distance between all individuals at different times of day, and overall, in each month (computed from ten days of observation, during which each individual, including the infant, was the subject for two days); (4) the proportion of observations in which each individual was in the same tree as the subject; (5) the distance of each individual from the subject throughout the alert period, and when in night positions, and (6) the mean distance from the subject of each individual during each maintenance activity – rest, feed, and travel.

Group Cohesion

Synchrony in siamang group behaviour has already been indicated by showing the consistently short time interval between the first and last group member to change maintenance activities (Chapter 4). It can be demonstrated

Table 5.1. Seasonal and individual variation in proportion of daytime all group members were engaged in the same activity

Month	Same activity				Different activity,		
	all		all but one		two or more		
	$\bar{x}$	R	$\bar{x}$	R	$\bar{x}$	R	n
October	76	65–89	11	3–17	13		10
December	74	64–83	13	4–24	13		10
January	71	61–80	15	6–19	14		10
March	70	56–80	17	11–25	13		10
Total	73		14		13		40
Individual							
Male	75	65–89	10	4–19	15		8
Female	71	61–80	14	8–25	15		8
Subadult	73	64–80	15	7–24	12		8
Juvenile	71	62–82	15	3–22	14		8
Infant	72	65–80	15	8–19	13		8

by time sampling that all group members were engaged in the same activity in the same tree for a high proportion of daily observations. This shows the high degree of cohesion within the family group, in terms of activity and space. It also confirms the validity of applying a concept of bouts of activity to the siamang group as a whole, as well as to the individual.

During 40 days of systematic observations spread over four months, the whole group was engaged in the same activity for 73% of the time. This shows a remarkable synchrony considering that there were five individuals (table 5.1), and that in only 14% of observations did one siamang differ in its activity; thus, more than one differed in 13%. Since the average time between first and last to start a new bout of feeding or travel was about 8 min, this 27% of the 10-min samples probably included the transition from one group activity to another on many occasions.

There was a remarkable consistency from one month to the next, and between individuals. There was also little variation in the daily pattern between months (fig. 5.1). Synchrony between individual activity patterns decreased steadily from dawn, when it was complete for more than 80% of observations, until the early afternoon when all animals were engaged in the same activity for only about 60% of observations (fig. 5.2); this pattern was more marked in the dry season.

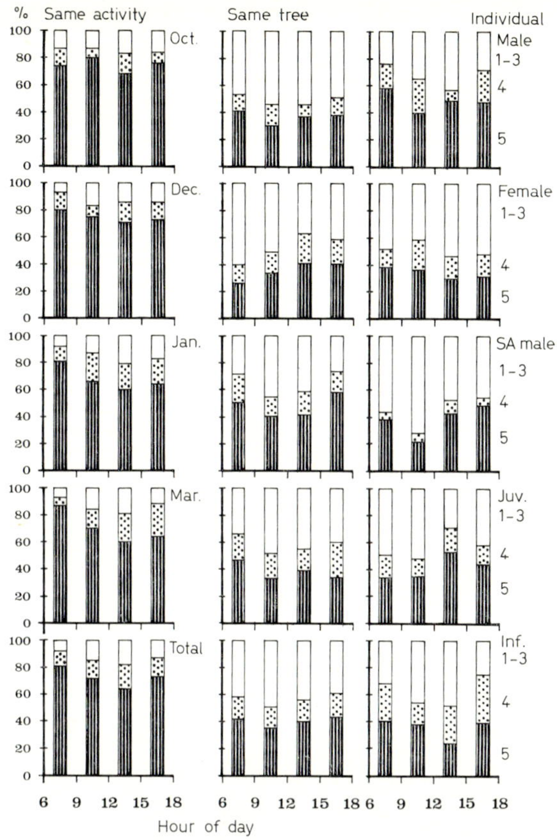

a, b

Fig. 5.1. Daily pattern (3-hourly) of proportion of 10-min observations for whole group (5), all but one (4) and three or less of siamang group TS1 in (a) the same activity in each month, and (b) the same tree in each month and in relation to each individual.

On average the whole group was in the same tree for only 40% of time (fig. 5.2). Two or more were in different trees for 43% of time. If one considers that the group was travelling for nearly 15% of time (and not far separated although in different trees), and that the climbers on which the group often fed were distributed over more than one tree, these scores do not seriously detract from the high degree of spatial cohesion postulated. It was mentioned in Chapter 2, in connection with night sleeping positions, that individuals were often closer in the terminal branches of adjacent trees, than if they were in the same tree. It will be shown that the actual distance between individuals was never very large.

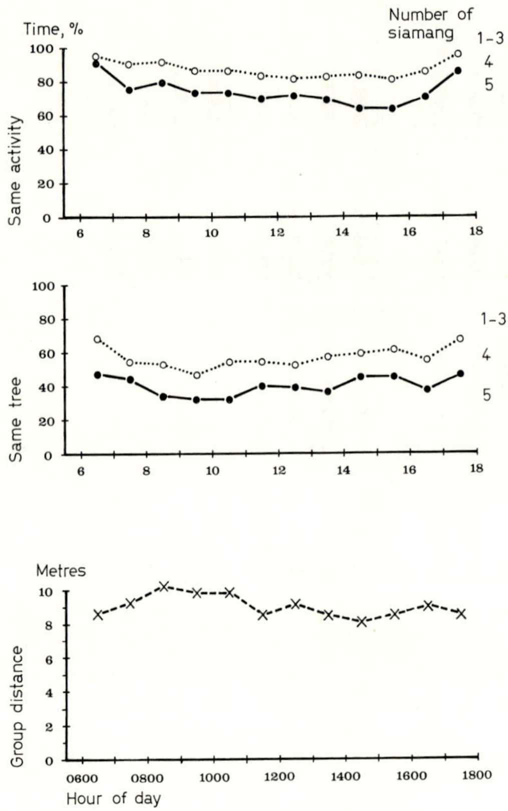

Fig. 5.2. Daily pattern (hourly) of proportion of time that the whole group, all but one, and three or less of TS1 were in the same activity, and the same tree during four months at Kuala Lompat, and also the mean distance between individuals in each hour of the day.

Table 5.2. Seasonal and individual variation in proportion of daytime all group members are in the same tree

Month	Same tree				Different tree,	
	all		all but one		two or more	
	$\bar{x}$	R	$\bar{x}$	R	$\bar{x}$	n
October	37	21 – 61	14	4 – 26	49	10
December	36	21 – 51	17	11 – 26	47	10
January	48	18 – 65	17	7 – 54	25	10
March	28	28 – 49	21	1 – 37	51	10
Total	40		17		43	
Individual						
Male	49	38 – 65	19	12 – 28	32	8
Female	34	21 – 43	17	4 – 37	49	8
Subadult	38	21 – 51	8	1 – 17	54	8
Juvenile	42	28 – 61	15	10 – 24	43	8
Infant	36	18 – 60	26	11 – 54	38	8

The group was more often in the same tree in the wet season than in the dry season, but the January scores were unusually high. The daily range of observations was much larger than for 'same activity'. By summing the scores for the two consecutive days of observation of each individual in each month, the overall pattern was obtained (fig. 5.2). It was the adult male who most often (49%) had all other group members in the same tree with him; the female and infant had the lowest scores (table 5.2). But it was the subadult male who most often had only one, two, or no others with him; he was either with all the group or few of them.

The significance of these patterns will become clearer as the chapter develops. If the group was together in the same tree, it made little difference which was the subject; thus the difference between individuals was mainly due to daily variations in group behaviour accentuated by the small sample. Again, analyses of the actual distances between individuals will help to resolve this problem.

The daily pattern of congregation in the same tree was more variable between months than was synchrony of activity. The group was most widely dispersed in mid-morning (fig. 5.1, 3-hourly pattern; fig. 5.2, overall hourly pattern). For monthly variations in the daily patterns of this and other parameters, and the significance of differences see CHIVERS [1972b], Appendices X and XI.

Table 5.3. Monthly variation in group spread (in metres), and variation in spread of group from each individual in TS 1 siamang group at Kuala Lompat

Month	Behaviour									
	total		night		rest		feed		travel	
	$\bar{x}$	rank	$\bar{x}$	rank	$\bar{x}$	rank	$\bar{x}$	rank	$\bar{x}$	rank
October	10.3	4	6.9	3	5.3	4	7.1	3	9.6	4
December	8.6	2	8.4	4	4.3	1	7.6	4	8.3	2
January	8.4	1	5.6	2	4.3	1	5.9	1	7.9	1
March	8.8	3	5.1	1	4.4	3	6.3	2	8.7	3
Individual										
Male	8.4	1	8.3	3	6.3	3	7.6	2	9.8	2
Female	8.7	3	8.9	4	4.1	1	7.4	1	11.7	4
Subadult	10.7	5	12.8	5	7.2	5	9.9	5	12.2	5
Juvenile	8.9	4	6.4	2	5.3	2	8.1	4	11.3	3
Infant	8.6	2	5.9	1	6.3	3	7.8	3	9.7	1

Group Spread

Having considered the cohesion of a siamang family group in terms of the extent to which individuals do the same activity in the same place, attention is now turned to the dispersion of the group in terms of the distances between individuals. One measure of the spread of the group is the mean distance between individuals throughout a number of days in which each animal has been the subject for the same proportion of time, i.e. for the same number of days. The mean group spread throughout the day declined from over 10 m in October to 8–9 m in subsequent months, being lowest at the turn of the year – end of the wet season (table 5.3). Also shown are the spread of the group in night positions, and during resting, feeding, and travel, with scores ranked, so that the months of closest group proximity for each category can be more readily discerned. Thus the group slept closer at night in March than in other months, whereas January was the month of closest proximity for maintenance activities.

The 'group' was closest to the male overall, but to the female during resting and feeding, and to the infant at night, and during travel. The subadult male was the most separated from the other individuals (table 5.3). Group spread was maximal in the morning, approaching 12 m on average around 0900 h in some months; it declined to as little as 6 m in mid-afternoon.

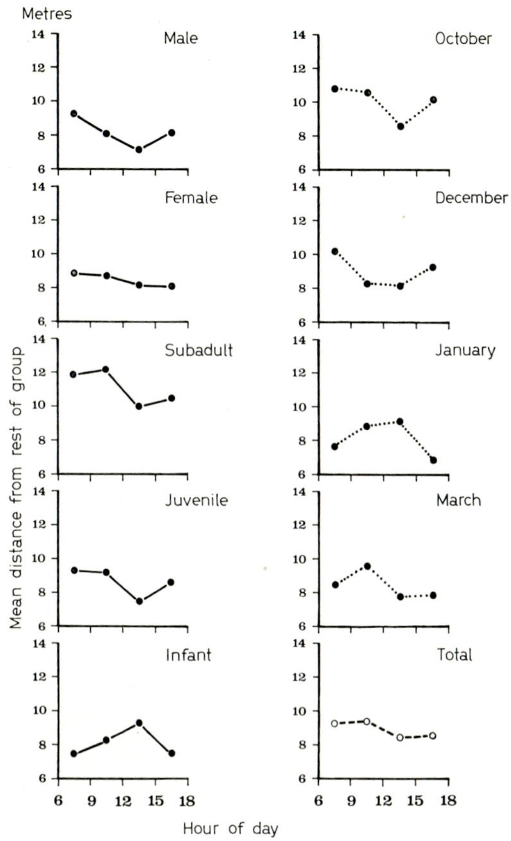

Fig. 5.3. Daily pattern (3-hourly) of mean group spread in each month, and the mean distance of all others from each individual of TS1 in the wet and dry seasons at Kuala Lompat.

The monthly (and individual) patterns were not significantly concorded because of an increase in spread in the late afternoon in October, a decrease in late morning in December, and an increase in the early afternoon in January (fig. 5.3). The individual patterns are equally variable with the spread lowest in early afternoon when resting was most frequent, and greatest in mid-morning when feeding and travel were maximal. The daily pattern for the infant was the converse.

Thus the daily pattern of group spread tended to approximate that of synchrony of activity, but with greater amplitude and more fluctuations (hence not correlated). It reciprocated the daily pattern of all animals being in the same tree with a significant negative correlation in most months (Spearman rank correlation coefficient). The daily pattern of group spread was negatively correlated with feeding in December, and with resting in March, and was positively correlated with travel in March (cf. Chapter 4). Thus there was a strong tendency for siamang to be closer during feeding and resting, and more dispersed during travel.

Table 5.4. Overall average distance in metres between individual siamang at Kuala Lompat in the wet and dry seasons during 40 days

Subject	Object						Mean distance of other siamang from subject, m
	M	F	SA	J	I	n	
Day							
Male		9.1	10.2	8.9	5.4	489	8.4
Female	8.0		10.2	9.0	7.6	495	8.7
Subadult	10.9	10.8		10.5	10.7	518	10.7
Juvenile	8.4	8.9	10.1		8.1	515	8.9
Infant	5.7	7.8	12.4	8.6		476	8.6
Total							8.9
Night							
Male		10.0	12.9	0.4	10.0	87	8.3
Female	11.1		13.6	11.0	0	75	8.9
Subadult	12.1	13.0		13.0	13.2	58	12.8
Juvenile	0.5	7.8	9.9		7.5	61	6.4
Infant	7.0	0	10.0	6.8		90	5.9
Total							8.3
							Mean
Proportion of time in same tree							
Male		72	61	72	85		72
Female	61		52	58	65		59
Subadult	53	54		48	50		51
Juvenile	67	65	54		65		63
Infant	85	75	53	69			70

Inter-Individual Distance

The analysis of distances between each pair of individuals in the siamang family group gives more precise information than group spread. In terms of proportion of time in the same tree, the male was closest to the infant (85%) and furthest from the subadult male (53 or 61%). The subadult was only in the same tree as each other individual about half the time; for other individuals scores were usually around two thirds (table 5.4). That reciprocal pairs do not equal each other could be because the sample was not large enough (5×8 days) to account for the considerable day-to-day variation in ranging behaviour and activity patterns.

Table 5.5. Mean distances between individual siamang at Kuala Lompat during the wet and dry seasons during resting, feeding and travelling

Subject	Object						Mean distance of other siamang from subject, m
	M	F	SA	J	I	n	
Rest							
Male		7.0	7.7	6.5	3.8	73	6.3
Female	3.3		4.7	5.2	3.2	54	4.1
Subadult	7.3	5.8		7.6	8.3	58	7.2
Juvenile	5.3	5.0	5.6		5.5	68	5.3
Infant	3.6	7.2	8.3	7.7		58	6.3
Total							5.9
Feed							
Male		7.5	9.8	8.1	4.9	219	7.6
Female	6.5		8.1	8.4	6.5	201	7.4
Subadult	10.4	9.2		10.3	9.7	231	9.9
Juvenile	7.5	8.1	9.7		7.1	210	8.1
Infant	6.7	6.2	10.8	7.6		183	7.8
Total							8.2
Travel							
Male		11.6	11.8	10.5	5.3	54	9.8
Female	12.1		13.6	10.0	11.1	54	11.7
Subadult	12.0	13.7		11.5	11.4	57	12.2
Juvenile	10.6	11.3	13.1		10.1	60	11.3
Infant	4.0	10.6	12.4	10.1		62	9.7
Total							10.8

Plate XVII. TS1 rest together in the early afternoon in a *Koompassia* (above) – male scratches, subadult, infant, juvenile, and female (from left to right). Group rest and groom on branches of a fig strangling a *Dipterocarpus* – juvenile self-grooms on left, subadult grooms male in centre, and female grooms hanging infant (middle). Later the male grooms the juvenile, the female grooms the subadult, and the infant plays (below).

Individual siamang were usually within 10 m of each other during the alert period, although it was not unusual for them to be side by side during rest bouts, and 15–25 m apart during travel. It was rare for any individual to be more than 30 m from the others in the usual course of group life. As has been indicated it was the subadult who was the greatest distance from the others, averaging about 11 m from each by day, and 13 m by night (table 5.4). The infant slept with the female at night, and the juvenile with the male, but during the day the infant was closer to the male (plate XVI).

Individuals were twice as close during resting than when travelling (plate XVII, cf. plate XV), with intermediate scores for feeding (table 5.5). The adults were closest in food trees, with the infant nearby and the older young at a greater distance. The subadult was markedly closer to all other individuals during rest bouts, and not far displaced during travel. The changes in spatial relations from one month to the next in overall daily distance, during feeding, resting, and travel, in night positions, and in the proportions of time in the same tree are set out in CHIVERS [1972b], Appendix X, 2–7.

Spacing of RS2 in Ulu Sempam

It was only possible to obtain a complete daily record (10-min sample) of spatial relations of RS2 on 16 days spread over February, March, and April 1970. because observation conditions were more difficult and the siamang were more shy in Ulu Sempam. Thus attention was focused on one animal for all but three days; this was the adult male, who appeared to occupy a central position in group activity (plate XVI). For direct comparison with the spacing behaviour of TS1 similar data were collected on 6 days in April 1970.

As at Kuala Lompat activity was less well synchronised in the early afternoon in Ulu Sempam, but at a lower level overall. But there was little difference in the scores for all but one, which reflects the greater independence of the RS2 subadult male (fig. 5.4). In contrast there was a higher frequency of all being in the same tree in Ulu Sempam, especially for all but the subadult male. This is confirmed by the lower mean distances of the female and infant from the male. But the daily pattern of a lower frequency of all, and all but one, in the same tree in mid-morning was identical with that at Kuala Lompat. Thus group spread was greatest during the morning.

The male infant in Ulu Sempam was more dependent on the male than the TS1 female infant, although the RS2 subadult male was more independent of the rest of the group than his TS1 counterpart. In both areas the subadult

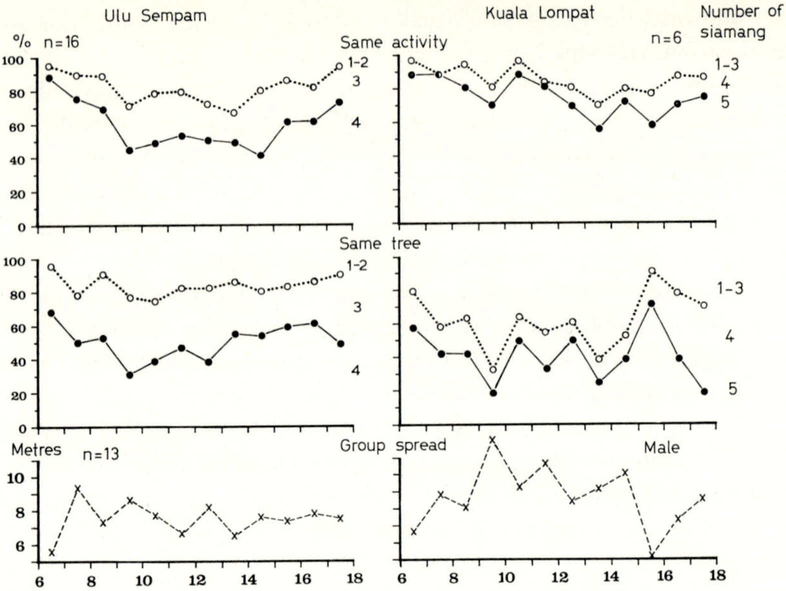

Fig. 5.4. Daily patterns of proportion of time that the whole group (RS2 = 4, TS1 = 5), all but one, and 3 or less (TS1), 2 or less (RS2) were in the same activity, and the same tree, and also the mean distance of rest of group from male in both study areas in 1970.

male and female were further from the male in mid-morning (especially) and mid-afternoon. At these times the infant was closer to the male than at other times of day (fig. 5.5). There was a tendency in Ulu Sempam for the female to be close to the male when the subadult was more distant, and vice versa; this is only partly supported by the data. At Kuala Lompat the morning dispersion was marked, and the afternoon scatter was followed by a clustering of the group. The amplitude of the female cycle was not as great as that of the subadult, reflecting her tendency to remain closer to the male. This differential phasic separation from the male probably reflects the differences in time spent in each maintenance activity, especially feeding (Chapter 4).

Spacing and Weather

During the 46 days at Kuala Lompat when spacing data were collected there were only seven really hot and sunny days, and five cloudy and wet days. The daily patterns of rest and feed (fig. 5.6) were highly correlated, but those of frequency of same activity and position, and group spread, were not

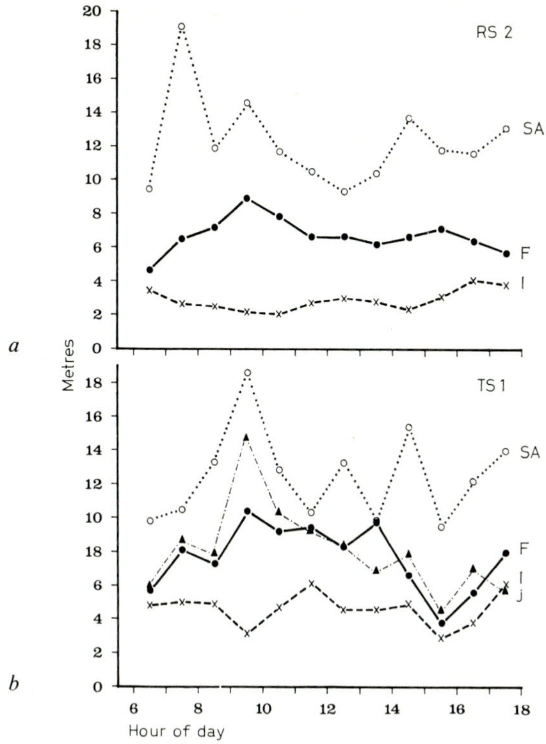

Fig. 5.5 Daily pattern (hourly) of the mean distance of each group member from the adult male in (a) RS2 during February, March, and April in Ulu Sempam, and (b) TS1 in April and May 1970.

(Spearman rank correlation coefficient). Although there were no significant differences in the overall scores for synchrony of activity (CHIVERS [1972b], Appendix XI) activity was less well synchronised in mid-morning, as well as mid-afternoon, on cloudy and wet days.

On hot and sunny days there was a marked decrease in the frequency of individuals being in the same activity and position around midday; there was a significantly lower frequency for all animals being in the same tree on hot and sunny days (fig. 5.6). The group was more dispersed after noon on hot and sunny days, and less dispersed on cloudy and wet days with peaks in the early morning, at noon, and in mid-afternoon, i.e. when feeding. The decrease in synchrony on hot and sunny days clearly associated with a slower transfer of the group from resting to feeding (fig. 5.6). There was more feeding and travel on hot and sunny days.

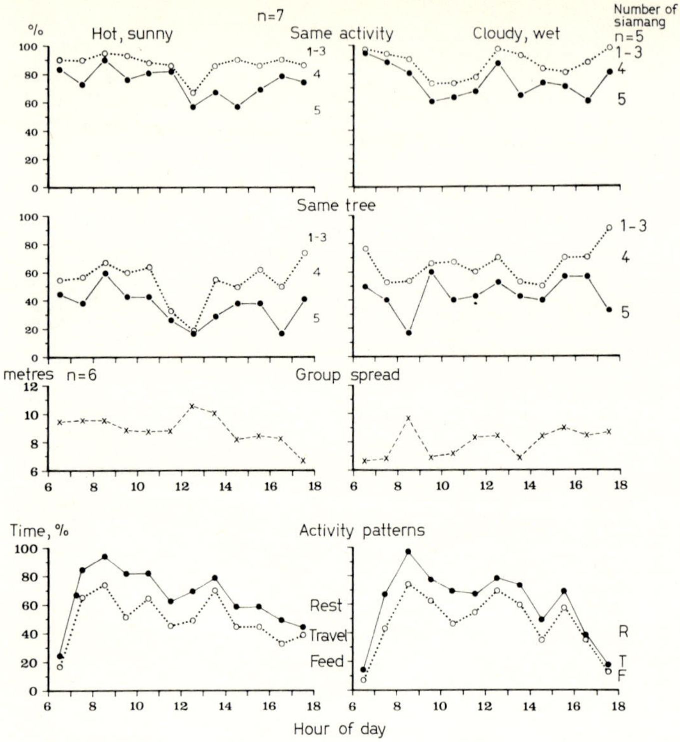

Fig. 5.6. Differences in daily patterns (hourly) of synchronised activity, congregation in the same tree, group spread, and maintenance activities on hot and sunny days and cloudy and wet days at Kuala Lompat.

Interim Discussion

It is useful at this point to discuss those aspects of spatial relations that might have a special bearing on the social organisation of the siamang family group. The high degree of cohesion was demonstrated by all animals being in the same tree for nearly 50%, and in the same activity for 75% of the alert period at an average separation of less than 10 m. Each pair of animals was in the same tree for more than half the day. Although sleeping with the female at night, the infant was nearer to the male during the day. The subadult male was spatially the most peripheral member of the group.

The adult pair in the white-handed gibbon group is recognised as co-dominant [Carpenter, 1940; Ellefson, 1967], so that the various responsibilities of group leadership and organisation are shared between them. Spacing

Month	Group spread	Average distance from male, m			
		group	female	SA male	infant
Kuala Lompat, TS1					
October 1969	10.3	11.3	11.3	12.4	3.9
December 1969	8.6	7.9	9.1	9.8	8.1
January 1970	8.4	7.2	8.2	8.7	6.7
March 1970	8.8	7.4	5.9	11.4	5.9
April 1970		9.1	8.0	14.0	4.5
Ulu Sempam, RS2					
February 1970		9.2	9.8	19.2	1.7
March 1970		7.6	6.9	12.1	3.0
April 1970		7.0	8.7	11.6	3.1

a

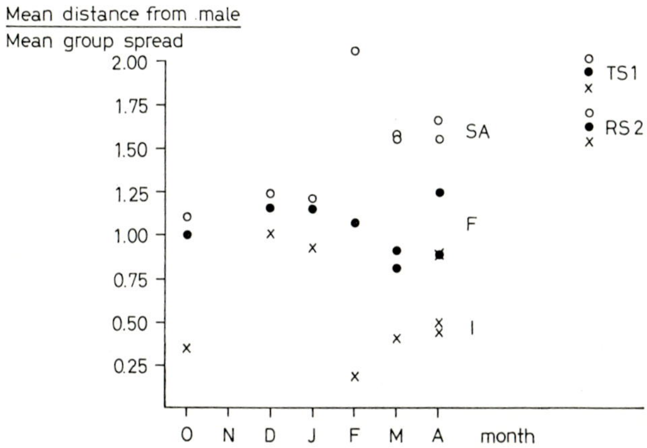

b

Fig. 5.7. Changing relations within the siamang family group: (a) mean group spread, and mean distances from the adult male of the rest of the group, the adult female, the subadult male, and the infant in those months in which spatial relations were investigated at Kuala Lompat (TS1) and in Ulu Sempam (RS2); and (b) monthly plots of the mean distance of each group member from the adult male as a proportion of the mean group spread.

within the siamang group suggested a similar pattern, with individuals being closer to the male for some activities, and to the female for others. The impression was formed that while the adult male was the focus for many group activities, it was the female who led the group about its day range. The subsequent sections of this chapter will attempt to resolve these problems, and to determine the social position of each individual, especially the subadult male, within the group.

The other major feature of social organisation to be considered is the changes that were recorded through the seasons, and from one year to the next. To what extent is there a behavioural cycle based on reproductive behaviour, in addition to the shorter annual cycle of environmental seasonality [CHIVERS, 1971b]? The spacing data showed an increased cohesion of the group in the wet season. But in the following dispersion in the dry season, the subadult male became more peripheral than expected, and in both areas the distance between male and female decreased (fig. 5.7). The significance of these behaviours will be investigated below, when it is pertinent to inquire which individual was responsible for increasing, maintaining, or decreasing distance during interactions between each pair. But first there will be a description of group progressions, which will include the sequence of individuals starting and ceasing daily activities.

Group Progressions

Start and Finish of Daily Activities

Although the times of first movement, leaving and entering night trees, and reaching night position were accurately determined for the group on all days of study (Chapter 4), the sequence and timing for each individual in each parameter were only accurately measured on 90 days for starting activity, and on 113 days for ceasing activity (61 and 76% of 148 days, respectively). The infant is usually omitted at this stage of the argument because of its low level of independence during most of the study. It can usually be regarded as effectively part of the male during day time; in this way the relations between fully independent individuals can be more clearly ascertained.

The adult male was usually the first to move in the night sleeping tree each morning, but on 26% days it was the subadult male; the female or subadult male were usually last (table 5.6). It was more often the female who left the night tree first, thereby initiating the first progression of the day; the juvenile was second, the male third and the subadult male last (within 15 min on average, see Chapter 4). They were more consistent in their pattern of leaving the night tree than of first movement.

Towards the end of the day the female was usually the first to enter the next night tree, the male was usually second (or third) and the subadult male last. The juvenile was second, third, or last on 92% days; his tendency to be third was more marked for reaching the night sleeping position. Although the female was first into the tree on 50% days, she was second, third, or last to her sleeping position on 61% days (table 5.6).

Table 5.6. Sequence of starting and finishing daily activity at Kuala Lompat

Parameter	Number of days	Order	Individual, % days			
			male	female	subadult	juvenile
First movement	90	1st	46	16	26	13
		2nd	38	18	7	38
		3rd	12	31	24	32
		4th	4	36	43	17
Leave night tree	90	1st	20	39	24	17
		2nd	19	28	10	43
		3rd	40	20	11	29
		4th	21	13	54	11
Enter night tree	113	1st	18	50	25	8
		2nd	41	26	4	29
		3rd	35	12	16	37
		4th	7	12	55	26
Reach night position	113	1st	28	38	27	6
		2nd	46	22	4	28
		3rd	19	19	17	44
		4th	6	20	52	21

Travel Order

The siamang group usually moved through the forest in single file along the same arboreal pathway (plate XV), so that the first and last were within 5 min or 45 m of each other (see Chapter 4 and above, respectively). This illustrates the slow speed of group progressions. It will be shown that the order in which the group travels from one food source to the next is of special interest in elucidating patterns of social organisation [KUMMER, 1968]. The position of each animal was even more consistent than for starting and ceasing activity. Because of the variable timing of group progressions, and difficult observational conditions, data were collected when conditions permitted. Most sequences recorded represent the initial order of progression, although additional records were made on some occasions when the group was still travelling after 15 min.

Observations were made on 1,257 occasions at Kuala Lompat, fairly evenly spread over the seasons, and 249 times in Ulu Sempam (table 5.7). At Kuala Lompat the female was leading in 65% of travel bouts, with the juvenile second (or third), the male third (or second or last), and the subadult male usually last. In Ulu Sempam, in the dry season, the pattern of female first, subadult male last, and male in the middle was even more distinctive, but early in the study the more usual order was male, subadult, and female. This discrepancy can be explained by the female becoming habituated to observation more rapidly; thus the male leadership with the subadult close behind represented flight from the observers (cf. KAWABE, 1970]. The wet season scores indicate that the subadult was the last to become fully habituated.

The tendency for each individual to occupy a certain position in the group

Table 5.7. Individual performance during the progressions of siamang family groups RS2 in Ulu Sempam, and TS1 at Kuala Lompat

Group and individual	(1) Travel order					(2) Travel sequence, followed						
	1	2	3	4	n	M	F	SA	J	total %	n	
RS2	June – September 1969											Follows
Male	48	28	24				33	67		20		
Female	44	12	44			23		77		29		
SA male	8	60	32			70	30			51		
					25	42	22	36			45	
	February – April 1970											
Male	23	62	15				75	25		38		
Female	57	24	19			72		28		21		
Subadult	20	15	66			72	28			41		
					172	44	40	15			304	
TS1	June – September 1969											Follows
Male	11	21	35	33			37	23	40	30		
Female	59	24	13	3		23		33	44	12		
SA male	17	18	23	42		44	29		27	27		
Juvenile	13	36	29	22		28	45	27		31		
					402	23	33	19	25			
	February – May 1970											
Male	4	39	40	17			58	12	30	33		
Female	71	19	8	2		9		38	53	9		
SA male	14	11	14	61		47	10		43	29		
Juvenile	11	32	38	19		47	36	17		30		
					464	28	33	12	26		1,296	

(1) represents percentage of observations for each individual in each position, and (2) represents percentage of total travel pairs in which each individual follows each other, and the percentage of following observations in which each group member is followed (and vice versa).

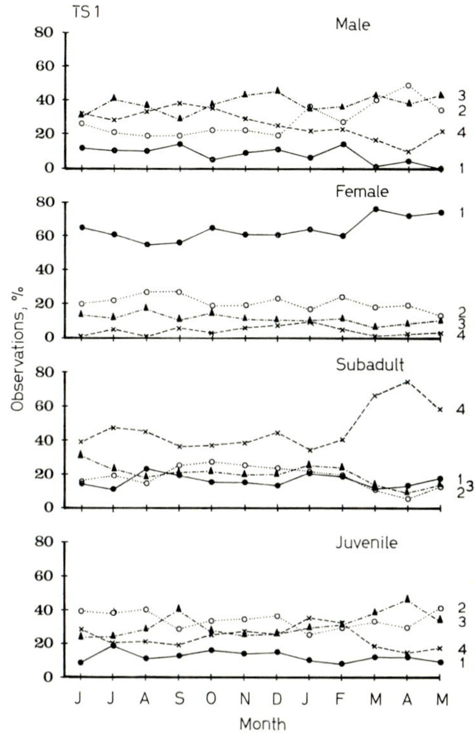

Fig. 5.8. Monthly variation in travel order in a siamang family group, TS1. Proportion of observations in which each independent group member was recorded in each position in group progressions.

travel order (fig. 5.8) was highly significant in all cases (χ^2 test). The adult male tended to travel nearer the front in the dry season, when the subadult was particularly conspicuous as the last in the order. Both the male and juvenile were more variable in their performance in the wet season, but the female was completely consistent throughout. The suggestion that the adult male siamang usually leads group progressions [KAWABE, 1970; KOYAMA, 1971] is not supported by the evidence presented here.

Travel Sequence

The distinctive order of travel observed in both siamang family groups can represent (1) individual preference and/or (2) attraction or repulsion between individuals. Although the problems are too complex to elucidate fully here, some useful indications can be given. The tendency for the female to initiate group progressions can largely be explained by intrinsic motivation. There were no indications that the behaviour of other individuals prompted her departure from a food or rest tree, but it is possible that her movements represented avoidance of the others (see Discussion). On the other hand, although an intrinsic component may be involved in the subadult male's behaviour, it will be shown that the aggressive behaviour of the adult male towards the end of the study was an important factor in his tendency to travel in the rear of group progressions. This suggests that his tendency to follow the male or

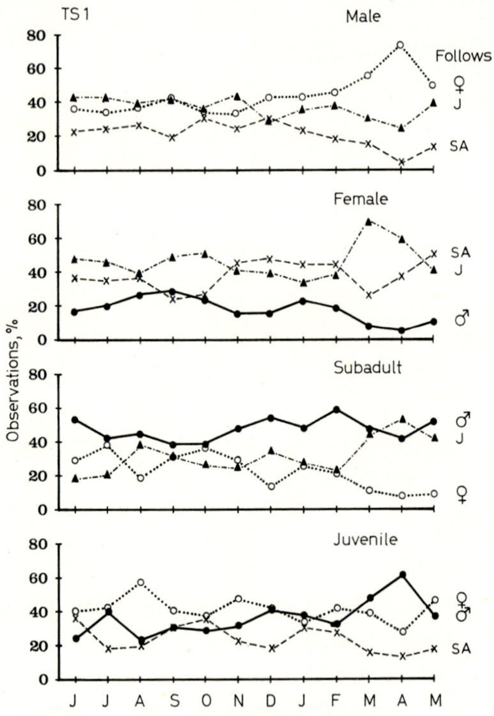

Fig. 5.9. Monthly variation in travel sequence in a siamang family group, TS1. Proportion of observations in which each independent group member was recorded following each other in group progressions.

juvenile did not indicate any special affinity to these individuals (table 5.7).

Similarly, the juvenile appeared to be showing a preference to travel near the front of the group, irrespective of who was leading. He followed the female and the male significantly more than the subadult (χ^2 test). The tendency of the male to follow the female was accentuated in the dry season (fig. 5.9; plate XV). When the female was not leading she tended to follow the younger animals that were, and not the adult male; in fact, she showed less tendency to follow him in the 1970 dry season than at other times of year. The juvenile followed the male more than the female towards the end of the study (a significant change from the intermediate season), for which there is as yet no clear explanation. In the wet season the juvenile was not following any individual in particular; neither was the adult male. The subadult male followed the juvenile rather than the male in the dry season, probably because the juvenile was travelling farther back in the order than usual.

The performance of each individual was highly consistent throughout the months of the study (fig. 5.9). In Ulu Sempam the female followed the male rather than the subadult, the male followed the female (more often), and the subadult male followed the adult male, once fully habituated (table 5.7). But it seems that the tendency of an individual to travel in a particular part of the group progression was usually stronger than a tendency to follow a particular individual.

Infant Behaviour

Having established the patterns of behaviour of independent animals, the behaviour of infants in group progressions can now be investigated. This assumes special significance as during the course of the study in both areas the infants were slowly gaining independence with some dramatic changes occurring. At the beginning of the study, when observations were few, the infant was apparently carried all the time by the adult female. In March 1969 the TS1 male was seen to take the infant (estimated to be eight to ten months old) from the female twice when she was mildly alarmed. Thereafter, he carried the infant throughout the day, except when it moved on its own (plates XIV and XV). This transfer, observed at the same time in Ulu Sempam, and the subsequent gradual gaining of independence by the infant with lower frequency of carriage by the male are clearly shown in figure 5.10. Once the transfer was complete, the adult female did not avoid the immediate proximity of the male and infant as she had done before July.

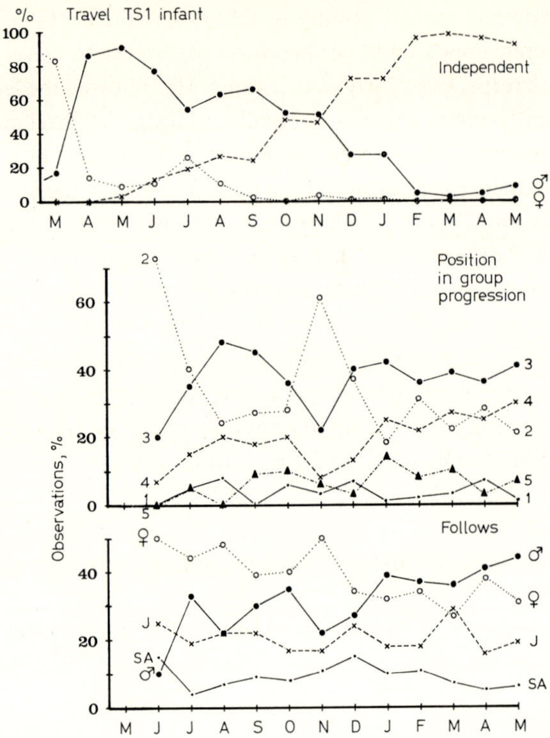

Fig. 5.10. Monthly variation in infant behaviour during group progressions, TS1. Proportion of observations: (1) carried by adult female, (2) carried by adult male, and (3) independent; (4) in each position in travel order, and (5) following each other group member.

Thus the proportion of observed travel bouts when the infant was carried by the male declined from 78% during the 1969 dry season at Kuala Lompat to 4% one year later, while independent travel increased from 2 to 96% (table 5.8). By the beginning of the third year of life the infant only required assistance for the more difficult crossings between trees. Although the transfer occurred at the same time in Ulu Sempam the infant male was much slower in gaining locomotor independence, being carried by the male for 76% of the time in the 1970 dry season. He was independent for 38% of observations in April 1970; this delay may have been caused initially by the slower habituation and the increased wariness of the group, and also by the greater problems of moving between trees in hilly terrain.

When moving independently the infant female at Kuala Lompat travelled

Table 5.8. Seasonal variation in infant siamang's position during group travel

Season	n	Carried by		Independent, % observations									
		male, %	female, %	order						sequence			
				total	1st	2nd	3rd	4th	5th	M	F	SA	J
TS1													
Dry													
Feb. – May 1969	46	78	20	2	0	100	0	0	0	0	100	0	0
Intermediate													
June – Sept. 1969	413	66	14	20	4	38	39	16	4	25	45	8	22
Wet													
Oct. – Jan. 1969	368	39	1	60	4	32	37	18	8	32	37	11	19
Dry													
Feb. – May 1970	490	4	0	96	4	25	38	26	7	39	32	7	21
Total (June – May)													
2nd year of life	1,271	35	5	61	4	29	38	23	7	38	38	8	21
RS2													
Dry													
Jan. – May 1969	7	71	19	0									
Intermediate													
June – Sept. 1969	25	88	4	8	0	50	50	0		50	0	50	
Wet													
Oct. – Jan. 1969	52	94	2	4	0	50	0	50		50	50	0	
Dry													
Feb. – April 1970	179	76	1	23	7	38	48	7		59	31	10	
Total (May – April)													
2nd year of life	256	81	1	18	7	39	46	9		58	30	12	

near the centre of the group, only being first or last on 11% occasions (table 5.8). In June and November she was closely following the female near the front (fig. 5.10). But overall the infant's position each month was highly concordant and her tendency to be third was significant; the infant was second or third on more than two thirds of occasions in the intermediate and wet seasons, with a higher frequency for fourth position (26%) in the 1970 dry season (table 5.8).

It appeared that, although gaining independence from the male, the infant increasingly followed the male rather than the female as time passed (fig. 5.10), but no significant change could be detected (except when comparing the seasonal scores for frequency of following the male). The infant followed the male and female on more than one third observations, the juvenile on about one fifth, and the subadult on less than one tenth. There was a high degree of concordance between months, especially in relation to the latter two individuals.

The infant was usually carried out of the night tree by the adult female, having passed the night with her, but the frequency declined from 78 to 30% days through the seasons of the study (table 5.9). Initially the infant was transferred to the male during or after the first group progression of the day; subsequently, the infant more often waited for, or moved towards, the adult male in the night sleeping tree. Detailed information was only obtained on 86 days.

Table 5.9. Infant behaviour while the siamang group at Kuala Lompat (TS 1) started and finished daily activities

Season	Percent of days with male	with female	Ratio M:F	Number of days
Infant entered night tree				
Intermediate	22	78	0.3	36
Wet	70	30	2.3	20
Dry	70	30	2.3	30
Overall	50	50	1.0	86
Infant entered night tree				
Intermediate	83	17	4.7	40
Wet	74	26	2.8	23
Dry	57	43	1.3	46
Overall	70	30	2.3	109
Percent of days on which infant reached night position (with female)				
1st	2nd	3rd	4th	5th
3	21	24	25	27

The infant entered the next night tree on or with the adult male on 70% of the 109 days when information was available. The increasing frequency of entry with the female as the study passed (up to 43% in the 1970 dry season) indicated the increasing tendency to follow the female rather than the male later in the day (table 5.9); the female never carried the infant into the night tree at this time. Unlike the male, the female invariably expected the infant to follow if she moved; from March 1969 she rarely assisted the infant at any time of day.

Although rarely first to reach her night sleeping position, the infant showed no particular timing in reaching the female relative to the others (table 5.9). It appeared to depend on the distance at which she was deposited from the female by the male, that is the distance between the adults' night positions. For except in the early months of the study, and quite often in Ulu Sempam, the adult male made no effort to take the infant to the female. The infant would make its way round the terminal branches to the female; for, unlike the others, it was unable to negotiate the larger branches, which afforded a more direct route.

Grooming Behaviour

Grooming was the most frequent behaviour observed during the relatively few social interactions involving contact between individuals. Although siamang scratched most parts of their body thoroughly, especially before a change in activity, self-grooming was rare. At Kuala Lompat it occurred for only 2% of 5,387 min of grooming time, as 8% of 1,124 bouts, during 148 days of observation; in Ulu Sempam for 2.5% of 656 min and 6% of 175 grooming bouts. Grooming was distinguished from scratching by the slower more carefully controlled movements of usually both hands in examining the skin and hair roots, with the occasional removal of foreign bodies; these were usually eaten. Thus while siamang generally removed sources of irritation by scratching, the more precise cleansing action of grooming occurred in a social context during rest bouts.

It was possible to record almost all occurrences of grooming within the group during the study period, but detailed information on grooming bouts was primarily collected in relation to the subject of detailed observation (Chapter 4). But because grooming only occupied about one eighth of the alert period, and individuals tended to congregate together, a full analysis was by no means restricted to the subject and his interactions (plate XVII). A grooming bout refers to one pair of individuals; on the rare occasions when tripartite relations were formed they were treated as two pairs (the third potential pair not usually interacting).

A grooming bout commenced when one animal approached another and remained close to it for a certain time interval, during which grooming occurred. The starting times of some bouts had to be calculated in retrospect, because grooming did not always commence when the individuals came together; grooming was not necessarily reciprocated. There were distinct, but sometimes subtle, movements which brought two siamang into close association, even among others in the kind of huddle that was characteristic

Plate XVIII. Social grooming. Adult male (RS2) grooms the side of the subadult male's head (upper left), then the subadult grooms the male's axilla, and finally the male grooms the subadult's genital region. The male and female rest together with the infant between them (lower left); then the male grooms the infant and elicits an open-mouth threat from the female (lower right).

of some group resting bouts. These changes involved both movement in space, and the quality of that movement in adopting certain postures.

Once (or as) the association was initiated one of the pair would 'present' for grooming; this involved facing away from the other siamang in a sitting or lying posture, with the spine curved concavely, and some part close to the other. Observation conditions did not permit a detailed analysis of parts groomed, but grooming usually started around the lower trunk, working up to the less accessible parts of the head and limbs in particular (plate XVIII). The groomer would eventually turn away and 'present'; he might resume grooming if this evoked no response. Grooming was terminated when individuals ceased 'presenting', or when these presentations brought no further response. The grooming bout was terminated when the two animals moved apart – however subtly.

Thus there were two main measures of a grooming bout – the time the two individuals were together, and the time for which grooming actually occurred. These are referred to as pair association time and grooming time respectively, and were determined on the basis of a 1-min sample of behaviour. The former is specifically associated with grooming (see above) – such relations were almost never developed without grooming occurring at least briefly; the latter represents cumulative time, rather than consecutive minutes, especially with reference to individual behaviour.

In this section the following aspects of grooming will be investigated: (1) the timing, frequency and duration of grooming bouts; (2) monthly and seasonal variation in pair association time (PAT) and grooming time, and their relationship with resting time, and (3) individual participation in grooming according to age and sex in terms of (a) initiation of pair association and grooming, (b) number of bouts, (c) daily pattern of involvement, and (d) number of minutes of grooming and being groomed.

The emphasis will be on the behaviour of TS1 at Kuala Lompat, although some useful comparisons can be made with RS2 in the 1970 dry season.

Group Behaviour

At Kuala Lompat grooming bouts occupied 15% of daily activity time (the alert period) – an average of about 97 min/day. Grooming time and PAT were cumulative in the sense that more than one pair of animals might be grooming at any one time; also, an individual might be involved in more than one bout at any time, but this was rare. There were 7.5 bouts/day, each

Table 5.10. Comparison of daily grooming behaviour at Kuala Lompat (TS1) and Ulu Sempam (RS2)

	Kuala Lompat, TS1				Ulu Sempam,
	Apr. – Aug. 1969	Sept. – Dec. 1969	Jan. – May 1970	total	Feb. – April 1970
Bouts/day	7.1	8.0	7.6	7.5	6.0
Min/day = PAT	91	131	77	97	71
$\frac{\text{PAT}}{\text{DAT}}$, %	15	21	12	15	12
$\frac{\text{Groom time}}{\text{PAT}}$, %	43	38	45	42	32
Bout duration, min	12.8	16.4	10.1	12.9	11.5
Grooming bouts, %					
0600 – 0900	8	14	7	8	30
0900 – 1200	26	20	11	19	33
1200 – 1500	47	42	41	40	24
1500 – 1800	30	24	42	33	14
Pair association time, %					
0600 – 0900	6	10	3	7	34
0900 – 1200	24	17	9	17	33
1200 – 1500	38	46	50	43	19
1500 – 1800	32	26	39	32	11

lasting about 13 min on average; individuals were actually grooming for 42% PAT (this does not include the 10% of PAT in which no grooming details were obtained). In Ulu Sempam there were only 6 bouts/day lasting 11.5 min; grooming occupied 12% of the alert period – i.e. 71 min/day on average – and only 32% PAT (table 5.10). These results are similar to those for TS1 at the same time.

The most consistent parameter through the months of study was the proportion of PAT during which grooming actually occurred (fig. 5.11). Otherwise, the frequency and duration of bouts – and, therefore, of PAT as a proportion of the alert period – tended to increase in the wet season. There was a highly significant monthly variation in the number of grooming bouts (χ^2 test, 7.5 bouts-day expected, $\chi^2 = 52.00$, d.f. = 13, p = 0.001 –), with marked increases in October and December, and a decrease in April 1969. There was no significant difference between the three 'seasons' (n = 1,101, $\chi^2 = 2.46$, d.f. = 2, p = 0.30 – 0.20). The scores in Ulu Sempam approximate those at Kuala Lompat more closely than they did for maintenance activity patterns (fig. 5.11, see also Chapter 4).

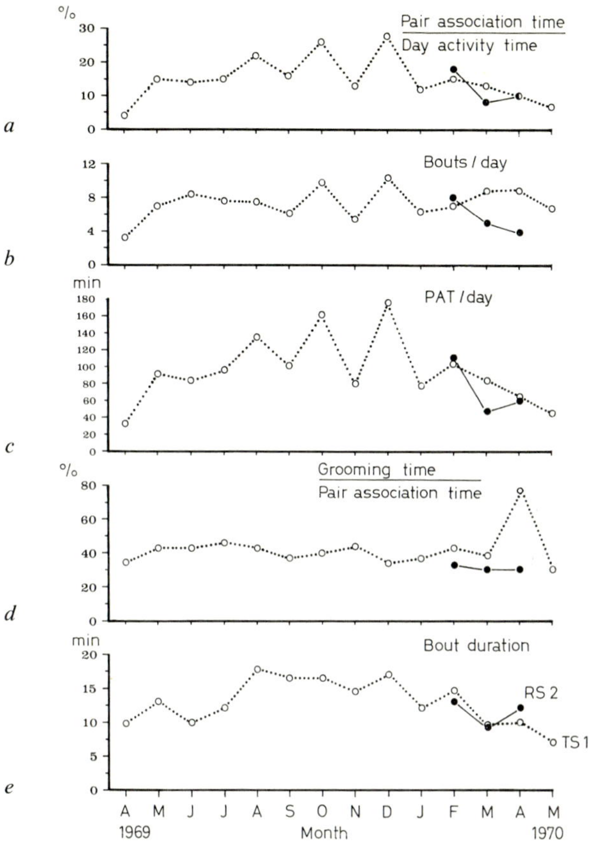

Fig. 5.11. Monthly variation in grooming behaviour: (a) pair association time (PAT) as a percentage of alert period; (b) number of bouts/day; (c) PAT/day; (d) grooming time as a percentage of PAT, and (e) mean bout duration for TS1 (and RS2).

The most striking differences between the two siamang groups were in the daily patterns of grooming behaviour (table 5.10). In Ulu Sempam nearly two thirds of grooming bouts and PAT were before noon, whereas that proportion at Kuala Lompat occurred after noon (comparing RS2 with TS1 dry season and TS1 overall, $\chi^2 = 117.35$ and 95.63, respectively). At Kuala Lompat there was a secondary grooming peak in mid-morning (fig. 5.12), which was most pronounced in the early months of the study. Grooming was more concentrated in the afternoon in the dry season.

The highly significant concordance of daily patterns of grooming time between 10-day observation periods was least marked in the wet season;

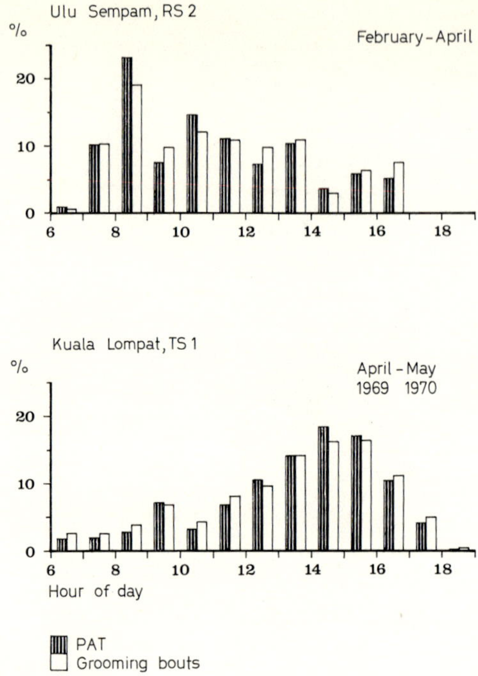

Fig. 5.12. Daily patterns (hourly) of grooming in Ulu Sempam in the 1970 dry season, and at Kuala Lompat from April 1969 to May 1970 inclusive, in terms of PAT and frequency of grooming bouts as a proportion of daily totals.

there was also significant concordance between seasons (W = 0.86 and 0.87 for daily patterns of grooming time and bouts, respectively). Overall, more than one third of PAT and grooming bouts occurred from 1400 to 1600 h at Kuala Lompat (fig. 5.12). By contrast, in Ulu Sempam one fifth of bouts, and nearly one quarter of PAT were between 0800 and 0900 h, with more than one third from 1000 to 1400 h. But this unusual pattern was mainly produced by the February behaviour, which has already been suggested to be atypical (Chapters 2 and 3).

Reference to the daily patterns of maintenance activities in each month in Ulu Sempam (Appendix D) shows that at the peak hour of grooming in February there was also a marked increase in resting following the first feed. Thus in February resting was fairly evenly divided between morning and afternoon, but with twice as much PAT in the morning. In April, resting was

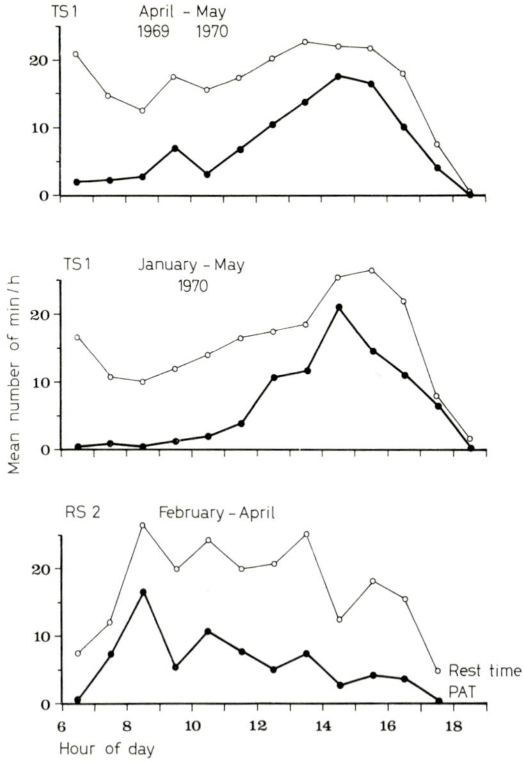

Fig. 5.13. Daily patterns of resting and PAT: April 1969 to May 1970, and January to May 1970 inclusive for TS1 at Kuala Lompat, and the 1970 dry season for RS2 in Ulu Sempam.

still well spread, and grooming was more dispersed through the day, although PAT was halved. But in each month at least 1 h showed a high intensity of grooming activity – 1000, 0700, and 1600 h in February, March, and April, respectively.

The daily patterns of resting and PAT were highly correlated for both RS2 and TS1 in the 1970 dry season (r_S = 0.82 and 0.80, respectively). But comparing the daily patterns of PAT and grooming time in these samples produced no significant correlation (r_S = –0.30). Although PAT was clearly correlated with resting time through the day (fig. 5.13), PAT was far more concentrated in a particular part of the day in both study areas. This is more clearly demonstrated by expressing PAT as a proportion of resting time for each hour of the day for RS2 and TS1 in the dry season, and for TS1 overall (fig. 5.14).

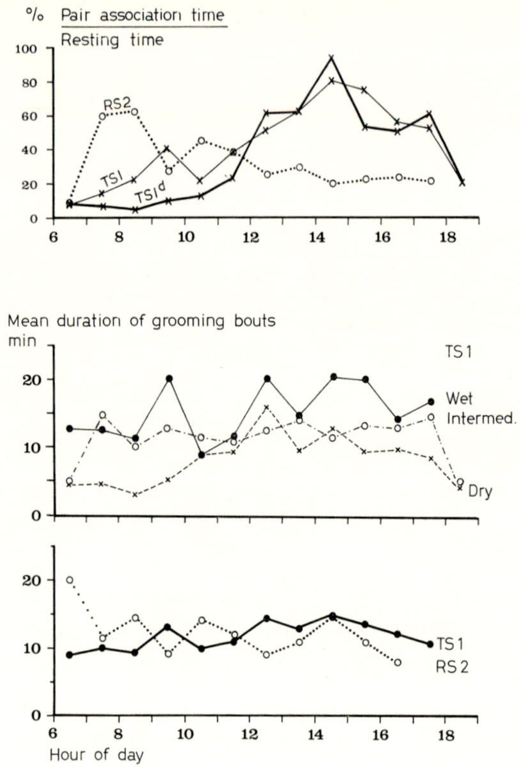

Fig. 5.14. Grooming behaviour: daily pattern of PAT as a percentage of resting time from April 1969 to May 1970, and from January to May 1970 at Kuala Lompat, and also in the 1970 dry season in Ulu Sempam; seasonal variation in the mean duration of grooming bouts in each hour of the day, and mean duration of grooming bouts in each hour of the day for TS1 overall, and for RS2 in 1970 dry season.

The daily patterns of (1) resting, (2) association, and (3) resting time as a proportion of PAT in each month of the study at Kuala Lompat are illustrated in Appendix D. Although there was little variation overall in the duration of grooming bouts at different times of the day, a seasonal breakdown indicates a tendency for grooming bouts to be longer in those hours when they were most frequent (fig. 5.14).

Individual Behaviour

The behaviour of individuals during grooming bouts varied markedly according to the age of the participants. Bouts were clearly distinguished as involving mature animals only (including subadults), or mature and young, or, rarely, immature animals only. The pairings to be considered are: (1) male and female; (2) male and subadult (male); (3) female and subadult; (4 – 9) each mature animal with juvenile and infant, and (10) juvenile and infant. Pairings involving juveniles apply only to Kuala Lompat. Discontinuous observations in Ulu Gombak indicated that subadult participation in grooming does not vary according to sex. There follows a description of the initiation, timing, and performance of each individual in these two types of grooming bout.

1. Initiation of Pair Association and Grooming

In pairings involving mature animals only, one individual approached the other to initiate pair association as often as vice versa. The main exception was the subadult male at Kuala Lompat, who approached the male three times more often. And in Ulu Sempam the female approached the male nearly twice as often as he approached her (table 5.11). The commonest association was between male and subadult, the least frequent between female and subadult (especially in Ulu Sempam). At Kuala Lompat the only variations between seasons were that in the dry season, on fewer occasions, the frequency of subadult approaches to female was markedly increased. The most dramatic change was the switch from an intermediate season pattern of female approaches to the male to a dry season pattern of male approaches to female, with an increased frequency of interaction.

At Kuala Lompat the adult male usually initiated grooming, but in Ulu Sempam the subadult usually started grooming the male; there was no marked difference in female and subadult interactions (table 5.11). Although the male initiated pair association in the dry season, he still started grooming; on the few occasions when female and subadult groomed in the dry season, it was the female rather than the subadult, which initiated grooming. Thus there was a weak tendency overall for the approached animal to start grooming.

The pattern was more distinctive when immature animals were involved. It was almost invariably the young animal which approached the older one, and the mature animal which started grooming (table 5.11). For example, although the juvenile made 89% of the 157 approaches which brought the adult male and juvenile together, the male nevertheless started grooming in 90% of these grooming bouts. If the two are classified as approacher and approached, the former started grooming in 16% of bouts.

Table 5.11. Variation in grooming behaviour between siamang individuals with regard to initiation of pair association and grooming at Kuala Lompat (TS1) and in Ulu Sempam (RS2)

Pair	n	Bouts, %									
		initiates pair association				initiates grooming				initiates grooming	
										approacher	approached
TS1											
M–F	159	M	50	F	50	M	68	F	32	48	52
M–SA	209	M	24	SA	76	M	68	SA	32	44	56
F–SA	90	F	40	SA	60	F	43	SA	57	49	51
Mature	458									46	54
M–J	157	M	11	J	89	M	90	J	10	16	84
M–I	41	M	8	I	92	M	100	I	0	7	93
F–J	89	F	4	J	96	F	89	J	11	13	87
F–I	73	F	3	I	97	F	100	I	0	3	97
SA–J	45	SA	15	J	85	SA	87	J	13	27	73
SA–I	22	SA	9	I	91	SA	95	I	5	14	86
J–I	9	J	11	I	89	J	100	I	0	0	100
Mature-immature	427									13	87
RS2											
M–F	32	M	36	F	64	M	58	F	42	52	48
M–SA	36	M	46	SA	54	M	27	SA	73	39	61
F–SA	2	F	50	SA	50	F	50	SA	50	100	0
Mature	70									50	50
M–I	12	M	0	I	100	M	100	I	0	0	100
F–I	12	F	0	I	100	F	100	I	0	0	100
SA–I	5	SA	0	I	100	SA	100	I	0	0	100
Mature-immature	29		0		100		100		0	0	100
TS1 Seasonal changes											
M–F											
Int.	43	M	30	F	70	M	63	F	37	44	56
Wet	43		42		58		70		30	38	62
Dry	73		66		34		70		30	56	44
M–SA											
Int.	88	M	24	SA	76	M	66	SA	34	41	59
Wet	78		27		73		69		31	50	50
Dry	43		21		79		67		33	37	63
F–SA											
Int.	35	F	43	SA	57	F	37	SA	63	50	50
Wet	34		44		56		41		59	44	56
Dry	18		28		72		56		44	56	44

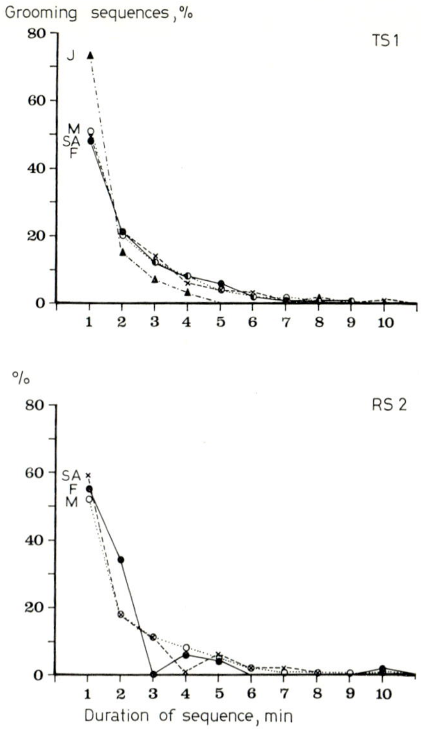

Fig. 5.15. Distribution of duration of grooming sequences by each individual of TS1 at Kuala Lompat, and RS2 in Ulu Sempam.

2. Duration and Frequency of Grooming Bouts and Sequences

80% of grooming sequences at Kuala Lompat (94% in Ulu Sempam) were terminated within 5 min of the start – 50% in less than 2 min; individuals showed almost identical patterns (fig. 5.15). Only the TS1 juvenile differed in its shorter grooming sequences. In only 13 out of 2,348 observations of TS1, and 1 out of 333 in RS2, did sequences exceed 10 min; the longest recorded was 20 min. Each sequence usually involved continuous grooming, with occasional pauses during which the groomer scratched himself and/or looked round the forest.

Overall scores for each individual showed that in terms of number of bouts, and PAT in Ulu Sempam, grooming frequency increased steadily with age (assuming the female to be younger than the male). But in terms of grooming and groomed time the males (adult and subadult) obtained the highest scores (table 5.12). Only the male markedly exceeded the expected score of 20% interactions (25% for RS2), and juveniles and infants showed little involvement. The reciprocal nature of this behaviour is demonstrated, although the adult male usually groomed more than he was groomed, and the converse was true for the RS2 female.

Table 5.12. Individual behaviour during grooming bouts and daily pattern of grooming for each siamang pair

Percent	Male	Female	SA male	Juvenile	Infant	Total, min
TS1						
Bouts	32	24	21	15	7	1,124 (83)[1]
PAT	33	25	28	11	3	12,854 (242)
Grooming	45	24	29	3	0	5,387 (126)
Groomed	27	23	27	17	6	
RS2						
Bouts	44	25	23		9	175 (11)
PAT	46	26	25		3	2,070 (16)
Grooming	52	14	34		1	656 (17)
Groomed	39	26	26		10	

Pair	Bouts, %									
	Kuala Lompat, TS1					Ulu Sempam, RS2				
	n	0600	0900	1200	1500	n	0600	0900	1200	1500
M–F	218	13	19	43	25	64	28	34	19	19
		32		68			62		38	
M–SA	261	10	24	36	30	65	31	35	31	3
		34		66			66		34	
F–SA	104	4	26	46	24	4	75	25	0	0
		30		70						
Mature	583	10	22	41	28	133	31	35	25	11
		32		69			66		36	
M–J	161	6	10	29	55					
		16		84						
M–I	40	18	20	32	30	12	17	17	42	25
		38		62			34		67	
F–J	95	3	18	49	30					
		21		79						
F–I	71	8	13	26	51	12	25	25	0	50
		21		78			50		50	
SA–J	48	4	14	50	28					
		18		78						
Mature-immature	415	7	15	36	43	24	24	21	24	31
		22		79			45		55	

1 Self-grooming (scores): expected score for each individual – TS1 = 20%; RS2 = 25%.

a b c d

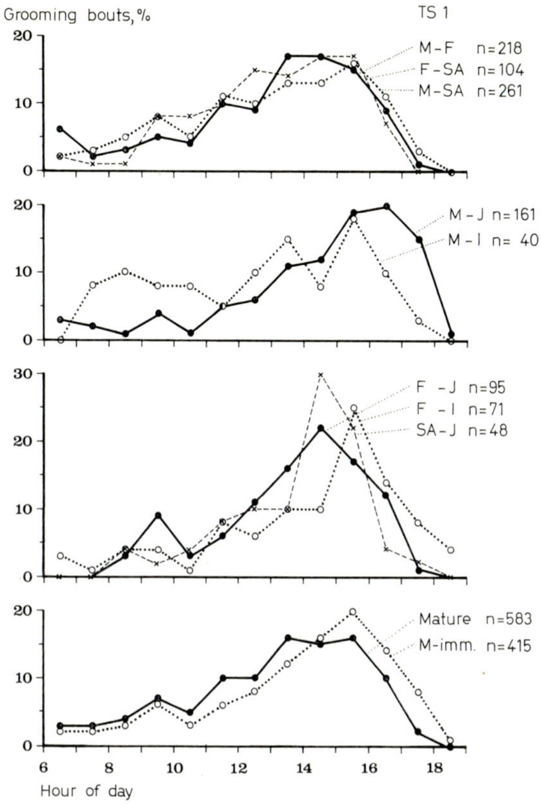

Fig. 5.16. Daily pattern of grooming for each pair of individuals in TS1 at Kuala Lompat: between (a) mature siamang; (b) male and immature; (c) female, subadult and immature, and (d) all mature and all mature-immature pairings.

3. Daily Patterns for Each Pairing

Examination of the daily pattern of each grooming pair revealed a low involvement of young animals in grooming (especially in the morning) until the late afternoon, whereas the peak of mature siamang grooming occurred in the early afternoon at Kuala Lompat (table 5.12). The pattern for young animals in Ulu Sempam was similar, although grooming was more evenly spread through the day. As has been indicated, the grooming peak for mature animals was in mid-morning. The hourly patterns for the combinations of mature animals at Kuala Lompat (fig. 5.16) were highly concordant (Kendall coefficient of concordance, $W = 0.87$, $p = 0.01 - 0.001$). While the daily patterns of grooming by male and female, and male and subadult, were

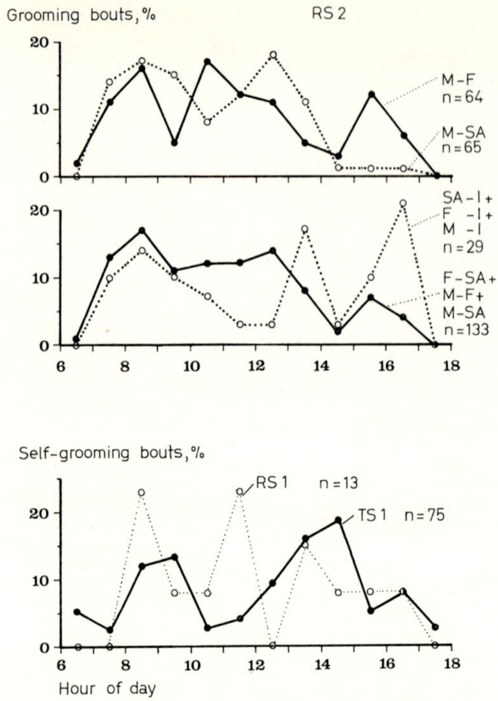

Fig. 5.17. Daily patterns of grooming between each pair of individuals of RS2 in Ulu Sempam, and self-grooming in both study areas.

highly correlated (Spearman rank correlation coefficient, $r_S = 0.82$), those of male and female, and male and juvenile, were not ($r_S = 0.53$). Neither was there a significant correlation between the patterns of male grooming juvenile and infant ($r_S = 0.30$). But although grooming between older and young animals tended to occur later in the day than that between mature animals, the daily patterns were correlated at Kuala Lompat ($r_S = 0.75$), but not in Ulu Sempam ($r_S = 0.36$). Furthermore, in RS2 (fig. 5.17) the male tended in groom with female and subadult alternately, so that hourly patterns were not closely correlated ($r_S = 0.53$).

Thus the male-female peaks in Ulu Sempam were from 0700 to 1000 and at 1200 h, and the male-subadult peak from 0700 to 0900, and at 1000 and 1500 h. The infant was also groomed during the morning rest bouts, but even more so during the afternoon. The TS1 peak of male-infant grooming preceded that of male-juvenile; the latter coincided with the two settling together for the night. Subadult-infant grooming coincided with the male-female peak (fig. 5.16) around 1400 h, which was followed by male-subadult and female-infant peaks an hour or so later. Grooming of the subadult by both adults, and male-infant and female-juvenile grooming made the major contributions to the overall morning grooming peak.

As has been indicated, the daily patterns of grooming for mature animals only, and for old and young together showed little correlation between the two groups, although the latter showed more similarities ($r_S = 0.02$ and 0.22, respectively). The infant was never seen grooming itself; self-grooming was not evenly distributed among the TS1 individuals. The female groomed herself much more than expected (40%, cf. 20%), and the juvenile much less (8% of self-grooming bouts). But although the overall scores were significantly different ($\chi^2 = 15.01$), the female did not self-groom significantly more than the male and subadult ($\chi^2 = 3.41$). The few instances of self-grooming in RS2 were fairly evenly divided between the three mature animals. The daily pattern of self-grooming approximated that of allo-grooming, but they were significantly different, in terms of hourly pattern ($r_S = 0.52$), and 3-hourly ($\chi^2 = 14.91$, d.f. $= 3$, $p = 0.01-0.001$).

4. Grooming Relationships between Each Individual

Grooming bouts between mature animals tended to be much longer on average than those involving young ones – respectively 16 and 8 min at Kuala Lompat (table 5.13), and 14 and 4 min in Ulu Sempam (table 5.14). At Kuala Lompat 58% bouts, and 72% PAT, involved mature pairs; in Ulu Sempam, where the ratio of mature to young animals was higher, 82% bouts and 94% PAT were between mature siamang. These two sets of facts alone indicate a major age difference in the significance of grooming behaviour. More light is shed on this by a detailed analysis of individual behaviour in each pairing.

At Kuala Lompat half the bouts involving mature animals only were between the adult male and subadult in the early months of the study, but by the 1970 dry season more than half of the shorter bouts were between male and female (table 5.13). Female and subadult pairings also decreased in frequency and duration towards the end of the study. But all other pairings (i.e. those involving young animals) were less variable between seasons. For pairings with high bout frequency, PAT was proportionately longer.

Final confirmation of the role of each age and sex class during grooming is provided by investigating the time for which each grooms and is groomed – as a percentage of total grooming time in the group, and of the individual's grooming or groomed time. Thus in the dry season in Ulu Sempam, the adult male groomed the adult female for 24% of group grooming time and 45% of the time in which he was actually grooming. He was groomed by her for only 8% of group grooming time, but this was 60% of the female's grooming time (table 5.14). The significance of this can be more clearly appreciated by noting that if grooming was randomly distributed through the group, expected scores would be 8.3% of total grooming time (%T), and 33% of individual grooming time (%t). Although the subadult groomed the male much more than expected and the female as expected, the male groomed both equally and almost exclusively.

Table 5.13. Distribution of bouts and PAT between siamang pairs at Kuala Lompat (TS 1) with mean bout duration

Pair	Season											
	intermediate, April – Aug.			wet, Sept. – Dec.			dry, Jan. – May			total, Apr. 69 – May 70		
	b	PA	d	b	PA	d	b	PA	d	b	PA	d
M – F	14	14.7	13.7	15	20.0	19.5	33	36.3	9.6	21	22.2	13.3
M – SA male	26	39.8	19.7	29	36.8	18.9	18	23.8	11.2	25	34.6	17.2
F – SA male	12	17.1	18.1	12	17.1	21.4	6	10.2	17.0	10	15.4	18.5
Mature-mature	53	72	17.7	57	74	19.6	57	70	10.6	58	72	15.7
M – J	19	14.4	9.7	13	11.0	13.1	14	10.8	6.7	16	12.2	8.6
M – I	4	0.4	1.2	5	1.2	3.4	4	2.1	5.2	5	1.1	3.1
F – J	11	7.3	8.5	8	9.1	17.0	7	2.9	3.7	9	6.8	9.6
F – I	4	1.3	4.2	5	2.7	7.6	11	5.2	4.3	7	2.8	5.1
SA male – J	7	3.8	7.0	3	1.9	9.4	4	3.3	8.1	5	2.9	7.8
SA male – I	1	0.8		2	0.2	1.0	3	3.2	8.5	2	1.1	6.9
J – I	trace[1]			1	0.1		1	2.1		1	0.7	9.2
Mature-immature	47	28	7.8	43	26	9.2	41	26	5.8	43	27	7.8
Sample size												
b = % bouts	355			325			371			1,033		
PA = % PAT, min	4,640			4,880			3,200			12,720		
d = average bout duration, min	13.1			15.0			8.6			12.3		

1 Term used by meteorologists to denote very small observation.

Table 5.14. Distribution of bouts, PAT, and grooming time between siamang pairs in Ulu Sempam, with mean bout duration from February to April 1970 inclusive

Pair	Bouts, %	PAT, %	Bout duration
M – F	40	44	13.9
M – SA male	40	46	14.3
F – SA male	2	4	19.3
Mature-mature	82	94	14.3
n	135	1,926	
M – I	7	2	2.6
F – I	7	4	6.9
SA male – I	3	1	3.2
Mature-immature	18	6	4.4
n	29	128	
Total	164	2,054	

Groomer	Groomed: male		female		SA male		infant		total		
	%t	%T	%t	%T	%t	%T	%t	%T	(t)	(T)	%T
Male	2	0.8	45	23.5	47	24.3	7	3.5	340		51.8
Female	60	8.4	3	0.5	1	0.2	36	5.0	92		14.0
SA male	88	29.8	4	1.2	4	1.2	10	1.5	221		33.7
Infant	0	0	100	0.5	0	0	0	0	3		0.5
%T		39.0		25.5		25.6		10.0		656	

%t = 4 (3 excluding self-grooming) pairings: expected value = 25 (33)%; %T = 16 (12 excluding self-grooming) pairings: expected value = 6.3 (8.3)%.

Similarly, in TS1 one might expect each individual to groom each other for 25% time (excluding self-grooming), and to be grooming or groomed for 5% time (4% including self-grooming). But the adult male groomed the subadult in particular, and the female, more than expected (table 5.15). The grooming relationship between male and juvenile was closer to expectation, except that, although the juvenile groomed the male much more than other individuals it was a very small proportion of grooming time. Grooming behaviour between male and female did not exceed the expected by much, and that between female and subadult, and female and juvenile were close to expectation. But, again the juvenile made a very small contribution to the association. Seasonal changes confirm the shift of activity by the male from subadult to female. The female did not fully reciprocate these attentions, and the subadult seemed reluctant to relinquish them from the way he initiated, or tried to initiate, pair association.

Table 5.15. Individual grooming time in siamang group TS1 at Kuala Lompat (totals for all seasons, April 1969 to May 1970)

	Groomed											
	male		female		SA male		juvenile		infant		total	
	%t	%T	%t	%T	%t	%T	%t	%T	%t	%T	(t), (T) min	%T
M	1	0.5	29	13.0	45	20.2	21	9.6	3	1.4	2,403	44.6
F	35	8.5	5	1.1	23	5.5	21	5.0	15	3.5	1,288	23.9
SA	58	16.8	29	8.5	2	0.6	8	2.4	3	0.1	1,553	28.8
J	56	1.4	19	0.5	14	0.4	4	0.1	5	0.1	140	2.6
I	0	0	0	0	100	0.1	0	0	0	0	3	0.1
%T		27.6		23.1		26.7		16.7		5.9	5,387	

Expected values for each cell = 4% (5% excluding self-grooming); expected values for each individual = 20% (25% excluding self-grooming).

Thus grooming mainly involved mature animals, especially adult males and subadults, and occurred later in the day between bouts of feeding and travel while the group rested. There were, however, striking differences in these daily patterns between the study areas. In contrast, grooming involving young mainly occurred as the group settled for the night (this is obscured to some extent by the variable time at which the group settled, fig. 5.17); it was not much reciprocated, being directed to the young sleeping with each adult.

Play and Maturation

The stages of maturation in the siamang have been outlined (Chapter 1), and compared with those of the white-handed gibbon [Chivers, 1972a]. The most important features are that the infant is dependent on the female for the first year of life. The second stage of infancy begins at about this time, when the young siamang is beginning to feed itself and to stray further from the female in its locomotor experiments (plate XIX). The adult female shows increasing intolerance of the presence of the infant, and the adult male rapidly takes over its daily care. Thus it is while with the male that the infant gains independence of movement (plate XIX), acquires the necessary skills for feeding, and becomes socially integrated into the group.

Plate XIX. Males and infants. TS1 infant female (about 15 months old) eating the young leaves of a climber (upper left), and being carried by the adult male during a group progression four months later (lower left). The infant male of RS2 (about 20 months old) feeding in a fig tree (upper right), and eating young climber leaves near adult male (lower right).

The siamang is unusual among primates in that all group members, except for the adult pair in most cases, are separated by at least two to three years in age. The young therefore receive less stimulus to develop such skills as are achieved through play with peers. It was apparent that while this lack was partly overcome by increased play behaviour in older animals, the more important aspect was self-play – the infant to some extent manipulating parts of the substrate in a play context. Thus, while the older animals were resting or feeding, the infant would indulge in varying degrees of acrobatics – from hanging and swinging in one place to chasing around small circular pathways – and would bite, chew, or manipulate small twigs.

But play, even in a social context, was much more sedate than in other primates (e.g. long-tailed macaques – personal filmed observations). The adult female, and sometimes the adult male, would occasionally encourage the infant in its play movements by grappling with hands or mouth. The infant showed great interest in its surroundings, perhaps through lack of peers, and the RS2 male infant was seen swinging at an infant dusky leaf monkey on one occasion, and swinging on (and biting!) the tail of a female banded leaf monkey on another.

Play developed socially during the second year of life, not with the closest animal in age, but with the subadult. The close association with the adult male brought the infant into close contact with the subadult male during rest bouts. At first the males would groom, but when the female arrived (after a longer feeding bout) the male transferred his attention to her, leaving the subadult. The infant exploited this situation by jumping on the sitting or lying older animal, either swinging from his limbs or a nearby branch. The subadult male reciprocated by grappling with limbs and mouth, occasionally with a show of teeth, and the emission of soft grunts or growls; this would last for several minutes. The juvenile became increasingly involved with the infant in such behaviour as time passed, while the older animals were grooming together.

Little play was observed during the intermediate season, but the incidence increased in the sunny months before the rains came to apparently inhibit such behaviour (fig. 5.18). In the dry season play became far more frequent and each bout lasted longer. On several occasions all three younger animals played together, while the male sat by and watched; he sometimes intervened when the infant appeared distressed. These play bouts involved little movement, but during hanging and grappling with an occasional lunge, sometimes for as long as 30 min, strange grunts and growls were frequent. By the end of the study the juvenile and infant had been involved together in nearly half the bouts observed (table 5.16).

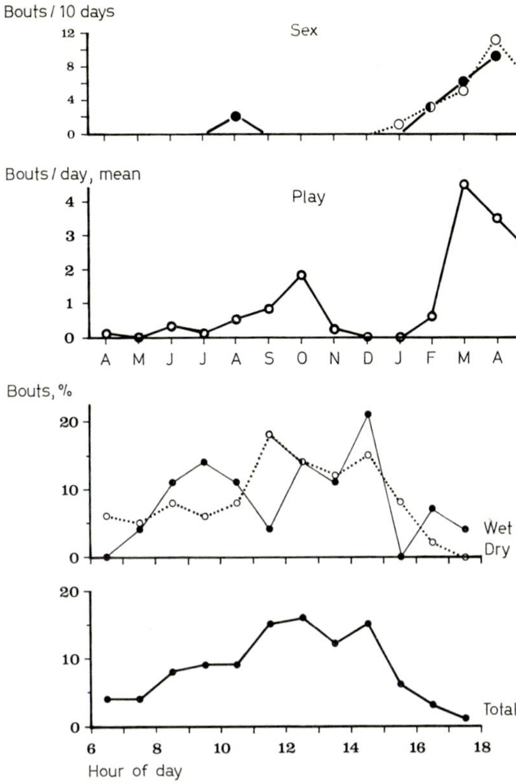

Fig. 5.18. Monthly variation in the frequency of sexual behaviour and play for TS1 at Kuala Lompat, and the daily patterns of play in the wet and dry seasons and overall at Kuala Lompat.

Table 5.16. Play behaviour at Kuala Lompat

Siamang pair	Total play bouts, %			
	intermediate season	wet season	dry season	total
M – I	20	0	4	4.2
M – J	0	0	2	1.4
F – J	0	0	1	0.7
SA – J	10	7	22	18.3
SA – I	40	43	22	27.0
J – I	30	50	50	48.6
Total, n	10	28	110	148

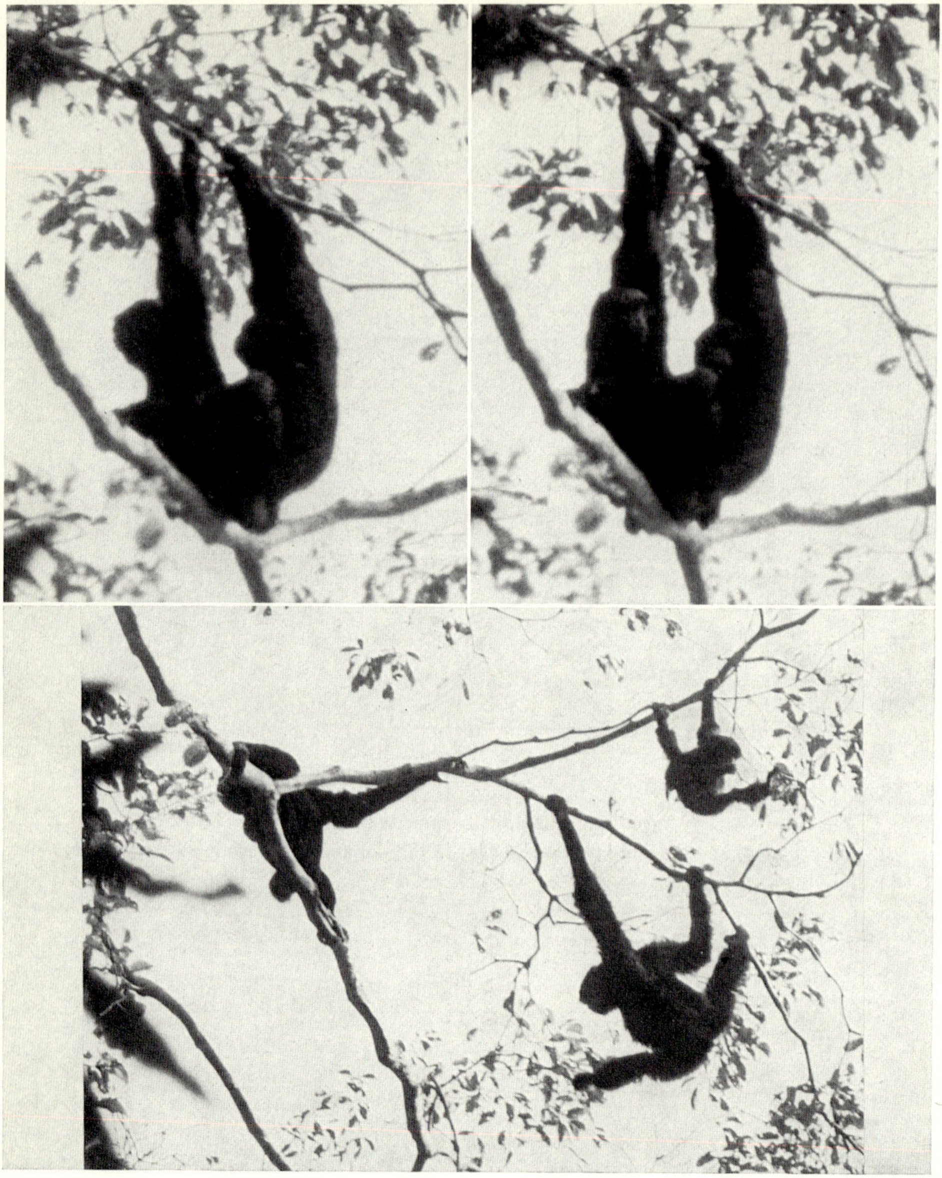

Plate XX. Sexual behaviour. Male and female (RS2) copulate dorso-ventrally in a partially suspended position, with the female occasionally turning towards the male (above). They separate (below) – the infant moves from behind them to hang feeding in the terminal branches of this fig tree, the female hangs preparatory to leaving the tree, and the male rests (note preputial tuft, diagnostic of males, but difficult to see in juvenile and infant males).

Play behaviour tended to occur during rest bouts, involving young animals while the older ones groomed, and thus the daily pattern approximated that of resting, with most play occurring in the early afternoon (fig. 5.18). But the hourly scores were not correlated with those for resting or grooming (r_S = 0.43 and 0.48, respectively). There was no significant difference between 3-hourly patterns in wet and dry seasons (χ^2 = 2.27). No quantitative data are available from Ulu Sempam where social play was less frequent.

Thus the infant became increasingly involved with other group members as it entered what was estimated to be its third year of life. As a juvenile it would receive less attention from the adults, at least during the day-time activities. The TS1 juvenile was even more peripheral than the infant in grooming bouts; it received little response from the adults to frequent presentations for grooming. But throughout the day it closely followed the leading member of the group, and was groomed by the adult male as they settled together to sleep through the night.

Sexual Behaviour

There are three reports of copulation in free-ranging hylobatids. In their studies of white-handed gibbons, CARPENTER [1940] observed two complete sequences and ELLEFSON [1967] gave a full account of sexual behaviour in one family group. From this ELLEFSON suggested that there is an average birth spacing of 2 – 2.5 years, and that mated pairs engage in sexual activities (inspection of genitalia, mounting, and copulation) for four to five months at these intervals. He saw 23 copulations on 18 different days over five months of discontinuous observation (on 32% of days), and suggested that conception occurred in the middle of this period. He indicated a clustering of copulations about an oestrus period; each averaged 30 pelvic thrusts during about 15 sec. Following these copulations, which occurred between December 1964 and March 1965, in which the male approached the crouching female from behind to mount her dorso-ventrally, an infant was born in August 1965. He observed little solicitation by the female, and that the male inspected the female's genitalia during oestrus. Only one observation of copulation in free-ranging siamang has been reported [KOYAMA, 1971].

In this study of the siamang no sexual behaviour was seen until January 1970, except for one definite, and one possible copulation between the male and female of RS2 in Ulu Sempam in August 1969. The occurrence of sexual behaviour was otherwise synchronised in both study areas, and coincided with the other changes in behaviour described – such as increased day range (Chapter 2), feeding (Chapter 3), play (see above), aggression (see below). In Ulu Sempam 18 bouts of sexual activity were observed during 29 days from February to April inclusive; at Kuala Lompat 26 bouts were observed in 49 days from January to May 1970 inclusive.

Bouts of sexual behaviour were classified as: (1) attempted copulation; (2) possible copulation, when observation conditions obscured the behaviour, and (3) completed copulation, when mounting was achieved and pelvic thrusts observed (plate XX). The former were most often observed at Kuala Lompat, when the male approached the female from behind in an odd shuffling gait, and attempted to mount her. He was thwarted in this by the female moving away – perhaps an indication that it is the female who plays the leading role in achieving a successful copulation. The adult male was observed to touch his penis and then lick that finger after some completed copulations. This is perhaps a reliable indication that ejaculation has occurred. All copulations observed were dorso-ventral, although one ventro-ventral copulation has been observed in the new group at Kuala Lompat [ALDRICH-BLAKE, personal commun.].

Sexual activity was increasing right up to the end of the study (fig. 5.18). The birth of an infant in TS1 at the end of January 1971 [HUNT, personal commun.] suggests that it was conceived in June – gestation period of 230 – 235 days or 7.8 months [HILL, 1967]. This helps to explain the female's avoidance behaviour earlier in the year. In both areas sexual bouts were evenly spread through each quarter of the day (table 5.17), although in Ulu Sempam there was a suggestion of peaks at 0800, 1100 and 1500 h, and at 0600, 1000, and 1400 h in Kuala Lompat (fig. 5.19). But these clusterings were not significant (Kolmogorov-Smirnov one-sample test, $D = 0.07$ and 0.11, $n = 20$ and 25, respectively). The RS2 male gave 89 pelvic thrusts per minute on average, or 1.5/sec. During the breeding period – as this limited time of sexual activity will be called – there was one bout of sexual activity every other day on average (table 5.17).

Gibbons have been shown to have a female reproductive cycle of about 28 – 30 days [POCOCK, 1905; MATTHEWS, 1946], and reference has been made to the changes in colour and turgidity of the female genitalia [CARPENTER, 1941]. But it is not clear whether the reddening and eversion occurred at menstruation or ovulation. In free-ranging siamang a comparable change was observed involving a colour change from black to white to red to white and then black again. The method practised here of comparing the behaviour of

Table 5.17. Summary of sexual behaviour in Ulu Sempam and at Kuala Lompat

	RS2, number of copulations				Days observed	TS1, number of copulations				Days observed
	A	P	C	t		A	P	C	t	
Time of year										
1969										
August	0	1	1	2	5	0	0	0	0	12
Sept. – Dec.	0	0	0	0	9	0	0	0	0	40
1970										
January	0	0	0	0	5	0	1	0	1	10
February	2	0	1	3	9	1	2	0	3	10
March	0	1	5	6	10	1	1	3	5	10
April	0	3	6	9	10	8	0	3	11	10
May						5	1	0	6	9
Time of day										
0600 – 0900	1	1	3	5		5	1	1	7	
0900 – 1200	1	2	2	5		4	1	2	7	
1200 – 1500	0	1	4	5		2	2	3	7	
1500 – 1800	0	1	4	5		4	1	0	5	
Total	2	5	13	20	29	15	5	6	26	49
					RS2			TS1		
Pelvic thrusts/min					89.1 (77 – 118)[1]					(104)
Pelvic thrusts/sec					1.5 (1.3 – 2.0)[1]					(1.7)
Breeding period										
Number of bouts/day					0.62			0.53		
Number of days/sex bout					1.61			1.88		
Days with sex bouts, %					52			43		
Overall (12 months), % days								15		

A = Attempted copulation; P = possible copulation; C = complete copulation.
1 Range.

widely separated siamang groups in contrasting habitats, did not facilitate the solution of these problems as continuous data were not collected over a long enough period.

It is possible that these colour changes and eversion were associated with ovulation as sexual activity appeared to be clustered around this time at Kuala Lompat (fig. 5.19). The clustering of more successful copulations in Ulu Sempam with no associated colour change, however, could provide stronger evidence that the changes were associated with menstruation – hence the avoidance behaviour shown by the TS1 female. This is one problem

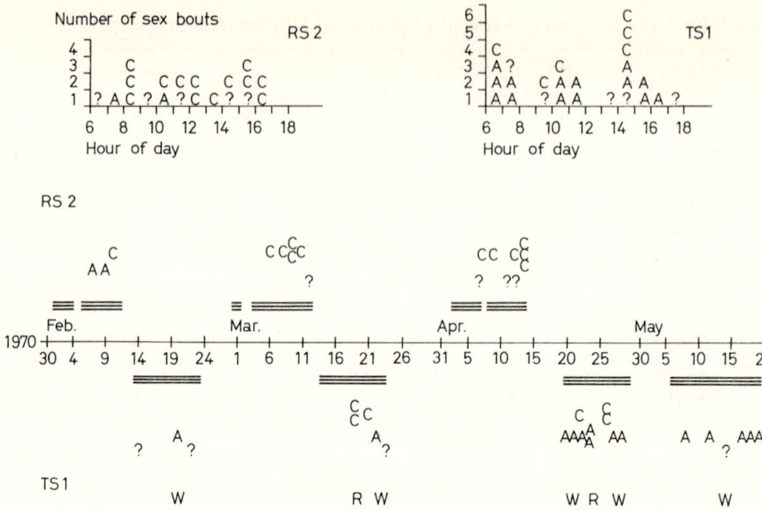

Fig. 5.19. Sexual behaviour in siamang showing daily patterns of copulation in both study areas – A = attempted, ? = possible, C = completed; and frequency of sexual behaviour during 1970 dry season, W = female labia white and everted, R = red and everted, otherwise indistinct. Triple lines indicate the days of observation in each study area.

that can be more successfully resolved physiologically in the laboratory. But it will require further field work to determine the environmental factors involved in the periodicity of reproductive behaviour.

Aggression and Group Formation

As play centred round the infant, so aggressive behaviour was directed mainly towards the subadult. The continuous vigilance shown by each siamang to other members of the group (plate XIV) has already been indicated as the main mode of communication *within* the group. Perhaps because of this, few facial expressions and gestures were seen, and few calls were heard. The structure of the habitat restricts sophisticated visual signals, and the siamang family group's way of life does not really require them.

Soft grunts were occasionally emitted in the initial stages of an alarm response to a source of potential danger outside the group. Although this developed into an aggressive group call (Chapter 6), it was important in communicating the disturbance to all group members. Squealing or bleating distress calls were given by the infant when isolated from the others, or attemping to follow the adult male. The juvenile or subadult usually stayed close to the infant, if the male wandered far from it. A squeal was the only other call heard within the group; it was usually emitted by the submissive, usually the younger, animal in an aggressive interaction.

Apart from an 'alert face', the only facial expression that was discerned in the black features of the siamang was an 'open-mouth threat' usually given with the teeth concealed (plate XVIII) [BALDWIN and TELEKI, in press]. But the teeth were sometimes exposed – as during play and yawning. The threat usually occurred when one individual was provoked by another; it represented a warning from the siamang (irrespective of age) that further approach – in terms of decrease in distance and the quality of movements – would be followed by a more aggressive response. It was often sufficient to cause the intruder into 'individual space' to retreat.

Older individuals showed varying degrees of tolerance to the close proximity of others according to their activity. Thus, tolerance was maximal during resting, although a third animal might not be welcomed if two were grooming together; adequate space was of greatest importance during feeding. It appeared to be the inability of the younger animal to predict these tolerance limits at a given time which resulted in aggressive interactions.

There were 1.5 aggressive interactions each day on average during the study at Kuala Lompat. It must be stressed that these 'attacks' were brief, with little forceful physical contact, and that they had little adverse effect on the close harmony of the siamang family group for most of the study. Few aggressive interactions were observed within RS2 in Ulu Sempam. This may have been because the subadult maintained a greater distance between himself and the adult male, thereby avoiding stressful situations (see above). The presence of siamang groups in adjacent parts of the forest perhaps reduced conflict within the group to produce closer cohesion between group members.

The frequency of aggressive interactions increased steadily from 1.31 to 1.70 as the study of TS1 developed, apart from a decrease at the end of the wet season (fig. 5.20); but there was no significant difference between seasonal frequencies ($\chi^2 = 2.40$). Overall they were evenly spread through the day, with no significant clustering in any part of the day (Kolmogorov-Smirnov one-sample test, D = 0.131, 0.097, and 0.111, n = 71, 59, and 83, for intermediate, wet, and dry seasons, respectively). There was no significant difference between daily patterns in each season (3-hourly pattern, $\chi^2 = 9.79$, d.f. = 6), although two thirds of bouts were after noon in the wet season, changing gradually to before noon in the dry season (fig. 5.20). The latter could indicate a need to resolve disputes earlier in the day in this more active period.

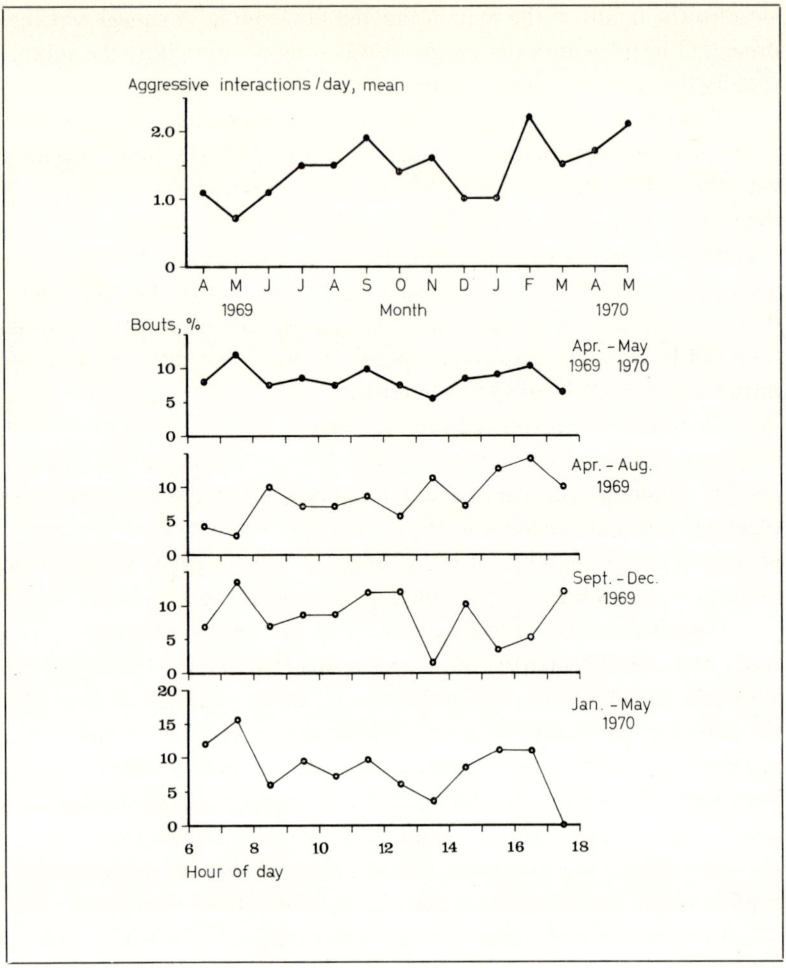

Fig. 5.20. Monthly variation in the frequency of aggressive encounters each day at Kuala Lompat, and the daily patterns of aggression in each season and overall.

As in grooming behaviour the male and subadult male were the main participants; nearly two thirds of observed interactions involved the male attacking the subadult (table 5.18). Most of the aggressive behaviour of the female and juvenile were also directed at the subadult. The juvenile never attacked an adult, and the only attack by the infant was on the female; otherwise, all possible combinations were observed. Some brief attacks were made following over-vigorous grooming.

Table 5.18. The direction of aggressive interactions, in a siamang family group at Kuala Lompat (TS 1)

Attacker	Attacked											
	male		female		SA male		juvenile		infant		total	
	%t	%T	%t	%T	%t	%T	%t	%T	%t	%T	t	%T
Male			12	7.5	75	48.0	13	8.1	1	0.6	111	64
Female	9	1.7			69	12.7	19	3.5	3	0.6	32	18
SA male	5	0.6	15	1.7			75	8.7	5	0.6	20	12
Juvenile	0	0	0	0	89	4.6			11	C.6	9	5
Infant	0	0	100	0.6	0	0	0	0			1	1
%T		2		10		65		20	2		173	

t = Number of bouts in which each individual is involved; %T = number of bouts in which each individual is involved as a percentage of total number of bouts.

The adult female was the second most aggressive animal, although four times less so than the adult male; the juvenile was the second most attacked individual. Thus for attacking the group was ranked: (1) male, (2) female, (3) subadult, (4) juvenile, (5) infant; and for being attacked: (1) subadult, (2) juvenile, (3) female, (4) male, (5) infant (table 5.18). The only noticeable changes in aggressive relations between individuals as the study progressed were that the male attacked the female most during the mating period, and that the juvenile was attacked more by the male as he got older.

Many aggressive bouts occurred in one of two situations. A younger animal (usually the subadult) intruded on the 'individual space' of an adult during a feeding bout while moving from one part of the tree to another; sometimes this intrusion was achieved by movements of the older animal. When this occurred the older animal (usually the adult male) lunged at the intruder, flailing arms and grappling with the mouth. The younger animal retreated rapidly squealing and eliminating, giving an 'open-mouth threat' if the adult followed.

The second situation was when male and subadult were grooming in the late afternoon. The juvenile approached them to take up his night sleeping position with the adult male. He slowly intruded upon the male and subadult, provoking the subadult in particular in his efforts to squeeze in closer to the male and thereby initiate grooming. It was difficult to determine the behaviour patterns involved, and their significance, because they were relatively brief and infrequent, and one participant almost invariably had his back to the observer. But it was clear that 'open-mouth threats' from the juvenile were followed by a lunge by the subadult. The male rapidly intervened by lunging and flapping at the subadult, who retreated flapping, squealing, and eliminating. In all cases the initial retreat was only a few feet, from which the animals involved confronted each other; after a pause the subadult moved right away – in this case to take up his night position.

For most of the study the subadult slept in the same tree as the others, albeit in a different part. By the end of the study, when he must have been eight to nine years old – socially as well as physically mature – he was more often sleeping in an adjacent tree. This was just one aspect of the change in relations that occurred at this time. The frequency of male-subadult aggressive interactions increased during the dry season until the end of April, when there were five such conflicts in one day. In the next and final month of the study a new facet of aggressive behaviour was observed. As the subadult was about to follow the rest of the group into a food tree, the male displayed by swinging vigorously around a small part of the tree. He ended up facing directly away from the subadult. This display was one of the components of inter-group calls and chases (Chapter 6).

Even though the force of the display was reduced by its redirected termination, the subadult usually stayed quietly where he was, or retreated into a nearby tree until such time as the group, particularly the male, moved on. In this way the spatial and temporal separation of the subadult from the group during the day range was rapidly increased. But there were other factors involved. The peak of sexual behaviour (see above) probably resulting in the male's intolerance of maturing male conspecifics was also correlated with a marked increase in group calling (Chapter 6). Many of these calling bouts were initiated by the subadult, who moved away from the group (sometimes more than 50 m towards the territorial boundary, and continued calling long after the adults had stopped and moved on with the young about their day range. Although he soon followed the others it was clear that these calls indicated the presence of a solitary male (both spatially and temporally separated from the adult pair), and resulted in the attraction of a young female to this part of the forest.

Such an animal was observed in the last month of the study, and within five months they had begun establishing a territory to the east of TS1 [Hunt, personal commun.] This is already providing excellent opportunities for further study of the processes of group formation and the establishment and maintenance of territories in siamang [Aldrich-Blake and Chivers, 1973].

Discussion

The siamang family fulfills completely the operational definition of a social group proposed by STRUHSAKER [1969]. All individuals were part of the same social network, having all non-aggressive interactions with other group members; all were associated, at least indirectly, in grooming, play, contact, agonistic, and sexual behaviour. Group progressions and reactions to conspecifics (and congeners) were highly synchronised, and individuals shared precisely the same area of habitat – the subadult was the only group member to deviate occasionally. One of the most conspicuous features was the marked spatial proximity of group members, which was not directly mediated by the environment – forest structure and food dispersion (see Chapter 3 and concluding Discussion).

The close spatial and behavioural integration of the siamang family group poses questions as to how this is achieved. The group was closest to the adult male overall, to the adult female during resting and feeding, and to the infant during travel and at night. These indicate the coordinating effect of the male, the food-seeking or leadership (about day range) responsibilities of the female, and the protection afforded to the infant at times of great vulnerability. On occasions when the infant was separated from both adults, it was noted in both areas that one individual remained very close to the infant until the adult returned.

The integration of group activities was most obviously achieved, in the absence of a repertoire of intra-group calls, gestures, and facial expressions, by each individual (especially the younger ones) watching the others continuously and imitating their actions [MENZEL, 1971]. The presence and diversity of subtle gestures have still to be determined, but the observer rapidly became aware of the intentions of the adults, and it became possible to predict changes in activity with some accuracy.

The movement of the female away from a resting huddle indicated that a group progression was about to begin; she would hang a few metres away from the others watching the forest ahead, and glancing back occasionally. Usually such movements could be predicted by a marked increase in scratching that spread from adults to young as they stirred from their resting and grooming; this scratching appeared to represent an inner conflict as to when and where to move next. That it was not confined to the female indicates the possible involvement of other individuals in determining the timing and direction of movements (see below). But it might merely represent the 'infectious' nature of such behaviour. Movement out of food trees was initiated most often by the adult female in a similar way.

Because they could find no clear evidence of dominance within a gibbon family group, CARPENTER [1940] and ELLEFSON [1967] referred to the adult male and female as co-dominant, pointing out that this did not necessarily reflect a stable relationship. Dominance was also described between groups (sometimes of different species) within food trees, but this relationship was also unstable, particularly between conspecifics. Hence the difficulty of applying a concept of dominance to small social groupings of this kind; it was not helpful to think in terms of one siamang dominating another, even in a specific context.

GARTLAN [1968] has outlined the overall problems of dominance, which involve the lack of correlation between ranks in the same social group for priority of access to need satisfying objects [CHANCE, 1956] such as food and copulation. GARTLAN [1968] and STRUHSAKER [1969] prefer to view individual behaviour in terms of social roles, which vary with age and sex, and even independently with time. This can be considered in terms of the THOMPSON [1969] enquiry as to what extent a group's pattern of organisation limits, or is determined by, the behaviour of each group member.

But GARTLAN views social roles in terms of adaptive significance, whereas BERNSTEIN [1971] uses the concept for 'activity profiles'. The latter is preferred here, so that each member of a siamang group can be assigned roles according to what they do (and when they do it), without probing too deeply into the complex reasons for performing those roles until more information is available. The change in spatial relations during the day, and over the seasons, with reference to the adult male and female, highlights the conflicting evidence as to which is the focus for group activities. The evidence suggests that each has a leading role in different activities, so that responsibilities are fairly evenly divided.

Thus the behaviour of the adults limits the activities of the young siamang, in terms of ranging patterns, or the occurrence and timing of social activities. Although the group was closer to the female during resting bouts, it was the male who usually started resting, and who acted as a focus for grooming. The adult, especially the male, tolerated the close proximity of others during resting, but maintained an increased distance between individuals during feeding. At other times the subadult male kept at a greater distance from the male, probably because of the brief but frequent encounters between them.

Several problems surround the female's role in leading the group about the territory in search of food. Most group members must be very familiar with the resources of their small territory, particularly with the distribution of food at a given time. It can be argued that individual leadership is not essential if the group travels as a coordinated unit through the experienced actions of the adults.

Nevertheless, one still has to explain why the adult female initiates and leads most group progressions. The 'shunting' mode of movement, whereby the female moves several metres and then waits until she is followed before moving further, is indicative of positive leadership. So is her behaviour when another individual starts travelling, and she follows rapidly to assume the lead or follow closely. These observations tend to refute the possibility that the female's position at the head of group progressions represents her attempts to avoid the others.

But the problem still remains as to what extent the adult male (or any other individual) influences the timing and direction of group progressions from behind [MENZEL, 1971]. The female's habit of waiting to be followed, and the way the group swung round to follow the male on occasions when he changed direction, indicated that the female's role in leading progressions might be to 'suggest' the timing and direction of movements for possible modification by the adult male. Thus one can summarise the daily behaviour of individuals in a siamang family group as a compromise between following the female and remaining near the male.

A reduction in the frequency of the whole group being in the same tree did not invariably mean an increase in group spread, because individuals were sometimes closer in the terminal branches of adjacent trees. The occurrence of more than one activity in each hour posed problems for showing correlations between daily patterns of synchronised activity, coordinated position, and group spread with resting, feeding, and travel. The infant was closest to the male at those times of day when the group was most dispersed; the infant strayed further from the male when the others were closer together.

The male-infant relationship is of special interest because – although some degree of paternal care has been described in several primate species [MITCHELL, 1969] and the male titi monkey carries the infant from birth (except during feeding by the female) [MASON, 1968] – no similar pattern of intensive care by adult males has been described among primates. It has been shown that such a transfer does not facilitate the birth of another infant; in fact it merely means that the male rather than the female is incapacitated. It is unlikely that the delay in onset of sexual behaviour was caused by the infant suckling from the female overnight until 18 months of age. Although signs of female reproductive cycling (and sexual behaviour) were not perceived until the infant was nearly two years old, it is possible that in a few months they have been missed in the weeks of observations in the other study area.

The daily care of the infant by the male seems most likely to be a factor of body size. The adult male is usually somewhat larger than the female, and it is possible that the growing infant presents too great a burden to the female in daily travel, particularly in view of her leadership responsibilities and the 'trial-and-error' tactics involved. The adult male can care for the infant from a safe central position in the group. The other advantage is that such an association increases the infant's acquaintance and social integration with other group members; not only with the male, but with the subadult (especially through the latter's association with the male). It is of interest that the last group member with whom the TS1 infant became acquainted was the juvenile. Apart from the advantages of playing with peers, it can be argued that the infant depends least on the juvenile for its survival – relying most on the adults. On the one occasion when the infant fell to the ground unharmed, it was the adult male who retrieved him, although the female approached showing consternation by her actions and soft grunts in reply to the squeals of the falling infant.

This male care of the siamang infant, observed in three well-separated areas, is strikingly different from the white-handed gibbon [ELLEFSON, 1967; personal observation of TG2]. Similar behaviour is reported for the siamang in Sumatra [MACKINNON, personal commun.; WILSON, personal commun.]. Interest in the infant by the adult male has been observed in captivity [FOX, personal commun.] but it does not develop as in the wild, because the female cannot escape from the infant as she does in the wild by crossing into another tree. Thus isolated, the infant turns to the nearest solicitous individual – the adult male – and so the relationship develops.

The other striking difference from the white-handed gibbon is the much greater tolerance shown by the siamang adults to the subadult. It was only in the final stages of peripheralisation that he was excluded from the immediate vicinity of the group. The adult female generally showed indifference to his presence, or avoided him. It was the adult male who both attracted and repelled the subadult. Observations in Ulu Gombak [personal observation] and Milwaukee Zoo [FOX, personal commun.] indicate that this relationship is not affected by sex. The female avoids, and the male grooms and scraps with a subadult female as much as with a subadult male.

The timing of grooming (and resting) appears to be related to two main factors. Grooming occurred after feeding bouts, either in the early morning in the sun (particularly after a wet night), or more often in the afternoon as the group became increasingly satiated and/or fatigued. The fact that actual grooming occupied only one third of pair association time enhances the social significance of the close proximity that occurred during grooming bouts. The increase in morning resting in RS2 ean be related to a period of vigilance near the territory boundary (Chapter 2). The increased grooming might reflect mounting tension amongst the participants; each grooming sequence was followed by a scanning of the canopy.

At Kuala Lompat 40% of self-grooming bouts involved the adult female, which may reflect a compensation for her low score for allo-grooming. The frequency of self-grooming decreased in the 1970 dry season when she was increasingly involved with the adult male. The social aspects of grooming are highlighted by the much greater involvement of mature individuals. There were indications that this provided an appeasement to actual or incipient antagonism.

Aggression between male and female increased during the dry season in parallel with increase in grooming and sexual behaviour. The main participants in grooming were also those most involved in aggression – the male and subadult. Each potential pairing was ranked for frequency of attacks and amount of grooming time in the same order; there was a highly significant correlation ($r_S = 0.68$, $N = 20$, $p = 0.001$). The daily patterns of aggression, grooming, and play were not correlated. The amount of grooming decreased with a decline in the frequency of aggressive encounters. If copulation is equated with aggression this explains the lower grooming scores between male and female outside the breeding period. The lower frequency of grooming between adults and young, with their distinctive daily pattern, retains the cleansing importance – the soothing aspects being derived from parental care rather than submission.

It becomes increasingly apparent that there is a 2- to 3-year behavioural cycle based on reproductive behaviour and the maturation of the young, in addition to the annual ecological cycle (Chapters 2 and 3), affecting the closely synchronised pattern of activities and social organisation of the siamang family group. Observations in Ulu Gombak confirmed the tendency for the final expulsion of the subadult to be associated with the conception or birth of a new group member, and for group formation to require a positive response from a young female to the signals of a young unattached male. The synchrony of reproductive behaviour in widely separated areas (and between several primate species) indicated the ultimate importance of long-term environmental cycles on social behaviour within the siamang family group.

Summary

The five siamang of TS1 were all in the same tree for 40% of 10-min observations between dawn and dusk on average; all but one were together for 17% of observations. RS2 were better coordinated spatially with the four individuals in the same tree for 50%, and all but one for 34% of the time. The groups were most dispersed in the late morning, and most concentrated in mid-afternoon, with hourly scores changing from 30 to 60%. These patterns of spatial coordination were confirmed by measuring the average distance between individuals over 10 days – mean group spread. The mean daily spread changed from about 12 to 6 m by early afternoon.

Thus the group was more concentrated and closer to the male during resting, and most dispersed during travel. The rest of the group were with the male for 49% of the time, but with the female for only 34% observations. The siamang family group was much better coordinated in terms of activity than of position, especially at Kuala Lompat. TS1 were all engaged in the same activity for 73%, and all but one for 13% of the time; the scores for RS2 were 59 and 23%, respectively. There was little difference with respect to each individual.

Considering the spatial relations of each individual in more detail, scores ranged from the male and infant averaging 6 m separation through the day, being in the same tree for 85% of observations on average, to the male and subadult, who were in the same tree for little more than half the daily observations, separated by 10–11 m on average. Scores ranged from zero during rest bouts up to 30 m during travel; on average individuals were 8–9 m apart during the day, but somewhat more at night, although the infant was with the female, and the juvenile with the male. On days of contrasting weather there was little difference in the synchrony of activity, but the group was more dispersed on hot and sunny days. Group spread was greatest in the early afternoon on hot and sunny days, and in mid-morning on cloudy and wet days.

The order of starting daily activities was usually male, juvenile, female, and subadult at Kuala Lompat, but the female left the night tree first, and the male third. The tree for the following night was entered first by the female followed by the male, with the juvenile and subadult later. During the day the usual order of group progression in travel bouts was female, juvenile, male (and infant), and subadult; when moving independently the infant travelled in the centre of the group. For most of the study she was carried about the day range by the adult male (following the rapid transfer from the female), with decreasing frequency as she developed the necessary locomotor skills.

The order of travel in RS2 was also female first, male (and infant) in the middle, and the subadult male last. It seemed that in both groups these positions represented individual preference, because tendencies to follow specific individuals were less conspicuous or consistent. The juvenile followed the leading animal (usually the female); the female followed the juvenile or subadult when either was in the lead; and the subadult followed the male, because these two were usually in the rear of the progression. In the dry season the male showed a clearer tendency specifically to follow the female, and although increasingly independent the infant was following the male more than the female.

Grooming was the most frequent of the conspicuous social interactions; little information could be collected on more subtle communication processes. Although the adult and subadult males were the most widely separated pair in terms of distance throughout the day, they were the main participants in grooming. At Kuala Lompat grooming occurred for about 42% of the time that pairs of siamang were associated, which was about 97 min /day,or 15% of daily activity time. Each grooming bout lasted for 13 min on average; scores were a little lower throughout in Ulu Sempam. But the main difference between the two groups was that 75% of grooming in TS1 occurred after noon, and 67% of grooming in RS2 was before noon.

Although daily patterns of grooming and resting were highly correlated, grooming was much more concentrated in a specific part of the day. This was confirmed by scoring PAT as a proportion of resting time in each hour; scores increased from 10 to 70% at peak grooming times. At Kuala Lompat 58% of bouts and 72% of PAT were between mature animals only; those involving young animals were at least half the duration, and tended to occur at 'off-peak' grooming times (when bouts were shorter) – especially as the group settled for the night. Thus participation in grooming increased with age, which highlights the social significance of this essentially cleansing behaviour.

There was weak tendency for the approached animal to initiate grooming; in bouts between mature and young, however, the young did the approaching, and the older animal almost all the grooming. Otherwise grooming was more fully reciprocated, and patterns were subject to individual variations. Grooming sequences tended to be short – 80% (94% for RS2) were concluded within 5 min, and only 0.05% exceded 10 min. In both groups the adult males did much more grooming than any other, but were not groomed much more than expected; the RS2 female was groomed more than the male. The TS1 male and subadult groomed each other much more than any other pairing until the dry season, when the male was repelling the subadult and was much more active in seeking out the female. Those individuals most involved in aggressive encounters (including copulation), were also most involved in grooming.

Play was initially a solitary activity which developed into a social activity during the infant's second year of life. It was important for developing a better coordination of movements, particularly for acquiring feeding skills, and for social integration. Although sleeping with the female at night, the infant slowly gained its independence from the adult male during the day. The subadult (mainly) and subsequently the juvenile became increasingly involved with the infant in the increased frequency of play bouts in the dry season, while the adults groomed.

Sexual behaviour was not observed until the 1970 dry season, when it was fairly frequent in both study areas. Bouts were evenly spread through the day, and associated with the seasonal increase in ranging, feeding, play, calling, and aggression. The close proximity of others was tolerated much less during feeding. Although the subadult relationship with the rest of the group (especially the male) was very harmonious, apart from the occasional brief scrap, antagonism was finally increased to such a level that the subadult was rapidly separated from the others in spatial and temporal terms. But this coincided with subadult initiative in pair formation, which was achieved within a few months. Thus by the middle of 1970 the 8- to 9-year-old subadult male was becoming an adult, the 2-year-old infant was entering the independent juvenile stage, and the juvenile male (about 5 years old) was in the middle of the relatively static (socially) juvenile stage (3 – 6 years).

Chapter 6: Calling Behaviour

Introduction

The most dramatic event in siamang behaviour is when group members join together in a bout of calling. The sound carries several kilometres, and at intervals the tree is shaken by the vigorous display movements of the older individuals. The pattern of these group calls, in terms of the contributions made by individuals of each sex and different ages, is investigated herein. A fuller analysis of individual vocalisations, including sonographic studies, will be the subject of a future publication.

There follows a description of the main features of calling behaviour in the three study areas[1], and a detailed analysis of the patterns of calling by a sample of groups from each area. Particular attention is given to variations in the frequency, timing, and structure of group calls, the time calling starts each day, and the daily proportion of the local population calling. Finally the significance of these concerted group calls in spacing, conflict, and alarm situations is discussed; reference is made to the importance of such calls within the family group.

The significance of siamang calling behaviour is more clearly appreciated when compared with that of the white-handed gibbon (hereafter referred to as gibbon) which was present in good numbers in all three study areas (Chapter 7). This provides an opportunity to initiate inter-specific comparisons in the concluding chapters which are being pursued and developed as part of a continuing programme of field work.

1 The location and identity of social groups of siamang and gibbons referred to are given in figures 6.2 – 6.5; group size (where known) is given in Chapter 2 under 'territory size'.

Plate XXI. Group calling behaviour. TS1 male screaming, with open mouth and fully inflated throat sac, while female on right barks rhythmically; the juvenile sits between them in *Koompassia excelsa* barking with his left hand before his mouth (above). At the climax of the female bark series, the male swings vigorously around screaming, followed by the bark-chattering female – note protruded lips of juvenile and the subadult male on the far side of the tree (below).

Pattern of Group Calls

Group calls, or calling bouts, are as discrete as bouts of maintenance activities (Chapter 4). As in travel bouts, the start and finish are very closely synchronised among group members, and there are pauses in calling similar to the pauses in movement that occur during group progressions. Both species exhibit distinct sexual dimorphism in calling; additional information on calls in gibbon family groups [CARPENTER, 1940; ELLEFSON, 1967] is presented to facilitate an inter-specific comparison.

Siamang

Apart from grunts and squeals given in intra-group situations (Chapter 5), bouts of group calling involve three loud vocalisations – the boom, bark, and scream. These basal calls may be classified according to the ways in which they are used as follows:

1. Bark: (a) single (connected into regular series by female); (b) chatter (starts bout – usually male); (c) bark-chatter (follows female bark series and male scream).

2. Boom: (a) diphasic between bark series (more pronounced in male); (b) monophasic (usually, invariably precedes bark).

3. Scream: male only, during and at end of female bark series.

Group calls were frequently made from the emergent level of the forest canopy. It was very difficult, therefore, to determine precisely what factors elicited these calls, and to follow the behaviour of each individual. But the following description was constructed during the study:

From a distance the start of a calling bout was heralded by a staccato chatter of about 2 sec duration, repeated a few times (usually by the adult male). At close quarters it was sometimes possible to detect quiet grunting by the female. This alarm call was then transformed into one of threat – irregularly repeated barks. These calls were soon taken up by the rest of the group as they climbed higher, also starting to boom. Booms were loudest in males, resulting from the passage of air into, and across the entrance of, the inflated throat sac; this was under pressure because the mouth was almost closed. Barks and screams followed immediately when the mouth opened under continued expiration.

The call bout continued with the female launching into a double series of rhythmic barks; the first series was longer than the second, but both increased in rate and pitch. Sometimes during the longer series, but invariably at the climax of each as the female culminated in a bark-chatter, the male gave a deafening scream, which also tailed off into a bark-chatter. At this point most of the group moved from their sitting (or hanging) positions to swing vigorously in small circles, shaking the branches as they moved (plate XXI). Thus the force of this vocal display was reinforced by the visual display of movement, and by the pink colour of the inflated throat sac against the otherwise black pelage. Young animals contributed to this calling in varying degrees showing a certain lack of timing and quality. Each cycle of calling during the bout was followed by quiet, while animals rested giving occasional booms, barks, bark-chatters, and even screams, and repeated about five times at 3-min intervals (see below).

1. Male Calls

Apart from the sexual specificity of some vocalisations, the most conspicuous variable in male calling was the timing (and rate) of screams during the calling bout. The timing has already been referred to in relation to the female bark series, which will be more fully investigated below. The rate of screaming remained relatively constant at about three per minute irrespective of the number of males (adult and subadult) in the group (table 6.1).

The decrease in rate of screaming during some conflicts indicates a non-aggressive function for this vocalisation. Observations on captive animals revealed the necessity for both sexes to be present to complete group calls. In the absence of one sex, the other tended to be erratic in its incomplete performance (table 6.1). Simulated female calls elicited a wider variety of better-timed calls from a captive male. Females in particular attempted to reproduce the male part in his absence – but with little success.

2. Female Calls

It was the timing and frequency of female bark series, which provided the greatest scope for investigating the patterns of vocalisations during a group calling bout, group variation, and the identity of groups. The start of each bark series was timed in relation to the start of the calling bout, and the number of barks in each part of the series were counted. A note was made of the rate and pitch of delivery.

There was a warming-up period at the start of each bout before females fully adopted their characteristic pattern in a double bark series (table 6.2). In 45% of bouts there were 4, 5, or 6 bark series, but as many as 12 bark series were recorded in a call bout and the average was about six. The RS1 female

Table 6.1. Rate of screaming by male siamang

Location	Group	Number of screams/min		Female bark series
		per group	per male	
Captive				
Seremban[1]	lone male captive	1.3	1.3	
Temerloh[2]	lone male captive	1.8	1.8	
Frankfurt	M, M, F, F (SA)	5.4	2.7	26 – 7
Rotterdam	M, F	3.6	3.6	14 – 6
Kuala Lumpur	F, F (SA)	–	–	8 – 17 – 10
Wild				
Ulu Sempam	M, M, F RS2	3.0	1.5	12 – 7
	M, M, F RS1	2.4	2.4	18 – 6
	RS1 v. RS2 RS1	3.4	3.4	
	RS2	3.4	1.7	13 – 7
	RS4 v. RS2	1.3	1.3	
Ulu Gombak	M, F, F GS16B	3.0	3.0	17 – 6
	GS16B v. S16A	2.3	2.3	
	M, M, F GS6	3.8	1.9	18 – 8
	GS16C v. S6	2.3	1.5	

M = male; F = female; SA = subadult.
1 Tape recording by Lord Medway.
2 Includes response to female calls given by D. J. C.

consistently gave fewer in shorter call bouts, averaging less than five, whereas the RS4 female averaged nearly eight bark series per bout, these differences were not significant by the Kolmogorov-Smirnov two-sample test.

Bark series started about 2.5 min apart (table 6.2) with some females starting a cycle every 2 min on average (RS4A), and others every 4 min (RS12A). Each bark series was composed of 15 barks followed immediately after male screams by about seven more on average. But in some groups, especially at Kuala Lompat, adult females consistently gave up to 20 barks followed by eight or nine in each series. The distribution of scores for groups with contrasting patterns (fig. 6.1) showed that female bark series were significantly longer in RS3, RS11, RS11, and TS5 than in RS2, RS2, RS3, and TS1, respectively (Chivers [1972b] Appendix XIII. 2, Kolmogorov-Smirnov two-sample test). This applies to the first phase; females were more similar in the number of barks in the second phase of the series – only RS3 gave significantly more barks.

Table 6.2. Siamang male and female call characteristics

Group	Call duration	Male, average number of screams per min	Female, average number of bark series	Average number of barks in each part of bark series	Mean interval between bark series, min
RS1	12.7	2.4	4.7[1]	18−6	2.7
RS2	18.1	3.0[2]	6.3	12−7	2.9
RS1B	17.0		5.9	14−7	2.9
RS3	16.2		6.9	18−9	2.4
RS4	22.0		7.8	14−7	2.8
RS4A	13.5		6.2	15−9	2.2
RS11	23.3	3.3	6.9	13−7	3.4
RS12A	21.0	3.6[2]	5.6	15−7	3.8
RS12	15.2	2.5	6.6	15−8	2.3
Ulu Sempam	17.6	3.0	6.2	15−8 581[3]	2.8 128[3]
TS1	10.2	3.7[2]	4.6	15−7	2.2
TS5	14.6		5.9	20−7	2.5
TS4	13.3		5.8	19−9	2.3
TS6	10.0		4.8	21−8	2.1
Kuala Lompat	12.1	3.7	5.2	16−7 277[3]	2.3 76[3]
GS6	17.8	3.8	8.0	18−8	2.2
GS16B	14.1	3.0	4.8	17−6	2.9
GS16C	14.0		5.3	15−7	2.6
Ulu Gombak	15.0		5.7	17−7 83[3]	2.6 16[3]
Total	15.5	3.1	5.8	15.5−7.5	2.7

1 Two females call in group (one subadult).
2 Two males in group (one subadult).
3 Sample size.

In a small area, the difference in the pattern of female barking was usually distinctive between a few groups; this was emphasised by differences in rate and pitch. Thus the RS2 female barked more slowly at a lower pitch, giving 12 barks followed by seven on average, whereas the RS1 females barked more rapidly at a higher pitch giving 18 barks followed by six on average. The presence of two females at the same location was the most distinctive feature; one of them – thought to be the subadult – barked even faster, higher, and for longer than the other. Similarly, RS2 and TS1 could be distinguished from neighbouring groups by the presence of two males.

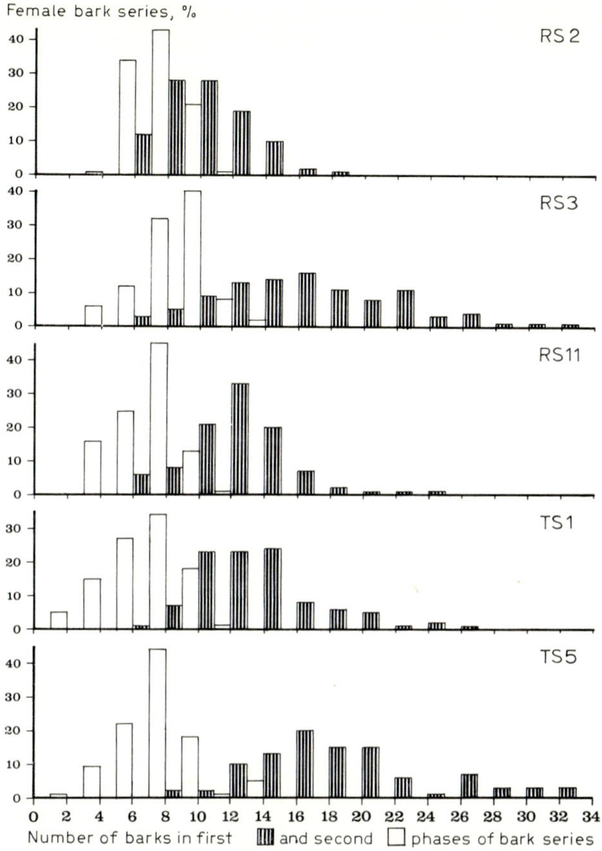

Fig. 6.1. Frequency of barks in the first and second phases of the bark series of female siamang for RS2, RS3, and RS11 in Ulu Sempam, and TS1 and TS5 at Kuala Lompat.

The intervals between bark series, and the number of barks in each, showed some variation at different stages of the call bout. Once the females had warmed up to their usual pattern, their performance was usually consistent until the final stages of the bout when bark series were often longer, but less frequent – every 3–4 min, rather than 2–4 min. The siamang at Kuala Lompat maintained a consistently high frequency of barking throughout the bout.

3. Inter-Group Conflicts

Interactions between groups across territorial boundaries have already been described (Chapter 2). Reference was made to the importance of calling prior to chasing on some occasions. It was difficult to detect significant differences in calling, whether a siamang group was confronting another or calling from the centre of its territory. In the few confrontations observed there was no real indication of increased intensity of barking by the female, or of screaming by the male (cf. tables 6.1 and 6.2).

4. Alarm

Group calling was conspicuously different in situations of real or potential danger. It was never possible to determine the source of alarm on the few occasions when this occurred, but it was likely to be a large cat, raptorial bird, or a snake [ELLEFSON, 1967]. The group was alerted by soft persistent grunts by the adult female (usually); these increased into barks with the rest of the group joining in. Individuals moved less than in other group calling bouts, tending to sit and watch the presumed source of alarm (usually down towards the ground).

There was no female bark series, and only occasional male screams; the call bout was composed mainly of booms, barks, and bark-chatters, which lacked the continuity of group calls in spacing and defence situations. To the human observer it was a very unpleasant noise, and perhaps had a similar deterrent effect to whatever animal may have been threatening the group. It should be noted that the response to man, in the early stages of the study, involved silent retreat rather than noisy confrontation.

This kind of group call, which probably had less significance in inter-group communication, accounted for most calls heard after midday and during the hours of darkness. Contrary to LIM BOO LIAT [1969] evening calling was rare during this two-year study.

White-Handed Gibbon

The group calls of gibbons showed similar sex-specific roles to those described for the siamang. The most distinctive feature of bouts was the female 'great call' [ELLEFSON, 1967; type I vocalisation, CARPENTER, 1940] repeated at regular intervals, against a background of male 'hoots' [CARPENTER, 1940, type II vocalisation]. It appeared to be mainly the male who gave wailing 'hoots' [ELLEFSON, 1967] becoming silent during the female 'great call'. Call bouts in which no female great calls were given tended to occur at the start of the day (see below). Calling prior to the first feed of the day appeared to involve the male only, and it was not until later that the gibbon group called together, exhibiting the full variety of calls.

ELLEFSON [1967] has reported that, following these group calls from a central part of the territory on 85% of mornings, the group proceeds towards any neighbouring group close to their territorial boundary. There followed conflicts in mid-morning every other day on average. In the present study these were indicated by bouts of calling in which male-type vocalisations predominated. But it was at this time that the female was also heard emitting 'hoots'.

The early morning male-only calls were considerably longer than subsequent group calls. The difference was less marked in Ulu Sempam, where they were more frequent:

		n	%	Duration, min
Ulu Sempam	male only	63	38	30.8
	male and female (g/c)	101	62	21.7
Kuala Lompat	male only	44	17	26.5
	male and female (g/c)	192	83	12.9

Overall, 27% of bouts were identified as male only, not only because no great calls were heard, but because there was frequently only one individual calling at each location. The female's role in these bouts was not determined, but it seems unlikely that she was always silent. The early morning calls lasted for 29 min on average, compared with 16 min for those that followed. ELLEFSON [1967] describes bouts as lasting 13 min with great calls of about 20 sec every 1–2 min.

Although the calls of individual gibbons were not as distinctive as in siamang, it was the female great call that had the greatest potential for identifying groups, and changes in patterns of calling. TG1 contained adult and subadult females, whose great calls were synchronised. Their calling bouts were about 3 min shorter than those of TG2 on average, containing three or four great calls spaced about 3 min apart (table 6.3). TG2 (adult male, adult female, subadult male, juvenile, and infant carried by female throughout) more closely reflected the overall bimodal pattern in both study areas, whereby 30% bouts contained only one or two great calls, and about 40% contained four or five. The TG2 female gave great calls every 4 min on average.

In Ulu Sempam call bouts were 5 min longer on average, but there were only four great calls per bout, spaced 5 min apart (table 6.3). In Ulu Gombak GG5 resembled TG2, whereas GG10 resembled RG8. Differences were not very great, although in Ulu Sempam the number of great calls/bout varied between groups from three to seven on average, with intervals between them of 3–8 min. RG3A was unusual in the way it consistently gave a great call within the first minute of the bout starting, and the RG1A female on average uttered no great call until 5 min from the start.

TG1 and TG2 had almost identical patterns of calling, in terms of the timing of great calls during the bout, with small variations from one season

Table 6.3. Gibbon group calls and female 'great call' characteristics

Group	Season	Call duration, min	Average number of 'great calls'	Mean interval between 'great calls'	Mean duration 'great call' sec	n
T G1	April–Aug. 1969	12.5	3.2	3.9	20.7	
	Sept.–Dec. 1969	12.5	3.4	3.7		
	Jan.–March 1970	7.1	3.0	2.4	22.1	
	April–May 1970	14.3	4.1	3.5		
	n = 78	11.0	3.4	3.2	20.3	29
T G2	April–Aug. 1969	15.8	3.8	4.2	23.2	
	Sept.–Dec. 1969	15.0	4.0	3.8	21.5	
	Jan.–March 1970	12.2	3.7	3.3	19.6	
	April–May 1970	11.8	3.7	3.2	16.8	
	n = 107	14.0	3.8	3.7	19.9	60
T G1A	4	14.0	5.3	2.6	26.5	3
G3	7	15.6	4.0	3.9		
G4	9	12.6	4.0	3.2		
G5	4	14.3	5.0	2.9		
G6	11	20.3	5.3	3.8		
Kuala Lompat	35	16.0	4.7	3.4	21.1	102
R G1	6	16.0	5.0	3.2		
G1A	3	19.7	4.7	4.2		
G3	16	20.6	5.3	3.9	22.3	4
G3A	4	24.5	6.8	3.6		
G7A	26	27.2	3.6	7.6	18.3	13
G7	16	19.1	3.1	6.2	22.3	4
G8	22	19.3	3.6	5.4	19.9	21
G9	3	22.7	4.0	5.7		
Ulu Sempam	96	21.8	4.2	5.2	19.7	44
GG5	22	13.9	3.5	4.0	20.6	9
G10	7	21.4	4.0	5.4		

to the next. There was also little difference in the duration of the great call. 69% of TG1 calls timed lasted for 19 – 22 sec inclusive, whereas 69% TG2 great calls lasted for 20 – 23 sec (Appendix F). 62% of RG7A great calls lasted for only 18 sec, whereas 52% RG8 calls were of 20 or 21 sec. The latter difference only was significant (Kolmogorov-Smirnov two-sample test, $n = 34$, $D = 0.71$, $p = 0.001 -$). The difference in duration of great calls is considered further during comparisons between white- and black-handed gibbons (Appendix F).

The relation between group call duration and number of female great calls was more pronounced, and less variable, than that for siamang – especially in Ulu Sempam.

Group Calling Behaviour

Having described the characteristics of calling bouts in each species, attention can now be focused on the calling behaviour of social groups, in terms of the frequency, timing, and duration of group calls. This will be followed by a description of the patterns of calling in local populations of each species at different times of year.

The study of calling behaviour was started in the first two months of the project during 49 days on and around Ulu Rumput. This is a forested hill behind the University of Malaya Field Centre in Ulu Gombak (Chapter 1; fig. 6.2). A further 74 days (3 – 4 in each month) were spent higher up the valley making behavioural observations and recording group calls from the roadside (fig. 6.3). Calling by hylobatids was recorded on 165 days in Ulu Sempam (fig. 6.5), and on 203 days at Kuala Lompat (fig. 6.4) from January 1969.

The following information was collected each day for the two species: (1) the time the first group started calling; (2) the onset time, and duration (where possible), of subsequent group calls; (3) the number of hours in which calling occurred, and (4) the number of groups calling (noting if any called more than once).

Analysis was concentrated on those groups closest to the study groups: TS1 at Kuala Lompat, RS2 in Ulu Sempam, GS1 and GS2 on Ulu Rumput, and the GS16 complex in Ulu Gombak. Recording conditions were more variable in the mountains – in some directions calls carried for several kilometres, but a group over a ridge could sometimes not be heard at quite close range.

Table 6.4. Siamang and gibbon group calling characteristics around the Ulu Gombak Field Studies Centre in June and July 1968

	Siamang					Gibbon						
	GS1	GS2	GS3	GS4	total (121 calls, 49 days)	GG1	GG2	GG3	GG4	total (99 calls, 47 days)		
Call days												
Call days, %	64	59	57	51	58	70	55	34	45	51		
Call bouts per call day	1.19	1.17	1.00	1.19	1.14	1.30	1.13	1.23	1.24	1.23		
Call bouts per day	0.76	0.69	0.57	0.43	0.61	0.92	0.49	0.28	0.36	0.51		
Percentage of days with two or more call bouts	19	24	14	8	17	39	15	31	10	25		
Call time												
Calls, % at												
0500−0900	41	35	29	19	33	60	70	69	59	64		
0900−1200	54	56	61	76	60	26	30	31	41	30		
1200−1800	5	9	11	5	8	14	0	0	0	6		
Call duration												
Duration, min	20.1	21.6	21.1	23.1	21.3	21.2	20.2	21.2	29.1	22.4		
Number	37	36	27	21	121	43	26	13	17	99		
Percentage of calls lasting												
0−10 min	11	11	14	10	12	21	27	15	6	19		
0−20 min	49	53	33	38	45	33	31	38	35	33		
0−30 min	38	22	37	24	31	22	31	23	24	25		
0−40 min	0	6	11	24	8	19	4	23	12	14		
0−15 min	3	8	0	5	4	5	0	0	18	5		
0−60 min	0	0	4	0	1	0	8	0	0	2		
0−70 min	0	0	0	0	0	0	0	0	6	1		
Call weather												
Predominant hours, %												
Sunny	48	46	42	44	46	50	55	58	68	56	55[1]	39[2]
Cloudy	36	46	47	41	42	38	32	33	26	34	31[1]	36[2]
Variable	11	8	11	16	11	11	10	8	6	9	11[1]	12[2]
Raining	2	0	0	0	1	2	3	0	0	1	3[1]	13[2]

1 July, mornings and afternoons.
2 June, mornings only.

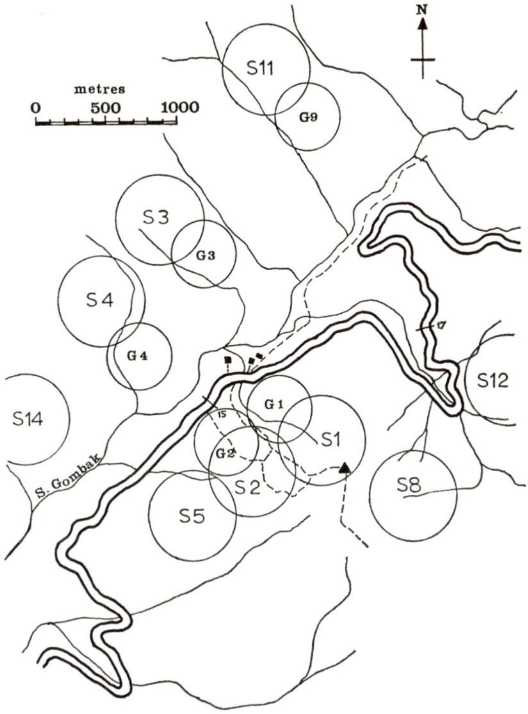

Fig. 6.2. Location and identity of siamang (S) and gibbon (G) family groups on and around Ulu Rumput, by Sungei Gombak and the University of Malaya Field Centre. Circles indicate centre of activity, rather than territorial limits of groups.

Ulu Gombak (G)

Ulu Rumput rises from 850 to 2,000 feet above sea level. In June and July 1968 there were three groups of siamang and two groups of gibbon resident on the north and west slopes (fig. 6.2). Immediately opposite the Field Centre were two other groups of both species. The mornings tended to be sunny, with clouds (and sometimes rain) moving into the valley during the afternoon. In both species less than 10% of group calls occurred after noon. A greater proportion of gibbon calls were given in hours of sunshine, which included the 10% of observations of clear skies before sunrise (table 6.4).

The groups all exhibited distinctive species-specific patterns. Thus every siamang group called on nearly 60% of days, whereas the frequency of calling by gibbon groups varied from 34 to 70% of the 49 days (differences can partly be explained by reduced accuracy for those groups further from the observer). Gibbon groups more often called twice in one day (table 6.3). Two thirds of gibbon group calls were before 0900 h, whereas 60% of siamang calls were from 0900 to 1200 h. In both species calling bouts lasted for just over 20 min on average, but gibbons had a higher proportion of both shorter and longer bouts – with only 58% between 11 and 30 min, compared with 76% for the siamang.

GS4 calls were more concentrated between 0900 and 1200 h than the others, and GS2 call bouts tended to be shorter – although they had a higher proportion of long calls than other siamang groups. It seemed that most calling occurred on mornings which started cloudy but soon became hot and sunny. It should be noted that this was the time of most calling by both species, especially the siamang, during the two years of study (see below).

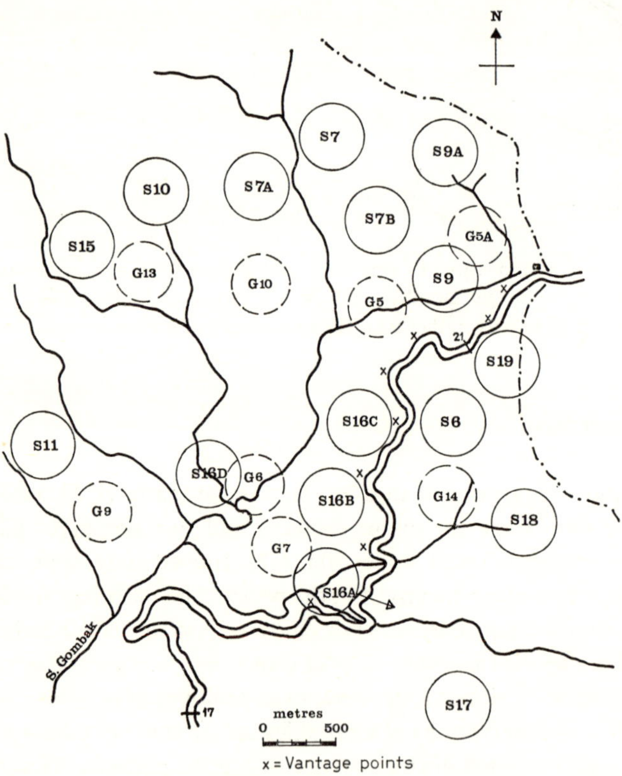

Fig. 6.3. Location and identity of siamang (S) and gibbon (G) family groups in and around the Jungle Reserve of Ulu Gombak near Kuala Lumpur, Selangor.

Observations were subsequently concentrated near the top of the valley. The road was patrolled between the 19th and 22nd milestones, from dawn until mid-afternoon, with long stops at each vantage point (fig. 6.3). There were four gibbon groups close to the road – GG5, GG5A, GG6, and GG7 – which were rarely seen. Two other groups were favourably placed for studies of calling behaviour – GG10 and GG13 – but GG14 was so inaccessible that their calling could only be heard from two places on the road.

There were three groups of siamang below the road – GS16A, GS16B, GS16C – and one above – GS6 (see Chapter 2 for compositions) – for which accurate data on calling, and behavioural observations, were obtained. Additional data were collected on 12 other groups with varying degrees of completeness. The frequency of calling in 1968 was similar to that lower down the valley, but the call bouts of both species were shorter (17.7 and 13.8 min for siamang and gibbon, respectively). The level of calling by siamang was as observed by McClure [1964] with a similar increase during the fruiting season. In both parts of the valley the frequency of siamang calls decreased in 1969 (55–22%), whereas gibbons continued to call as often as in 1968. At the top of the valley siamang calls were a little shorter (14.8 min), and gibbon calls longer (16.3 min), in 1969.

Kuala Lompat (T)

1. Siamang

Observations on calling indicated the presence of nine groups of each species within hearing range of the study area (TS1 territory, fig. 6.4). Each siamang group called on only 14–41% of days on average (table 6.5); the higher frequency of 58% for TS6 confirms that two groups were present on a similar compass bearing in this remote part of the area. On average groups called twice on only about one quarter of their call days, and the mean bout duration varied from 11 to 19 min.

TS1 rarely called until the last three months of the study, when they called frequently (Chapter 5); all but 8 of 44 call bouts were on 26 days (out of 41) in March, April, and May 1970. Calling by RS2 did not increase so markedly, but relations between the male and subadult were different. For TS1 the female and subadult both started 39% of bouts, and the male 23%. Their daily pattern of calling was negatively correlated with resting, not correlated with feeding, and positively correlated with travel (Spearman rank correlation coefficient, $r_S = -0.79$, 0.50, and 0.86, respectively).

Table 6.5. Group call characteristics of siamang during 12 months from May 1969 to April 1970 inclusive

Group	Days, %	Call bouts/ call day	Percentage of days with more than one call bouts	Duration min	n
Kuala Lompat (140 days)					
TS1	15	1.43	38	10.7	44
S2	20	1.32	25	17.6	7
S2A	19	1.27	15	18.9	10
S2B	14	1.21	21	12.4	10
S5	39	1.19	13	17.7	44
S7	41	1.09	9	16.0	15
S6 (2 groups)	58	1.35	30	13.7	12
Total	26	1.25	27	14.6	151
Ulu Sempam (130 days)					
RS1B	28	1.08	8	16.9	22
S1	25	1.06	6	13.5	33
S2	15	1.12	6	18.1	24
S3 (2 groups)	36	1.43	32	15.4	66
S4	9	1.17	8	16.7	17
S4A	19	1.12	12	17.5	29
S12	16	1.24	24	16.3	21
S12A	27	1.06	6	23.3	37
S11	24	1.16	16	21.8	26
Total	20	1.15	14	17.6	275
Ulu Gombak (36 days)					
S6	25	1.11	11	16.8	9
S16A	14	1.00	0		
S16B	36	1.00	0	14.2	17
S16C	25	1.00	0	14.1	9
Total	25			14.8	35

2. Gibbon

Groups called on 53% of days on average, but the higher frequencies for the closer groups can be regarded as more reliable (table 6.6). TG1 and TG2 called on 75% days with two or more bouts on more than one third of those days. The call bouts lasted for 16 min on average – only 1 min longer than siamang bouts. But gibbon groups were more variable in their performance; TG1 bouts averaged 14 min, whereas those of TG4 and TG6 averaged more than 20 min.

Fig. 6.4. Location and identity of siamang (S) and gibbon (G) family groups around the Kuala Lompat Ranger Post in the Krau Game Reserve, district of Temerloh in Central Pahang.

Call bouts were classified as male only, or as male and female (see above). Male only calls were more frequent – absolutely, and in terms of the hourly total – soon after dawn (fig. 6.7). Not only did gibbon groups call earlier than siamang, but they were more closely synchronised. Both TG1 and TG2 called on 57% of days, and only one of the two groups called on 29% of days; there were only 28 days when both were silent. Furthermore, the daily patterns of calling by each group were almost identical, with more than 75% of group calls between 0600 and 1100 h (fig. 6.7). Although this synchrony can partly be explained by the adjacency of their territories, no such explanation can be invoked for the siamang in Ulu Sempam (see below).

Table 6.6. Group call characteristics of white-handed gibbons during 12 months from May 1969 to April 1970 inclusive

Group	Days, %	Call bouts/ call day	Percentage of days with more than one call bouts	Duration min	n
Kuala Lompat (140 days)					
TG1	75	1.47	34	14.6	107
G2	75	1.49	42	15.8	152
G3	48	1.33	28	14.7	11
G4	51	1.40	49	22.2	18
G5	34	1.21	17	15.9	15
G6	34	1.15	15	21.1	14
Total	53	1.38	34	16.0	313
Ulu Sempam (130 days)					
RG1	39	1.10	15	29.1	28
G1A	31	1.22	18	32.3	33
G3	60	1.29	26	24.0	52
G3A	22	1.07	7	17.3	8
G9	40	1.31	27	20.4	10
G8	20	1.35	27	21.9	32
G7	66	1.51	40	26.5	34
G7A	54	1.26	21	28.7	38
Total	42	1.39	22	26.2	235
Ulu Gombak (36 days)					
GG5	81	1.66	63	14.8	30
GG10				22.1	7

Ulu Sempam (R)

The daily pattern of group calls showed that siamang called later, and gibbons earlier, than at Kuala Lompat (fig. 6.6). Thus the temporal separation of calling in the two species was increased. The peak of siamang calls was between 0900 and 1100 h, whereas 50% of gibbon group calls had been given by 0900 h. Data from nine siamang groups and eight gibbon groups are included in this analysis (fig. 6.5).

1. Siamang

Each siamang group called on between 15 and 30% of days (table 6.5). RS4 was a very quiet group, perhaps because of recent logging activity; RS3 calls can be attributed to two groups. The mean duration of call bouts varied from 13.5 min by RS1 to 23.3 min by RS12A. Groups called more than once on call days less often than at Kuala Lompat. The timing of group calls shows a bimodal pattern. Some groups called more often before 1000 h, others after that hour, irrespective of spatial arrangement (fig. 6.7, cf. fig. 6.5).

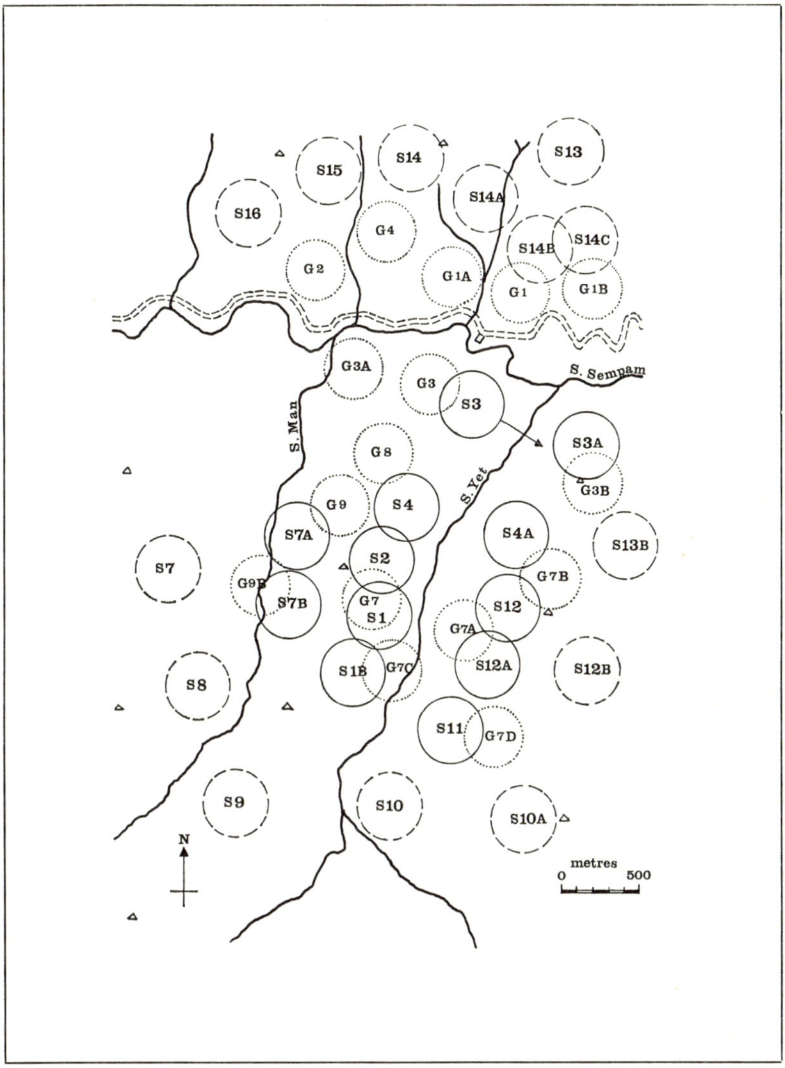

Fig. 6.5. Location and identity of siamang (S) and gibbon (G) family groups in Ulu Sempam, district of Raub in West Pahang.

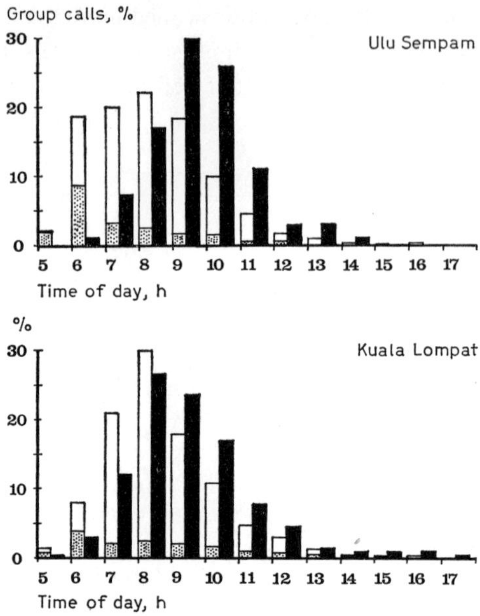

Fig. 6.6. Daily pattern of calling by groups of siamang (closed bar), and gibbons (open bar – male and female; stippled bar – male only) in Ulu Sempam and at Kuala Lompat during 17 months in 1969 and 1970.

The onset of calling by one siamang group appeared to inhibit other groups calling at that time, so that calling passed sequentially around the area. This contrasts with the chorus effect of simultaneously calling gibbon groups, which approximates to the calling pattern of howling monkeys [CHIVERS, 1969]. On 20 days when calls were long and frequent (containing 9–12 female bark series) an average of eight widely dispersed groups called. On the 28 days when there were only three to five bark series in shorter call bouts, nine groups called on average; these groups were more concentrated in a particular part of the area.

It appeared, therefore, that when many groups called the amount of calling by each was reduced through this inhibitory effect, which provided a greater opportunity for each group to 'listen' to others than occurs in many other species. This could be significant in maintaining the dispersion of groups. Although siamang attended to the calling of other groups, their subsequent actions rarely indicated a direct response. But it is difficult to interpret calling behaviour without sufficient observation of other aspects of behaviour of the groups concerned. Considering all groups involved on days when nine or more groups called, group calls were in fact longer (20.8 min) than on days when eight or fewer groups called (15.8 min). On days when groups gave many bark series (9 – 12) the duration of call bouts averaged at 17.4 min; but on days with few bark series (3 – 5) call bout duration was still as high as 16.4 min.

2. Gibbon

Although RG8 and RG9 were on the study area ridge, they were on the opposite side from RS2; thus more accurate information was on RG3 and the RG7 complex. The duration of bouts was longer than in other areas with some group calls averaging over 30 min (table 6.6). RG3 and RG7 call bouts lasted for 24.0 and 26.5 min, respectively, on average. It is clear that each group usually called on more than 50% of days, but possible that they did not call as often as at Kuala Lompat. And yet Ulu Sempam had the highest density of both species, with groups calling less often than in the other areas (see Discussion).

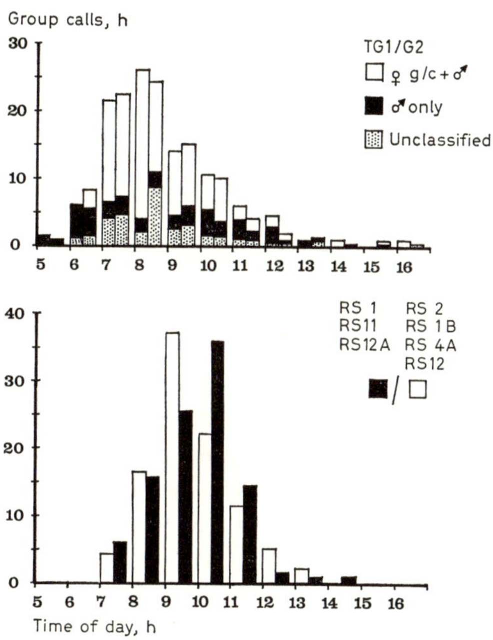

Fig. 6.7. Daily pattern (hourly) of call bouts by gibbon groups TG1 and TG2 at Kuala Lompat, and siamang groups RS2, RS1B, RS4A, RS12 and RS1, RS11, and RS12A in Ulu Sempam.

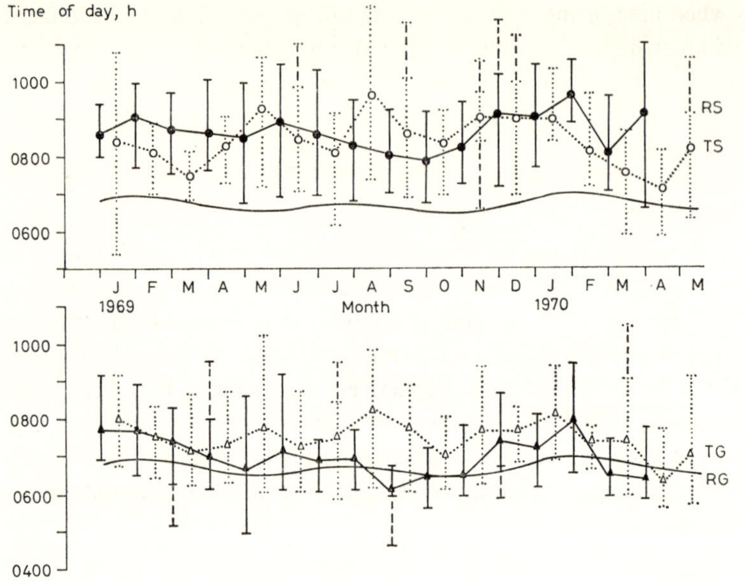

Fig. 6.8. Time of first calls (mean and range) for siamang (RS and TS) and gibbon (RG and TG) in each month at Ulu Sempam and Kuala Lompat, respectively. Broken line indicates extension of range for one observation only, curved line denotes sunrise.

Calling in Local Populations

Time of First Calls

Gibbon calls frequently started during darkness, sometimes as early as 0500 h; the average time of the first call bout starting (= onset of calling for the day) was 0656 in Ulu Sempam, and 0729 hours at Kuala Lompat (table 6.7). The wide dispersion of fewer groups in Ulu Gombak, and the hilly terrain, resulted in less accurate observations. Siamang calls were not often uttered before sunrise, and the average time of onset was between 0800 and 0900 h. Siamang, therefore, started calling later than gibbons, but differences were less marked at Kuala Lompat, and the time of first calls varied between 0430 h for gibbons, and 0530 for siamang, through until midday.

In both areas siamang calls were not heard on 6 – 7%, and gibbons on 0.5% of days; this includes days on which only one group called, since it was not possible on those days to refer to the onset of calling in the population. The greatest difference in the distribution of observations was between

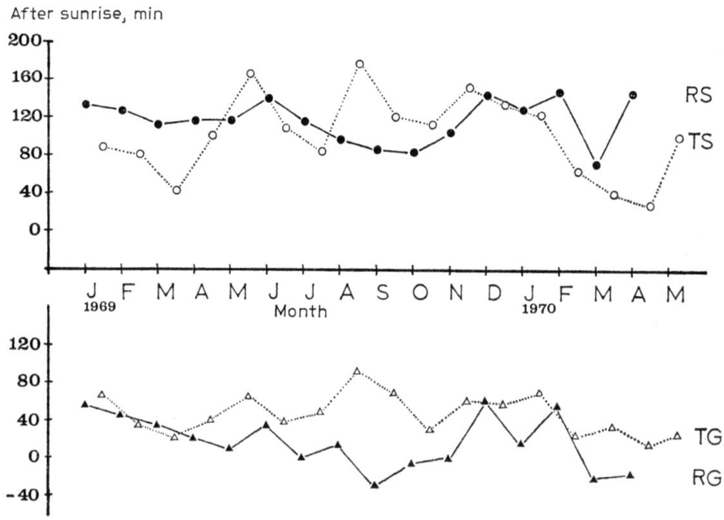

Fig. 6.9. Mean time of first calls from sunrise for siamang (RS and TS) and gibbon (RG and TG) in each month at Ulu Sempam and Kuala Lompat, respectively.

siamang and gibbons in Ulu Sempam, and the smallest between siamang in the two areas. But even the latter was significantly different (Kolmogorov-Smirnov two-sample test, D = 0.54 and 0.17, p = 0.001 – and 0.05 – 0.01, respectively). These differences in the timing of first calls were observed in most seasons; calls tended to start earlier in the wet season. Onset times were significantly different (Mann-Whitney U test) for siamang and gibbon in both study areas in the two months of greatest contrast – October and January or February (fig. 6.8).

The fluctuations in the time of first calls – represented by the means and ranges (fig. 6.8) – from one month to the next were highly correlated with sunrise for both species in Ulu Sempam (Spearman rank correlation coefficient, r_S = 0.60 and 0.64, respectively). At Kuala Lompat, the pattern for siamang approached a negative correlation. Correction of times of first calls from local time to sun time, however, showed little reduction in the spread of observations, except for gibbons at Kuala Lompat (fig. 6.9). But it can be deduced that the timing of first calls was not correlated for each species between the two study areas.

Table 6.7. Calling characteristics of gibbon and siamang groups in Ulu Sempam, Kuala Lompat, and Ulu Gombak during the 12 months from May 1969 to April 1970 inclusive

	Ulu Sempam (R)				Kuala Lompat (T)				Ulu Gombak (G)			
	siamang		gibbon		siamang		gibbon		siamang		gibbon	
	$\bar{x}$	n	$\bar{x}$	n	$\bar{x}$	n	$\bar{x}$	n	$\bar{x}$	n	$\bar{x}$	n
Onset time	0834	159	0656	170	0813	188	0729	198	0853	47	0804	47
Call hours per day	3.2		4.3		2.9		3.5		2.7		2.8	
Number of days		173				193				51		
Number of groups (estimated population)	26		16		12		9		16		7	
Group calls per day	4.8		4.6		3.4		3.9		4.1		2.0	
Daily proportion of population	0.20		0.31		0.30		0.44		0.26		0.29	
Call duration, min	16.2	439	29.0	217	14.3	174	16.0	310	16.6	97	13.6	71

Daily Number of Call Hours

Siamang group calls were uttered in 3 h of the day on average in all three areas; gibbon calls extended over a longer period – over 4 h on average at Kuala Lompat (table 6.7). Monthly variation in the mean number of hours daily with group calls indicated prolonged calling in the rainy season, with a decrease at the start and end of the dry season (fig. 6.10). This applied especially to both species in Ulu Sempam, and was weakest for gibbons at Kuala Lompat; the difference between the months of greatest contrast was significant for siamang in both study areas (Mann-Whitney U test, U = 12 and 17, p = 0.02 – for Ulu Sempam and Kuala Lompat, respectively).

Calling by both species occurred in up to 8 h daily with little seasonal variation. But calling was reduced from about 5 h daily in the wet season, to a bimodal pattern of 2 or 4 h in the dry season; this bimodality was weak but it appeared in both species in the two dry seasons in which observations were made. More conspicuous was the correlation between the time of first calls and the daily number of call hours. The earlier calling started the longer it continued; this relationship was weakest for the gibbons in Ulu Sempam. At Kuala Lompat gibbon groups called in 2 or 3 h on 50% of days, but in Ulu Sempam there were 4 or 5 h of calling on 50% of days. In both areas siamang called in 2 or 3 h on 40% of days (fig. 6.11).

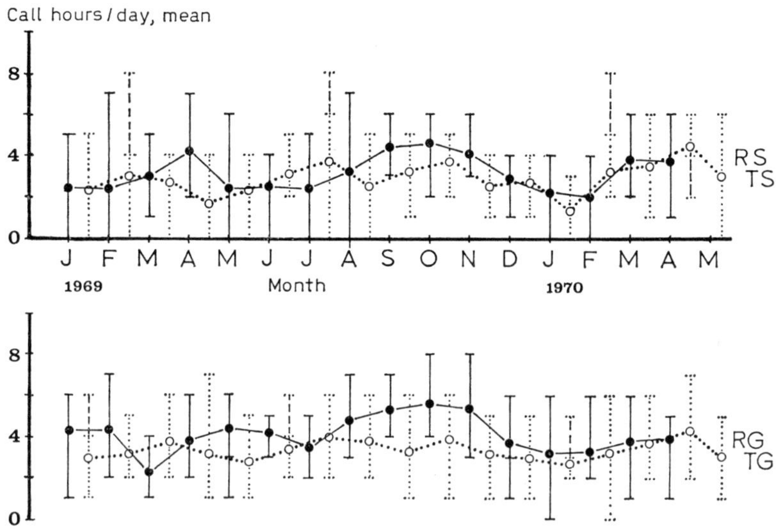

Fig. 6.10. Number of call hours-day (mean and range) for siamang (RS and TS) and gibbon (RG and TG) in each month at Ulu Sempam (closed circle) and Kuala Lompat, (open circle) respectively.

Daily Number of Group Calls

Nearly five groups of each species called each day in Ulu Sempam, and three to four of each on average at Kuala Lompat (table 6.7). It has been shown that it was reasonable to assume that each group (of siamang in particular) called only once each day, but that it was possible to accurately identify most calling groups (see above and Chapter 1). During the study it was determined that there were 26 groups of siamang and 16 groups of gibbon within hearing in Ulu Sempam (fig. 6.5); numbers were 12 and 9 at Kuala Lompat (fig. 6.4), and 16 and 7 in Ulu Gombak (fig. 6.2 and 6.3) for siamang and gibbons, respectively.

The significance of these densities will be discussed in Chapter 7, but for comparing calling between species and localities proportions of the local populations calling daily were calculated. In Ulu Sempam only one fifth of the siamang population called each day, i.e. one fifth of groups (on the assumption that all group members call); whereas at Kuala Lompat nearly one half of the gibbon population called daily on average (table 6.7). In Ulu Sempam on nearly two thirds of days less than 20% of siamang groups called, but at Kuala Lompat this low level of calling was recorded on less than 40% of days (fig. 6.11). For gibbons less than 40% of groups called on 80% of days, but at Kuala Lompat more than 40% of groups called on more than 50% of days; the composition of calling groups varied considerably from day to day.

The differences between seasons confirms the tendency for increased calling in the wet season, followed by a marked decrease in the dry season; the difference between the months of greatest contrast was significant for both species in Ulu Sempam (Mann-Whitney U test). Thus for siamang, the proportion of days on which less than 20% of the population called increased from 20 to nearly 80% in the dry season in Ulu Sempam. At Kuala Lompat the proportion for less than 40% of gibbon groups increased from 30 to 50% in the dry season.

Day-to-day fluctuation in the number of calling groups was considerable. But over 10 days of observation consistent patterns emerged in each month, in terms of both means and ranges of observations (fig. 6.12). Thus the wet season proportion of about one third of siamang groups calling daily declined

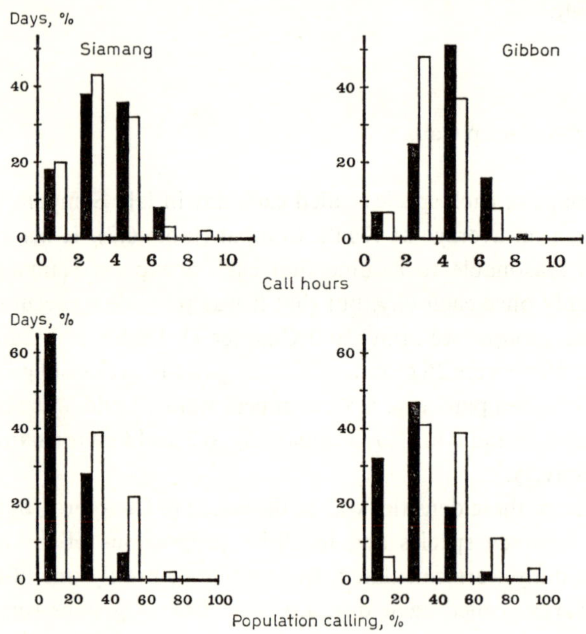

Fig. 6.11. Distribution of days according to number of hours in which calls were uttered and proportion of population calling daily, for siamang and gibbons, during 17 months in Ulu Sempam (closed bar) and at Kuala Lompat (open bar).

to about 10% in the dry season. Similarly, the proportion of gibbon groups calling daily declined from about 40 to 20% in Ulu Sempam in the dry season; there was no such pattern at Kuala Lompat.

In Ulu Gombak in 1968, at the height of an unusually heavy fruiting season, nearly two thirds of siamang groups called daily (fig. 6.13). This was followed by a decrease to less than 10% in the dry season month of March 1969. At other times 20 – 40% of siamang groups called daily, as in the other study areas. The gibbons displayed a similar pattern, but with smaller amplitude, and a peak of calling after the 1969 dry season when up to 50% of groups called daily.

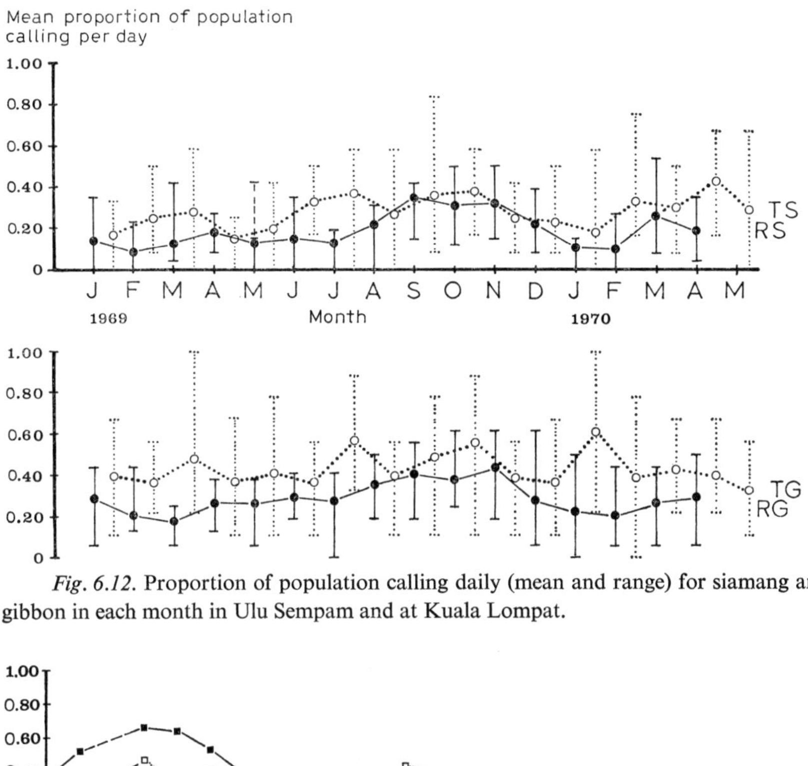

Fig. 6.12. Proportion of population calling daily (mean and range) for siamang and gibbon in each month in Ulu Sempam and at Kuala Lompat.

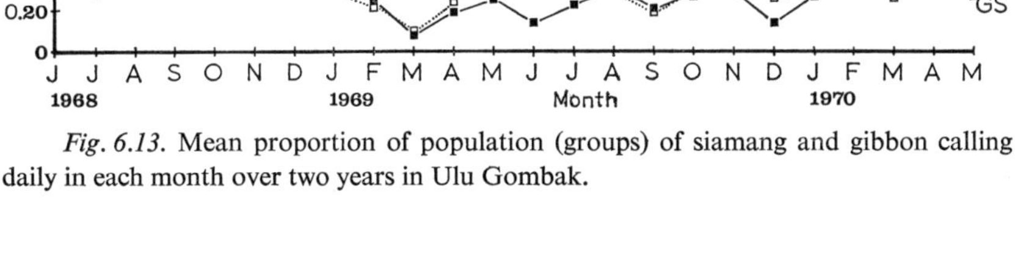

Fig. 6.13. Mean proportion of population (groups) of siamang and gibbon calling daily in each month over two years in Ulu Gombak.

Duration of Call Bouts

Siamang call bouts lasted for about 15 min; gibbon call bouts were markedly longer only in Ulu Sempam (table 6.7). 75% of siamang bouts in Ulu Sempam, and 86% at Kuala Lompat, lasted for less than 20 min; only 53% of gibbon bouts in Ulu Sempam, and 78% at Kuala Lompat, were as short. There was little seasonal variation in the duration of call bouts, although they tended to be shorter in the wet season.

The duration of call bouts fluctuated apparently at random from month to month, although siamang showed less variation, especially in Ulu Sempam (fig. 6.14). There was no correlation between monthly means for call bout duration and frequency, and no other indication of relationships to help explain the changing duration of call bouts. The varied duration of gibbon call bouts has already been related to the presence or absence of female great calls and the time of day; this may be related to a decrease or increase respectively in inter-group conflicts.

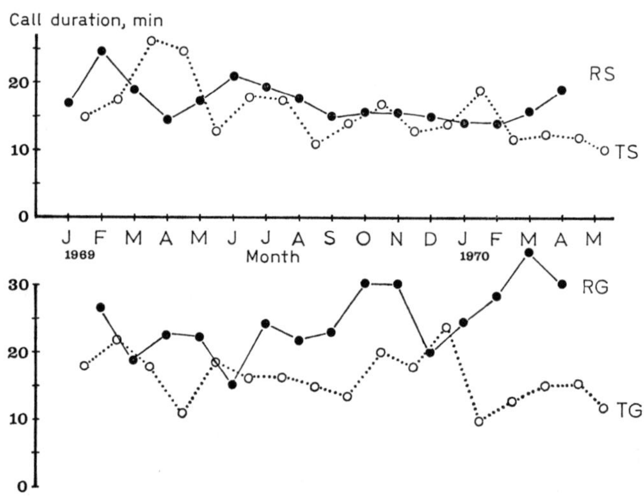

Fig. 6.14. Duration of call bouts (mean) for siamang (RS and TS) and gibbon (RG and TG) in each month in Ulu Sempam and at Kuala Lompat, respectively.

Table 6.8. Calling characteristics of gibbon and siamang under different weather conditions, and seasonal variation in calling

		Hot, sunny		Cloudy, wet	
		siamang	gibbon	siamang	gibbon
Onset time	R	0827	0646	0847	0649
	T	0751	0712	0904	0758
Total		0804	0701	0853	0751
Daily proportion of population	R	0.19	0.30	0.21	0.31
	T	0.32	0.44	0.26	0.40
Total		0.26	0.37	0.24	0.36
Call hours per day	R	3.1	4.2	3.0	4.4
	T	3.3	3.7	2.6	3.5
Total		3.2	3.9	2.8	4.0
Number of days	R		14		24
	T		21		17
Total			35		43

	Siamang		Gibbon	
	Ulu Sempam	Kuala Lompat	Ulu Sempam	Kuala Lompat
Number of group calls/day				
January – April 1969	2.57	2.13	3.86	4.15
May – August 1969	2.22	3.00	4.34	4.62
September – December 1969	3.60	2.90	5.68	4.91
January – May 1970	1.91	3.77	3.00	4.46
Number of days	165	203	165	203
Total	2.54	1.94	4.18	4.52
Number of groups	11	9	9	9 (7)
Number of calls/day/group	0.23	0.22	0.46	0.57

Calling and Weather

Calling in both species started nearly an hour earlier on hot, sunny days (table 6.8); this was only significant for the siamang at Kuala Lompat (Mann-Whitney U test, $N = 36$, $U = 75.5$, $p = 0.02 - 0.002$). There was no difference in the average onset time for gibbons in Ulu Sempam, and the siamang there started calling only 20 min later on cloudy and wet days. (The daily pattern of rainfall has been described in Chapter 4.) Otherwise there was virtually no difference in calling on days of contrasting weather.

Reference has already been made to seasonal variations in calling (indirectly related to weather) – to the increase in calling during the wet season, followed by a marked decline in the dry season. The increase was apparently related to increased food availability in good weather at the start of the rainy season; when the rains came they inhibited calling as expected, with the effect continuing through into the dry season.

Discussion

Loud calls play an important part in inter-group relations in many species of primate. Whereas terrestrial social groups mainly avoid, or interact with, conspecifics by visual signals [HALL, 1965; DEVORE and HALL, 1965] auditory signals assume great importance in forests, where visibility is restricted [JAY, 1965; CHALMERS, 1967; YOSHIBA, 1968]. The dawn chorus of calls by howling monkey groups is followed by mutual avoidance during the subsequent day ranges of each group [CHIVERS, 1969]. The two species of leaf monkey in Malaya, *Presbytis obscura* and *P. melalophos*, both have loud calls. The nasal 'chengkong' of the former was most often heard at dawn, and the harsh chatter of the latter before dawn and after dusk [CHIVERS, 1973].

But gibbons and siamang have particularly loud vocalisations and complicated bouts of calling. The ritualised patterns, with sex-specific interdependent utterances, indicate far more than simple functions in inter-group spacing, or group cohesion. The prominent part played by the adult female is also unusual among primates. This discussion will examine the variation in calling patterns in association with possible causes and functions of group calling bouts – both proximal and otherwise.

While it is possible to correlate the onset of calling in some species with increased light intensity at dawn, it appears that intrinsic factors play a more important part in those species which habitually call later in the day. Thus the main group calls of gibbon and siamang follow the first feed of the day, and sometimes a rest in the early morning sun. Calling may follow as an expression of individual well-being to reinforce group cohesion.

Once calling has started for the day, it is easier to see why other groups start, although siamang did not respond to the calls of neighbouring groups as rapidly as do other arboreal primates. When they did call they allowed the other group to complete the major part of its bout before starting, and thereby inhibiting calling by the first group. Thus siamang calls passed sequentially through the forest, whereas gibbon groups responded more quickly to their neighbours, so that several groups were often heard calling simultaneously.

Group calls of siamang were sometimes elicited by the activities of other animals – not necessarily potential predators. On some days calling throughout the study area was apparently initiated by one group calling in alarm at some danger, but on others siamang started calling because of the persistent proximity and calling of a gibbon group. It was interesting that the force of siamang calls was minimised by the group facing away from the gibbons. It therefore appeared that many calling bouts were started as an outlet for alarm or aggression, or both; this can be extended to an intra-specific (territorial) situation.

Whatever the proximal cause, such group calls early in the day would usually initiate calling by about one quarter of the siamang population, and over half the gibbons. The functions of these calls are even more complex than the causation, but there appear to be three main components: (1) group cohesion, specifically expressed in alarm and transformed to aggression; (2) dispersion, by advertisement of the group territory, and (3) territorial defence.

ELLEFSON [1967] has drawn attention to the special importance in small social groups of calls and gestures that may have a dual function in communication within and between groups. Alarm calls were initially audible only within the group, but subsequently the signal was transmitted to other groups, although of less relevance. The small size of the group also increases the frequency and significance of inter-group communication at the expense of calls and gestures within the group. The relative absence of the latter in the confined social network of the siamang group has already been described (Chapter 5).

It is clear that concerted group calls can have an important cohesive effect, and that the elaborate ritualisation may reflect this function. At the same time they provide a very clear signal to neighbouring groups. It is also clear in siamang, that in terms of defining the area within which conspecifics will not be tolerated, they do not need to reinforce this signal as gibbons apparently do, by frequent conflicts along territorial boundaries (Chapter 2).

The difference in calling between siamang and gibbon can largely be explained by the larger territory of the white-handed gibbon, which results from its more specialised diet. Thus gibbon family groups call earlier in the day, in chorus, more often and for longer than do siamang groups. Gibbon calls were also more clearly differentiated according to the distance of neighbouring groups, i.e. whether or not there was a confrontation across the territorial boundary.

MARLER [1968] has categorised signals between groups as (1) distance increasing, (2) distance maintaining, and (3) distance reducing. In many arboreal species close proximity is avoided by groups maintaining distance by auditory cues; this is true for the siamang, except that avoidance responses are stabilised by means of a permanent territory. Although gibbon territories appear to have equal permanence, gibbon calls often serve initially to decrease distance between adjacent groups. Visual signals then act with different calling patterns to increase distance between the groups concerned, although this has no lasting effect in terms of space. It does promote increased wariness by groups in the 'no-man's land', thereby reducing the likelihood of encroachment [ELLEFSON, 1967].

The gibbon also differs from the siamang in the way subadult gibbons are excluded from group calls. ELLEFSON asserts that the adults will not tolerate the close proximity of an apparently adult conspecific. But the participation of subadult females in gibbon group calls was established in both study areas, and it is only the absence of diagnostic male calls that prevents an equally confident assertion for subadult males. The value of sex-specific calls and individual idiosyncracies in identifying groups and determining their composition has already been indicated. The GS6 female has an odd wheeze, audible between booms and barks; the subadult male who was first observed in GS16C, was traced through GS16B, to GS16D by the structure of his scream. It is immediately obvious if two groups are interacting at the same location, because female calls will not be synchronised; it may subsequently be possible to determine the numbers of females and males present.

Finally, reference must be made to variations in calling patterns in each species during the study. Day-to-day fluctuations can be explained mainly in terms of intrinsic or other factors resulting in the onset of calling, and weather conditions. Gibbons were more consistent from one season to the next, for reasons advanced above. But siamang showed fluctuations that can be explained by one or more of the following: (1) food availability enhancing individual energy intake to surplus levels; (2) reproductive behaviour, with increased calling in consorting pairs and unmated individuals, and (3) territorial adjustments, resulting from disturbance, such as logging.

As the loggers worked along the ridge to the east of S. Yet in Ulu Sempam in the 1970 dry season they were greeted by increased calling from the siamang. The subsequent silence in similar circumstances elsewhere has given rise to theories among the local people of nomadism in siamang, but earlier selective logging activity on the study area ridge has indicated that affinity for their territory is too strong. Although RS2 had no infant in 1964–65 (unless it died young), they still remained; similarly the logging incursions in 1970 had no obvious effect on their territorial limits, although the group retreated from the immediate vicinity of the disturbance. There were 100 ha of forest available for each siamang group at Kuala Lompat, and only 67 ha in Ulu Sempam, but there was more calling in the lowland area, even when the gibbon population is taken into account.

The peak of calling in 1968 (fig. 6.13) appeared to be related to the unusually high abundance of food (Chapter 3), but in Ulu Gombak it seems likely that the increase in logging in surrounding valleys is responsible for the high density of siamang there (Chapter 7), and contributory to the increased calling as new territorial limits were established. In the 1970 dry season increased calling was clearly associated with breeding (Chapter 5). This most prominent aspect of siamang behaviour is therefore closely associated with many activities, in both intra- and inter-group contexts.

Summary

The concerted group calls, common to both siamang and gibbons, are important in communication between social groups. Siamang groups called later in the day, less often, and sequentially, with greater seasonal variations. Siamang calls started between 0700 and 0900 h on 71% of days, whereas gibbon calls started between 0600 and 0800 h on 69% of days – about 1 h earlier on average.

Siamang called in 2, 3, or 4 h of the morning on 65% of days, but gibbon calls were more frequent occurring in 2–5 h on 81% of days. The earlier calling started, the later it continued; calling started earlier in the wet season (which was partly correlated with sunrise), but otherwise calling tended to start earlier and be more frequent in sunny weather. About 25% of siamang, and more than 50% of gibbon groups in each locality called each day.

Group calls (bouts) in both species lasted for about 15 min, but in Ulu Sempam gibbon calls were longer, where a higher proportion were male only. 64% of gibbon calls occurred before 0900 h, and 60% of siamang calls were between 0900 and 1200 h in Ulu Gombak in 1968. Gibbon groups called twice in one day more often than siamang.

In both species sex-specific patterns of calling were derived from similar vocalisations. During siamang call bouts initiated by male chatters or female grunts, intensive series of rhythmic barks by the female culminated in movement bark-chatters by all, and screams by the male at 3-min intervals; they were interspersed by quiet periods with occasional booms and barks. On average there were six diphasic female bark series in a bout composed of 15 followed by seven barks. Females varied in pitch and rate of barking; males screamed three times/minute on average. There was little difference in calls prior to inter-group conflicts, but when alarmed siamang calls were composed of discordant barks, booms, and bark-chatters after prolonged grunting.

27% of gibbon call bouts, apparently involved male hooting only; they occurred mainly at dawn lasting about 29 min, which was 13 min longer than bouts in which female great calls, about four, were given at intervals of 3 – 5 min, each lasting for about 20 sec. Male hooting also predominated in inter-group conflicts.

There was a high degree of consistency in the calling behaviour of each species, and there were also some similarities between siamang and gibbon. Calls were initiated in response to environmental factors such as weather and the presence of other animals (not necessarily conspecifics), and by intrinsic factors as yet undetermined. Once calling had started each day, groups responded to each other, but not necessarily to those closest.

Apart from promoting group cohesion these group calls appeared to be important in maintaining an adequate dispersion, and in defending territorial boundaries. This was more elaborate in gibbons than in siamang, presumably because of the need to defend more specialised and widely dispersed food sources. Calling patterns varied according to food availability, the stage of the population's reproductive cycle, and the amount of disturbance in the area.

Chapter 7: Distribution and Density

Introduction

The distribution and density of siamang in the Malay Peninsula were assessed during the preliminary search for study areas (Chapter 1). The results are presented here to see how the behaviour observed in social groups (Chapters 2–5) might reflect high or low population density, and extensive or restricted distribution. Data are also given for the white-handed gibbon to continue the detailed comparison initiated in Chapter 6, and to see to what extent behavioural differences between the two species, especially in calling, can be explained by distribution and density, and by major ecological separation.

Following a description of the distribution of hylobatids in West Malaysia, in which a discussion of the status of the black-handed gibbon is introduced, the total populations of each species are estimated by several methods. Great reliance is placed on the frequency of calling in each locality visited (Appendices E and F).

Distribution

Siamang

During five months in the latter half of 1968, 101 localities were visited in all States of West Malaysia except for Perlis (fig. 1.3; Appendix A). Reliable records from other sources were included where they provided additional information of importance in determining the distribution of each species. Observations were concentrated around the periphery of the siamang's range, as the secondary aim to finding suitable study areas centrally was to define the limits of the range. There was no time to visit remote central areas or the interior of Trengganu. But whilst siamang are present in good numbers in and around the Taman Negara [WELLS, personal commun.], they appear to be absent from most of the East Range [Game Department, personal commun.].

Siamang were located mainly in hill forest, over about 1,000 ft elevation.

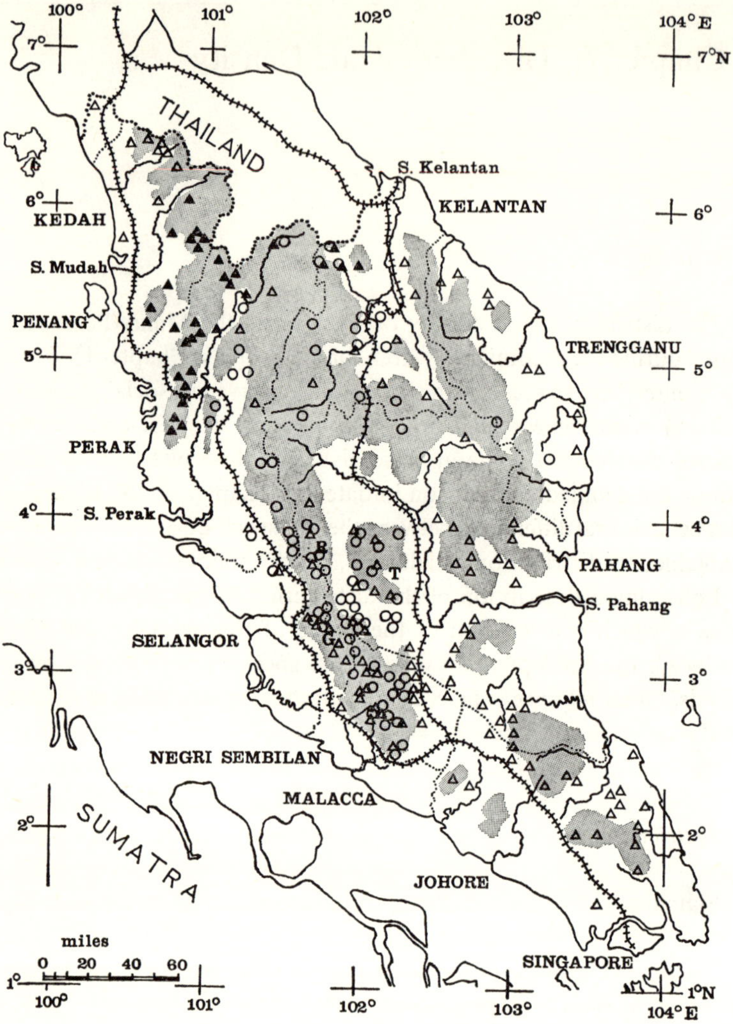

Fig. 7.1. Distribution of siamang (open circle), white-handed gibbon (open triangle), and black-handed gibbon (closed triangle) in West Malaysia. Also shows survey localities, and includes some records of Dr. J. Ellefson, Mr. D. Labang, Mr. Lim Boo Liat, Enche Mohamed Khan, Lord Medway, Mr. Cecil Rajah, Dr. D. Wells, Dr. T. C. Whitmore, and Mr. Yong Ghong Chong.

Fig. 7.2. Probable limits of distribution of the siamang (continuous line), the black-handed gibbon (dotted line), and the white-handed gibbon (dashed line) in the Malay Peninsula. Study areas: G = Ulu Gombak; R = Ulu Sempam; T = Kuala Lompat.

They occur throughout the main range of mountains (fig. 1.2), extending eastwards along the mountains of the Pahang-Kelantan border through the Taman Negara, into the south-west corner of Trengganu [WHITMORE, personal commun.]. They are absent from Johore, in the south of the country, to the north and west of the Perak river, and to the east of the Kelantan and Pahang rivers (fig. 7.1). The only lowland forest in which siamang were found was west of the main road and railway around the Perak-Selangor border, and south of Gunong Benom in central Penang (fig. 7.2).

Gibbons

1. White-Handed (lar)

Gibbons were located in almost every locality visited – in lowland or highland, inland or by the coast – where dry forest still stands. Thus *lar* occurs through most of the country (see below for exceptions), although sightings decreased further north up to the banks of Sungei Perak on the west, and S. Kelantan on the east (fig. 7.1). They also occur in Kedah, north of the Mudah river (see Appendix F for note on colour phases).

2. Black-Handed (agilis)

The Malayan sub-species of white-handed gibbon is separated from the Thai sub-species, by the black-handed gibbon. Originally known from Maxwell's Hill in north-west Perak, reports were received from the State Game Warden of Kelantan of their presence in the north-west of that State. These were confirmed by observations, especially of the distinctive calls (Appendix F). *Agilis* is distributed in dry forest from the west coast, between the Mudah and Perak rivers, towards the east coast, north of the Kelantan river and one of its tributaries – S. Pergau (fig. 7.1). On the east it extends an unknown distance into southern Thailand, although there are similar river systems which probably limit its distribution to south of the Perlis-Thai border.

Altitudinal Zonation

Each primate species occupies a particular ecological niche, which may in part be determined by the climatic effects at different altitude. The diversity of the Malayan rain forest may be such that the ecology of different primate species is provided irrespective of altitude. While elevation above sea level provides discrete measurements, great problems remain in indicating any biological significance. SOEPADMO [personal commun.] pointed out that there may be greater botanical differences between opposite sides of the same valley, than between opposite slopes of a mountain range. Nevertheless the Malayan forests can be classified by altitude (Chapter 1), and although all primate species occur at most elevations – in lower montane as well as lowland and hill dipterocarp forests – altitudinal zonation is not immediately obvious.

Siamang were recorded between 50 and 4,500 ft above sea level; the average of 39 observations was 1,655 ft. Gibbons were observed between 50 and 2,500 ft, at an average of 930 ft above sea level (n = 45).

Comparable figures for monkeys were:

	Range, ft	x, ft	n
Dusky leaf monkey	100 – 2,600	1,085	32
Banded leaf monkey	700 – 2,400	1,440	16
Long-tailed macaque	50 – 1,800	605	10
Pig-tailed macaque	50 – 2,400	1,360	8

These figures support the impression formed that siamang occur mainly at higher elevations than gibbons, and that banded leaf monkeys and pig-tailed macaques occur in higher and remoter forests than their congeners.

MEDWAY [1972a] observed gibbons up to 4,700 ft on Gunong Benom, siamang near the summit at 6,000 ft, and banded leaf monkeys right up to the final slopes of the summit at 6,500 ft. Dusky leaf monkeys were common from 700 to 2,800 ft above sea level, long-tailed macaques were seen at 700 ft, and pig-tailed macaques at 2,400 and 3,600 ft. Although these observations confirm the trends described, albeit extending the upper limits in several cases, MEDWAY [personal commun.] points out that dusky leaf monkeys occur at the summit of Bukit Fraser (4,777 ft above sea level), and pig-tailed macaques live in the mangrove swamps of the west coast – where the former also occur [POLUNIN, personal commun.].

The absence of altitudinal separation of species of Malayan primates is further indicated in the report of the 1905 expedition to Gunong Tahan (in the Taman Negara). Siamang were collected at 3,000 ft, a gibbon at 500 ft, dusky leaf monkeys at 200 and 500 ft, and banded leaf monkeys at 200 ft above sea level [BONHOTE, 1908]. All species were observed up to 3,000 ft, but no higher. Gibbons, dusky leaf monkeys and pig-tailed macaques figure in the list of mammals occurring in the lowlands of Pahang [KLOSS, 1911]. LIM BOO LIAT [1969] records pig-tailed macaques as rare in coastal forests, and the silver leaf monkey *(Presbytis cristata)* as confined to swamp forest on the west coast, but his data on forest primates inland are too incomplete for useful interpretation.

Nevertheless it is possible that each species is concentrated at certain elevations, but in the absence of more detailed quantitative evidence it appears

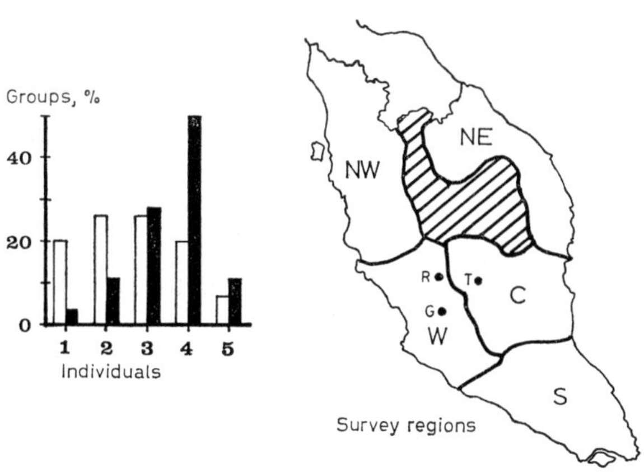

Fig. 7.3. Frequency of group size in siamang (closed bar) and gibbon (open bar, – one to five individuals), and survey regions showing location of study areas, in West Malaysia.

unlikely that the different ecologies of Malayan primates are clearly reflected by altitude. The situation will be better understood when the ecological requirements of each species have been determined.

Group Size

Returning to the distribution and density of hylobatids, 18 family groups of siamang were accurately counted, including those in the study areas. They were composed of 3.6 individuals on average (3.8 in the study areas); 45% of observations were of groups with four individuals (fig. 7.3). Only 15 groups of white-handed gibbon were accurately counted; the average size was 2.7 individuals (fig. 7.3). Nearly 50% of observations were of groups with two or three individuals, but observation conditions were such that these figures may still be an under-estimate. In the study areas, where groups of four or five were seen, the average group size was 4.0 (cf. 4.4, n = 21, CARPENTER [1940]; and 3.3, n = 28, ELLEFSON [1967].)

Population Density

It was very difficult to observe social groups of gibbons and siamang during the survey, even to ascertain group size (see above), because of (1) their elusive and arboreal habits, and (2) the structure of the tropical rain forest. But on the basis of prolonged observations in Ulu Gombak (Chapter 6), it was possible to estimate the number of groups – and thence the number of individuals – of each species in an area after a visit of only one day. This was because of the regular calling habits of each species, which was confirmed in Ulu Sempam and at Kuala Lompat. The decline in calling in 1969, however, precluded the use of the data from these main study areas in the estimates of population densities.

The 101 survey localities were classified according to the presence of: (1) no hylobatids, (2) many siamang, no gibbons, (3) many siamang, few gibbons, (4) many siamang, many gibbons, (5) few siamang, few gibbons, (6) few siamang, many gibbons, (7) no siamang, many gibbons, and (8) few siamang, few gibbons (table 7.1). The division between many and few was chosen as more or less than six groups of siamang calling daily in about 11 km^2 of forest, and more or less than eight groups of gibbon calling daily in about 17 km^2 of forest (Appendix E). Although siamang calls carried over a greater distance, and thus a larger area could be surveyed, in some parts of the country the survey was more mobile (so that a larger area was covered daily) – hence the larger mean area sampled for gibbons.

This crude analysis supplements the distribution map (fig. 7.1) in highlighting the gibbon strongholds of Johore, Pahang, and Trengganu, i.e. to a large extent where siamang are absent, and the large numbers of both species in the southern part of the Main Range – in Selangor and Negri Sembilan (table 7.1). Siamang were also abundant in the absence of gibbons in central Perak. The effect of the presence of one species on the other was investigated by comparing densities in areas with only one species and with both siamang and gibbons.

Methods

1. To count the number of groups of siamang and gibbon calling in one day, or part of one day, in many widely separated localities (Appendix E), and to extrapolate to an estimate of the population knowing (a) the patterns of calling behaviour from prolonged observations, and (b) regional variations in calling by each species.

2. To extrapolate from accurate densities of gibbon and siamang in the three study areas (Chapter 6); this is likely to produce an over-estimate, because of the reasons for choosing study areas.

In both methods it is necessary to determine the areas of forest sampled in study and survey areas (Appendix E), and the total area available to each species in the country from a knowledge of their distribution and of the areas of forest remaining.

Results

The decline in calling in 1969 was confirmed for all three study areas in terms of the mean proportion of the population of each species calling daily:

		Ulu Gombak	Ulu Sempam	Kuala Lompat
Siamang	1968	51	62[1]	64[1]
	1969	26	20	30
Gibbon	1968	38	44[1]	67[1]
	1969	29	31	44

1 One-day visit only.

Table 7.1. Distribution of siamangs and gibbons and relative frequency

State	S−	s−	Sg	SG
Selangor	Batang Kali FR			Ulu Gombak FR and ext. Semangkok FR Ulu Selangor FR Ulu Langta FR
Negri Sembilan	Serting FR ext.			Angsi FR Berumbun FR Teriang FR ext. Johol FR
Kelantan	Kg. Amai	Kemubu		
Terengganu				
Pahang, east and central			K. Lompat Bt. Patong	Lakum FR
Pahang, south and Johore				
Pahang, west		Ulu Teranum FR Ulu Kali	S. Sempam Bt. Tinggi FR Bt. Gelong	
Perak	Selim Hills FR Kledang Saiong FR Chior FR Piah FR ext.	Tapah Hills FR Korbu FR		
Kedah				

Sg = many siamangs, few gibbons; − G = no siamangs, many gibbons. No hylobatids recorded at: Serendah FR (Ulu Liam), Sel.; G. Bujang Melaka, Kampar, Perak; Sungkai, Perak; Bukit Legong FR, Sel.; Bt. Ulu Telang, Pahang; G. Tampin, NS; S. Lawin, Perak.

sg	sG	−G	−g
Serendah FR			
S.Lalang FR			
Bt.Belata FR			
Senaling Inas FR		Pasoh FR	Galla FR
Gapau FR			Jelei FR
			Gemas FR
S.Pergau	Kg.Kalai	Sokor Taki FR	S.Rek FR
Bt.Jual		Ulu Sat FR	
		Bt.Berangin	Bt.Bauk FR
		G.Tebu FR	Kg.Keruac
		Kg.Langkap	Ulu Chukai FR
		Kg.Sekayu	
Ulu Dong FR	G.Benom	S.Muring	Beserah FR
	Bt.Besar	Berkelah FR	Berkelah FR
		Lepar FR	S.Lembing
		Kg.Seri Jaya	
		Lesong FR	Palong FR
		Labis FR	Labis FR
		Tasek Bera	Lenggor FR
		Bt.Bertangga	Ulu Sedili FR
			Panti FR
			Endau-Kota Tinggi WLR
			G.Pulai (SWCA)
S.Sia FR			Tersang FR
Lentang FR			
S.Perting			
	Kg.Bersia	Larut Hills FR	G.Besout FR
		Bubu FR	Kinta Hills FR
		Lepang Nenering FR	Bintang Hijau FR
			Padang Chong FR
			Ijok FR
			Naga FR
		Bt.Perangin FR	S.Pedu
		G.Inas FR	Bt.Perak FR
		Ulu Muda FR	G.Jerai FR
		Bt.Sebelah FR	Bt.Relau FR
			G.Bongsu FR
			Rimba Telui FR
			Bt.Enggang FR

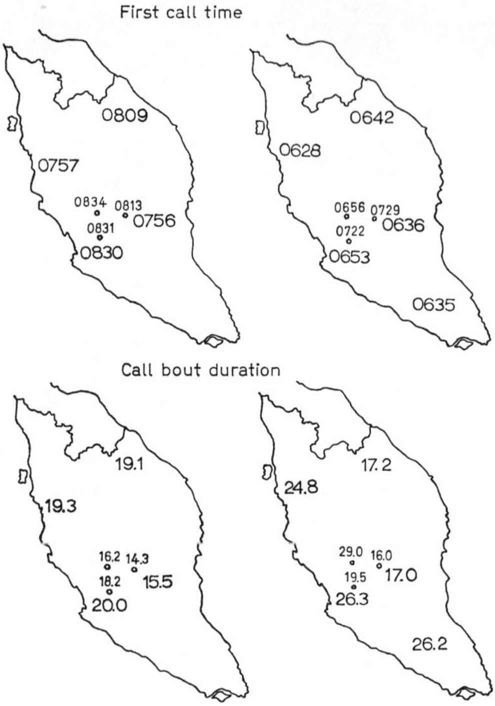

Fig. 7.4. Mean time of first call, and mean call bout duration (min), in each survey region and study area.

Hence the need to restrict estimates of population based on calling behaviour to data from 1968; the coincidence of this census with the marked increase in calling was extremely fortunate. With 51% of siamang groups, and 38% of gibbon groups, in an area (e.g. Ulu Gombak) calling daily (see CHIVERS [1972b], Appendix XV), nearly 80 and 60% of the respective populations of each species were heard after four days (fig. 7.6).

The total population size of siamangs and gibbons (white- and black-handed) in West Malaysia, was calculated on a regional basis, in terms of area and calling behaviour. The area available for each species in each region was calculated from the areas of forested land in each State (kindly supplied by Dr. LEE PENG CHOO of the Economic Planning Unit). For the gibbon this equalled the area of dry forest, but for the siamang only forest above 1,000 ft from sea level in those States, or parts of States, in which siamang occur, were included. The presence of the siamang below 1,000 ft in some localities is balanced by their absence over 4,000 ft in others.

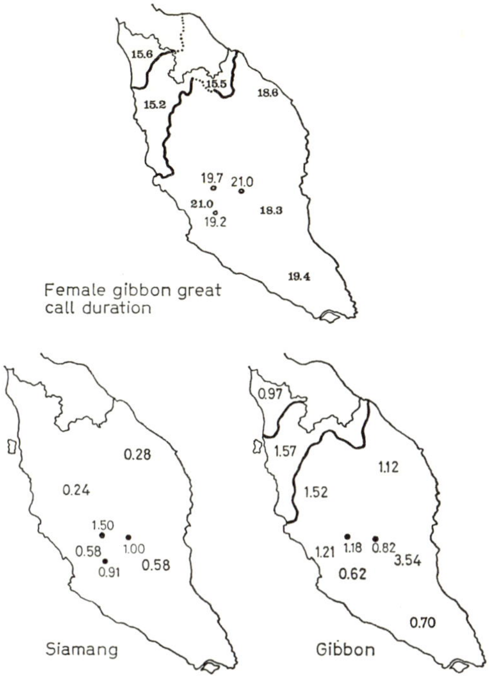

Fig. 7.5. Mean duration (sec) of great call of female gibbons in each survey region and study area, showing range of black-handed gibbon, and estimated density of siamang and gibbons in each survey region in 1968, and in each study area in 1969 (number of groups per km²).

The first major finding of the survey was that siamang are restricted to one fifth of the remaining Malaysian forest – 6.8% of this area was sampled (table 7.2), whereas only 1.8% of the more extensive range of white-handed gibbons, and 2.9% of the range of black-handed gibbons were sampled. From the number of groups heard calling, a one-day estimate of the population of each species in each locality was computed, and summed for each region.

Regional differences in calling behaviour indicated the proportion of the population calling daily on average; it was estimated to be higher in north-east and central regions (fig. 7.3) for both species (table 7.2). Siamang calls started earlier in the centre and north-east; gibbon calls started later in the west and north-east (fig. 7.4). Siamang call bouts were shorter in the central region, and gibbon calls were shorter in north-east and central regions (see also CHIVERS [1972b], Appendix XV). Gibbon groups called consistently earlier than siamang groups; but the duration of female gibbon great calls varied regionally (fig. 7.5), especially between *lar* and *agilis*.

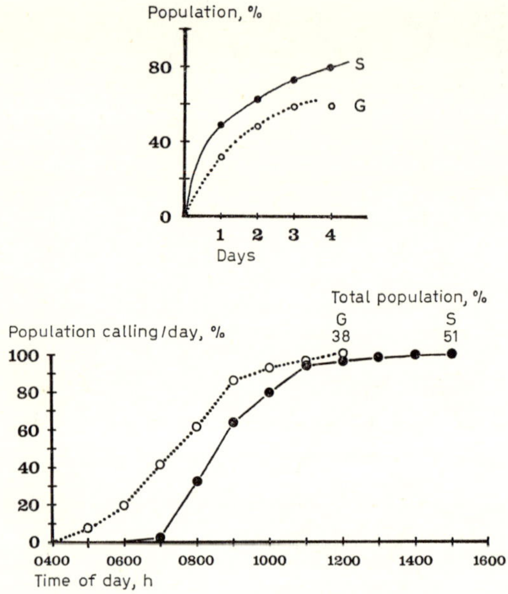

Fig. 7.6. Proportion of populations, in terms of groups, of siamang and white-handed gibbons heard calling after 1, 2, 3, and 4 days on average in Ulu Gombak in 1968, and proportion of population of each species which had called by each hour of the day in Ulu Gombak in 1968.

Thus the population of each survey locality, and for the region as a whole, was calculated – first as the number of groups, then as the number of individuals on the basis of there being 3.8 siamang and 4.0 gibbons per group on average (from the more reliable observations in the study areas; totals are also computed for an average gibbon group size of 2.7 to indicate the minimum level).

The density of siamang and gibbons, in terms of number of groups and individuals/km², in each region and study area illustrates the high density of siamang in the southern part of their range, and of gibbons in the centre of the country (fig. 7.5 and table 7.3). The one-day counts of groups were plotted in histograms according to density; there were 53 localities with gibbons only, then with siamang only and 35 with both species. There was no significant difference in the distribution of densities for both species, when occurring alone or with the other (fig. 7.7), Kolmogorov-Smirnov two-sample test.

In no area did the one-day count of siamang groups exceed a density of one group/km², whereas the density of gibbon groups exceeded this in over 40% of localitities irrespective of whether siamang were present or not; these differences were highly significant. The density of hylobatids was also significantly different between areas with only one species and areas with both. It must be remembered that the one-day counts represent only 50 and 40% of the siamang and gibbon populations respectively; thus all scores can be at least doubled to give real densities.

The estimates of total population size were finally computed in a variety of ways to account for variations in (1) calling, and (2) density in each region at different times.

Table 7.2. Regional distribution and size of hylobatid populations

Region	Survey area, km²	Area available, %	Groups	One-day estimate	Total population, %	Total estimate: survey area	Total estimate: total area	Individuals, total estimate	Locations
Siamang									
North-east	150	3.9	25	25	60	42	1,069	4,060	5
Central	129	3.5	34	44	60	72	2,094	7,970	6
South			0						
West	611	16.5	158	195	55	351	2,129	8,110	28
North-west	241	4.2	22	26	45	58	1,375	5,220	6
						523	6,667	25,360	
	1,130	6.8	239	290	51	568	8,370	31,750	45
					30	967	14,230	54,150	
Gibbon, white-handed (Malaya)									
North-east	160	0.8	83	79	43	175	21,820	87,280	10
Central	81	0.7	138	147	43	342	48,850	195,400	14
South	665	4.3	133	169	39	433	10,750	43,000	14
West	315	2.6	162	219	40	548	21,070	84,280	27
North-west	14	0.2	3	7	30	23	11,650	46,600	2
						1,500	114,140	456,560	
	1,244	1.8	519	621	38	1,634	90,800	363,200	67
					60	1,068	57,550	230,200	
Gibbon, black-handed									
North-east	42	3.2	23	22	45	49	1,528	6,112	4
North-west	133	2.2	72	97	40	242	11,030	44,120	15
							12,558	50,232	
	175	2.9	95	119	40	275	10,260	41,040	19
					60	199	6,875	27,500	
Gibbon, white-handed (Thailand)									
Kedah,	72	3.2	19	28	40	70	2,190	8,760	5
Perlis					60	47	1,458	5,832	

Table 7.3. Density of siamangs and gibbons in West Malaysia

Region	Siamang		Gibbon		
	mean number of groups/ km^2	mean number of siamang/ km^2 ($\bar{x} = 3.8$)	mean number of groups/ km^2	mean number of gibbons/km^2	
				$\bar{x} = 4.0$	$\bar{x} = 2.7$
North-east, 1968	0.28	1.06	1.12	4.48	3.02
Central	0.58	2.20	3.54	14.16	9.56
South			0.70	2.80	1.89
West	0.58	2.20	1.21	4.84	3.27
North-west	0.24	0.91	1.52	6.08	4.11
Overall survey	0.40	1.52	1.66	6.64	4.48
	0.51	1.94	1.32	5.28	3.57
Ulu Gombak, 1968	0.91	2.40	0.62	2.48	
Ulu Sempam, 1969	1.50	5.70	1.18	4.72	
Kuala Lompat, 1969	1.00	3.80	0.82	3.28	
Overall study	1.10	4.18	0.82	3.28	2.21
Black-handed gibbon			1.53	6.12	4.13
White-handed gibbon (Thai)			0.97	3.88	2.62

1. Siamang

If the pattern of calling in Ulu Gombak in 1968 was applicable to the country as a whole, a siamang population of 31,750 is indicated with 1.7 individuals/km^2 (table 7.4). But this is a locality of high density, with perhaps a corresponding increase in the frequency of calling. Correcting for this, according to regional variations in calling (table 7.2) suggests a population size of 25,630 individuals – 6,667 groups of 3.8 individuals. A much larger estimate was obtained on the basis of calling behaviour in the three main study areas in 1969; although the decrease in calling was corrected for, these localities are unrepresentative because of the high density of hylobatids.

Similarly, the original estimates are more than doubled by applying the study area density of 4.2 siamang/km^2 to the country as a whole. Results from the complete one-day estimates indicate a density of 0.22 groups/km^2 (0.84 individuals) and give a total of 28,260 siamang in West Malaysia. There were other indications that the populations in the survey localities were underestimated, because of the assumption that more than 50% of groups called daily on average; it may have been fortuitous that more than 60% were heard on the one-day visits to Ulu Sempam and Kuala Lompat. If the 1969 calling

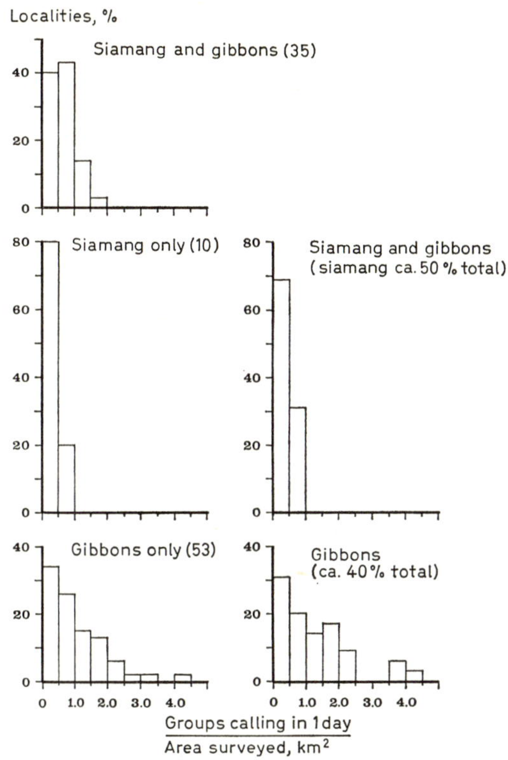

Fig. 7.7. Densities of hylobatid groups (after one day of calling) in localities with siamangs and/or gibbons.

peak in these areas, when 34% of groups called daily, were applicable to the country as a whole in 1968, a total population of about 48,000 is indicated.

It is with these reservations in mind that the siamang population of West Malaysia in 1968 is estimated to be 30,000 $\pm$ 5,000.

2. Gibbons

There was a greater margin of error in estimating the much larger population of gibbons, mainly because of difficulties in determining the levels of calling throughout the country; there was much greater regional variation than with siamang. From the known density of 3.3 gibbons/km² in the study areas, it is possible that the overall estimate of 5.8 gibbons/km² resulted from an underestimate of the proportion of the population calling daily, i.e. 60% might be a more likely proportion than 40%. This reduces a total estimate of over 400,000 to about 230,000.

Table 7.4. Total population estimates

Form of estimate	Siamang	Gibbon		
		white-handed		black-handed
Overall estimate from regional survey	31,750	363,200	245,300	41,040
Corrected for regional variation in proportion of groups calling per day	25,360	456,560	308,200	50,232
Corrected from behaviour in study areas	54,150	230,200	155,300	27,500
Overall estimate from study area density	70,650	224,800	151,650	23,680
Overall estimate from complete one-day estimates	28,260	268,700	180,800	28,120
Number of groups/km²	0.22	0.39		0.39
Mean group size	3.8	4.0	2.7	4.0
Final estimation	30,000 ±5,000	230,000 ±40,000		25,000 ±4,000

$$PS = \frac{nG}{pA} \times 10^4,$$

PS = population size; n = sum of one-day estimates of number of groups; G = mean group size; p = one-day estimated proportion of population; A = survey area as proportion of total available area.

The latter order of size is confirmed by estimates based on the density in the study areas, and in localities for which complete one-day estimates are available (table 7.4). It is therefore suggested that in 1968 there were about 225,000 white-handed gibbons south of the Perak and Kelantan rivers, 5,500 to the north of the Mudah river, and about 25,000 black-handed gibbons in Kedak, Perak, and Kelantan. This would be depressed to about 205,000 overall if 4.0 was an over-estimate of group size and ELLEFSON's estimate of 3.3 gibbons/group was more accurate.

Discussion

The limits of siamang distribution, and their high concentration in the south-west of the Malay Peninsula, support the contention that the larger dimensions of the Sumatran siamang indicate the geographical origin of the species [FRISCH, 1971]. The occurrence of fossil siamang in Java, and subfossils in Sumatra [HOOIJER, 1960] confirms the long history of the siamang in the southern part of the Sunda Shelf. It seems likely that the siamang evolved to its present form in this relatively isolated area, prior to the rapid divergence of gibbons elsewhere in Sundaland. This is in agreement with the assertion of GROVES [1972] that the siamang represents a conservative stage in hylobatid evolution, for which additional evidence was provided by a detailed study of dentition [FRISCH, 1965].

Thus, the Sumatran siamang would represent the 'mainland' form, somewhat surprisingly, and the Malayan siamang the 'island' form, in the zoological sense. This helps to explain the occurrence of siamang in the lowlands near the west coast, and their absence from most of the east coast. Barriers to their spread into the Peninsula from the south-west in relatively recent times have been the Perak river in the north, the Pahang river in the east, and Tasek Bera and the Muar river in the south; hylobatids are noted for their dislike of water. The greatest present-day concentration of siamang is in the southern part of the main range.

A more conventional explanation would suggest that siamang were part of a south-east migration of ancestral hylobatids, which then differentiated *in situ*, and are now represented by relict populations. But the current evidence conflicts with this; there is no explanation for their absence from the East Range in Trengganu. There are no obvious ecological differences in this area [WHITMORE, personal commun.] to explain why gibbons, with their more specialised diet, are so plentiful, and siamang so scarce. The greater seasonality in the north-east would be expected to mediate against a primate with dietary specialisations. It is also not clear why siamang are mainly confined to hill forest, although they are now seen to be more numerous in the lowlands than was originally thought.

The calls of the black-handed gibbon, especially those of males, were found to be very distinctive. This helped to demonstrate their distribution right across the Peninsula (as far as forest still stands) so that they effectively separate the white-handed gibbon in Thailand from that in Malaya. It is surprising that these relative distributions have not been described before [KHAN, 1970; LIM BOO LIAT, 1969]. The taxonomic status of *lar* and *agilis* has still be determined. GROVES [1972] puts both as sub-species of *Hylobates lar*, but he was not aware of the more extensive distribution of *agilis*, or in the difference in colouration between the two forms of *lar* (Appendix F). He further assumes that *lar* and *agilis* are sympatric in parts of their ranges and not reproductively isolated.

It has not yet been possible to penetrate the highland areas near the Thai border, where rivers no longer present a barrier, and where sympatry may occur. Although the density of *lar* decreased as one moved northwards towards *agilis*, and calling characteristics changed somewhat, there is no reason why sympatry should not occur. The distinctive morphology and calling of *agilis* indicate that where they meet each is likely to behave as good biological species. It remains to be seen whether *agilis* fills the ecological niche of the siamang, or *lar*, or both in Kedak, Perak, and Kelantan. The sympatry of siamang and *agilis* in Kelantan suggests the former to be unlikely – to the east of the Main Range at least.

In addition to seasonal changes in calling behaviour, there were indications that calling varied in frequency with changes in population density. Gibbons started calling earlier, with more longer group calls daily, in the south of the country. It was not possible to relate calling with behaviour, as in howling monkeys [CHIVERS, 1969]. This was because calling did not serve the same function of maintaining 'group space'. Hylobatid 'group space' was already maintained in the form of territories, and calling merely reinforced these stable boundaries.

Nevertheless, calling increased in response to disturbance (Chapter 6). This included reactions to conspecifics, which would be more frequent in areas of high density if competition for space was increased. This may be the explanation for the early vigilance and calling in certain localities. But it has been suggested that the increase in calling in 1968 was primarily in response to the temporary enrichment of the environment, with the coincidence of fruiting by trees of different floral cycles (Chapter 3). It could be that this increase in calling was associated with more frequent disputes over preferred food sources near territorial boundaries. Whatever the complex reasons for these phenomena, it was a most favourable time to conduct a census of hylobatids based on calling behaviour. It has been shown that calling patterns were more regular when the frequency of calling was increased.

It was surprising that siamang were at a lower density in smaller territories than gibbons, but nutritional studies are still necessary to confirm the dietary specialisations of gibbons. The lack of modification of the patterns of dispersion characteristic of each species in the presence of the other was also unexpected. This indicates the clear ecological separation of the two species, and that the carrying capacity of the environment, in terms of hylobatids alone, is well below saturation.

Such a census obviously has severe limitations, and a low level of accuracy. But it has great potential for determining numbers of elusive arboreal primates. It is now possible to give some indication of the number of siamang in the Malay Peninsula. While it merely confirmed the abundance of gibbons, the restricted distribution and relatively small population of siamang are of special interest. The ability of primates – of siamang in particular – to adapt to selectively logged forest was very encouraging for their future survival. But this in no way obviates their complete dependence on the rain forest habitat, which is increasingly threatened even in remote areas. The primate populations are rapidly being divided into small 'islands' by human activity (fig. 1.3); the effects of this have yet to be realised , but the consequences could be drastic from a biological viewpoint.

Summary

The search for suitable areas for the main study of siamang ecology and behaviour was elaborated into a census of hylobatids. By travelling as widely as possible through West Malaysia during five months in 1968, the distribution of each species was plotted and the density of groups was estimated according to the number of groups heard calling in one day. Total estimates were obtained by extrapolating from the scores for 6.8% of the range of siamang, and 1.8 and 2.9% of the ranges of white- and black-handed gibbons, respectively. The two main assumptions made concerned the size of groups and the proportion of the population calling daily.

Siamang occurred mainly on the west – in the Main Range over 1,000 ft above sea level – with an extension eastwards through the Taman Negara. They were absent from the south, north-west, and east of the country – occurring in only one fifth of the forested area. Black-handed gibbons were abundant between the Mudah and Perak rivers on the west, and north of the Kelantan and Pergau rivers on the east. White-handed gibbons occurred elsewhere where dry forest still stands – from the coast up into the mountains.

Although all primate species have been observed at most elevations, there was a tendency for siamang, pig-tailed macaques, and banded leaf monkeys to be observed more frequently in the higher and remoter forests, whereas gibbons, dusky leaf monkeys and long-tailed macaques occurred in the lowlands and on the lower slopes of mountain ranges. But such an analysis does not contribute much to defining the ecological separation of these species.

During the survey differences in calling behaviour between siamangs and gibbons were confirmed; gibbons called earlier, more often and for longer. The black-handed gibbon was clearly distinguished from the white-handed by the diphasic hoots of the male, and the shorter great call of the female. Calling was accentuated in the gibbons in the south, north-west, and west; siamang groups called earlier in the centre and north-west, and for longer in the west.

Siamang were found in 45 of 101 localities; gibbons in 67, and both species in 35 localities. There were estimated to be at least 1.52 siamang and 5.28 gibbons/km^2; but the density in the study areas was 4.18 and 3.28 individuals/km^2, respectively. After applying corrections for variations in group size and calling behaviour a total estimate of 30,000 $\pm$ 5,000 siamang was obtained for the Malay Peninsula. There were estimated to be about 25,000 *agilis* and 230,000 *lar* gibbons in West Malaysia.

Chapter 8: Concluding Discussion

Comparative Socio-Ecology

The study of forest primates has lagged behind that of more terrestrial species which live in open habitats. ALDRICH-BLAKE [1970] has outlined why in his analysis of problems of observation in forest. In general, poor visibility leads to bias because insufficient accurate data can be collected. But he indicates a uniformity in group size, range size, and population density among forest monkeys, although there is greater diversity in social structure. Consideration of these factors suggested that the best approach in defining the ecological and behavioural characteristics of forest primates was to start with small social groupings.

The small family groups of hylobatids, living in small territories, afford excellent opportunities to refine observation methods in forest. In studying the siamang it has been possible to overcome many of the problems described by ALDRICH-BLAKE – by observing individuals of a family group continuously from dawn to dusk over a long period. There are no obvious problems in modifying these methods to obtain comparable results for groups with up to 20 individuals. This would cover the majority of South-east Asian primates, and allow a fuller appreciation of primate socio-ecology in tropical forest than has hitherto been available.

The recent discovery of marked intra-specific variation in behaviour [see YOSHIBA, 1968, for example] has demanded an investigation of the adaptive significance to the habitat of contrasting social systems – both within and between species of primates [CROOK, 1970]. CROOK refers to direct environmental effects, the long-term acquisition of tradition, and natural selection of genetically controlled traits as involved in altering social structure, either individually or in combination.

A quantitative description of group social behaviour, in terms of activity profiles, has been developed by BERNSTEIN [1971]. The procedure is similar to that used in this study: (1) the identification of responses; (2) measurement of frequency and duration; (3) sequential analysis, and (4) synthesis into organising principles underlying the society. He points out that these profiles are sensitive to the age, size, sex, and condition of individuals, as well as to temperature, time of day, weather, and physical spatial features. Comparison of social responses, individual activity patterns, and sex differences across taxa produced phylogenetic differences that permitted the identification of congeners.

Comparison with non-primates is particularly useful in connection with social structure, feeding behaviour, bioenergetics, and breeding. Relevant information is provided by the study of birds, and economically important mammals. ELLEFSON [1967] drew attention to the CROOK [1965] review which showed that the majority of solitary or territorial birds subsisted on dispersed foods, whereas polygamous species were usually gregarious feeders on locally abundant foods. CROOK [1970] suggests that local concentrations of widely dispersed foods promote larger social groups with looser social organisation. In referring to similar social systems in other mammals, especially bovids, he associates a more even dispersion of food with one-male groups, and suggests that multi-male groups are an adaptation to an omnivorous, rather than a folivorous, diet.

With reference to ungulates, RODMAN [in prep.] points out that the energetic cost of producing offspring is small relative to the cost of maintaining the parents; but domestic cattle do not have to search for their food to the same extent as primates. He illustrates the difficulties of social coordination in sexually dimorphic species. He explains the absence of stable social groupings larger than an adult and dependent young in orang-utans, by the greater energy requirements of the adult male. These require him to be less selective in his diet, and to eat more, move less, and travel on the ground more than the adult female. RODMAN suggests increased breeding efficiency as the most important factor in the successful coordination of activities between male and female baboons and gorillas in their larger heterosexual groups.

In his review of nutritional effects on reproduction, MOUSTGAARD [1969] discusses the subnormal conception rate and impaired oestrus cycles in females, and cessation of spermatogenesis in males, that follow persistent underfeeding. Increased food intake prior to ovulation increases ovulation and birth rates in swine, rectifying the depressant effect of feed restrictions [LODGE and HARDY, 1968]. This process of 'flushing' has been developed in domestic animals, but it may be that similar phenomena occur naturally in primate populations.

Malayan Forest Primates

Attention has been drawn to the high density and diversity of primates in Malayan forests [BERNSTEIN, 1967; CHIVERS, 1973]. There was little stratification within the forest, and even less altitudinal zonation between species, but ecological separation could be detected in terms of diet and daily activity patterns, for example. While there is considerable dietary overlap, preliminary results indicate that the principal foods of each primate are not as important to other species. Thus, while all diurnal primates in Malaya consume figs, none do so much as the siamang.

Trees of two dominant species in lowland forest – *Parkia speciosa* and *Dillenia reticulata* – were principal food sources for banded and dusky leaf monkeys at Kuala Lompat, but the siamang was never seen feeding on the flowers or fruit of these species. HUNT [personal commun.] shows that where the two species of leaf monkey feed in the same trees, each often consumes different parts of the tree, or the same part in very different proportions. There were also foods eaten by siamang which were not apparently consumed by other primates.

The suggestion that forest ecosystems are relatively stable is somewhat misleading [CROOK, 1970], with regard to the great diversity of tree species in South-east Asian forests. There is evidence indicating that productivity in the forest as a whole is not constant [MACKINNON, 1971], and that in small areas, such as the home range or territory of a primate group, variation in the diversity and abundance of food can be marked – seasonally or in the long-term (Chapter 3). Throughout central Malaya in 1968 there was a high frequency of births in all primate species. This synchronised birth period coincided with the greatest abundance of food for five years, when the basal rate of fruit production was almost doubled [MEDWAY, 1972b].

This marked expansion of social groups was followed by a year of subdued behaviour, and then by a period of group fission. In siamang this involved only the departure of the eldest young, but in leaf monkeys small groups were formed [CHIVERS, 1973]. In view of the high density of primates in Malayan forests, and the apparent lack of marked ecological separation (see below), it is likely that primate populations in primary forest, where six species commonly occur, are near saturation and, therefore, very susceptible to fluctuations in food supply.

Although each species was not fully utilising the available resources at a given locality, in terms of space and food, collectively the pressures on each species may be very high. It certainly appeared that population expansion had been held in check since 1963 to permit the great number of births in the favourable year of 1968. Most siamang groups had new infants, and leaf monkey groups of nearly 20 individuals contained three to four infants (compared with zero to one, two years later).

The ranking of species by different environmental criteria indicates the complexities of ecological separation between Malayan primates in climax forest:

	Rank						
	1	2	3	4	5	6	
Altitudinal zonation	*S*	P	B	G	D	L	lowest
Forest stratification	G	D	*S*	B	L	P	lowest
Group size	L	P	B	D	G	*S*	smallest
Range size (actual)	P	L	G	*S*	B	D	smallest
Hectares/individual	G	*S*	P	L	B	D	smallest
Diet	G	L	P	B	*S*	D	
	frugivorous, omnivorous, folivorous						

The siamang (S) occurs at higher altitudes, in the middle part of the canopy, where it is mainly folivorous, and in very small groups in small territories – but with a low density of individuals – relative to other species. Congeners are most clearly indicated by group size, ranging and feeding behaviour, but less data are available for pig-tailed (P) and long-tailed (L) macaques, which spend more time on the ground than other species.

The predominantly folivorous species live in the smallest group ranges, especially the dusky leaf monkey (D). This species lives in small exclusive territories at much higher densities than siamang – less than 1 ha/monkey, compared to about 7 ha/siamang – whereas there is overlap between the larger home ranges of the less folivorous banded leaf monkey [personal observation, confirmed by Hunt, personal commun.]. Differences in ranging patterns can certainly compensate for similarities in diet [Chivers, 1968], and indicate those aspects of behaviour which are species-specific – these are mainly genetically determined, thereby reflecting phylogeny.

It is the aim of continuing research in the Krau Game Reserve, and elsewhere in Sundaland, to clarify these relationships through a more precise definition of the social organisation and ecological requirements of each species. Observations are concentrated on (1) ranging (and calling) behaviour, (2) diet and feeding (including nutrition), and (3) social organisation (with temporal variation).

The Siamang in Tropical Rain Forest

The discussion above has indicated that while the siamang is ecologically separated from most animal species by physiological mechanisms, behavioural mechanisms were conspicuous in their separation from more closely related species, i.e. other primates. It is perhaps relevant that siamang reacted much more intensely to gibbons feeding in the same tree, than to leaf monkeys, or giant squirrels, hornbills, or green pigeons.

This study has shown that, although living in monogamous family groups in small territories maintained by regular patrols and loud calls, siamang differ from gibbons in most other respects. Even their territories are smaller and less intensely maintained, and this appears to be related to their different diets. Siamang are primarily folivorous; although they consume considerable quantities of fruit, especially figs, for which they show strong preference, the bulk of their food intake is young leaves. Their diet is, therefore, less specialised than the gibbon's, and more easily obtained in a small area of rain forest.

These differences in diet and ranging behaviour are also apparently reflected by the closer integration of group members, whose activities are more closely synchronised, despite the absence of a conspicuous repertoire of calls, facial expressions, and gestures. This coordination is facilitated by the familiarity each group member has with its territory. The dispersion of individuals in small groups in non-overlapping ranges is obviously most effective for species with a varied diet, and possibly the frugivorous gibbon is exposed to greater pressures. But similar pressures may be exerted on the siamang in their efforts to maintain consistent proportions of each food category throughout the year.

One interesting assertion made by BERNSTEIN [1971] is that most members of a primate group spend 50–75% of time within body length of another. Although siamang individuals were consistently closer to each other than occurs in other primate species, they were not that close; individuals were probably within body length of another for less than 25% of the day. Social interactions were less frequent and more sedate than in species living in larger social groups; daily behaviour was more harmonious and 'predictable' from all points of view – even compared with gibbons.

It is not yet possible to confirm whether breeding is synchronised throughout the siamang population – at regular or irregular intervals – and whether it is a seasonal activity. MCCLURE [1964], supported by KOYAMA [1971], suggests that births occur mainly in the dry season (January to March) following mating during the preceding fruiting season (June to September). The observations of growing infants during 1968, and the birth of an infant to TS1 at the end of January 1971 [HUNT, personal commun.] support this, but indicate that it is not an annual event within a local population.

It is difficult to tell to what extent increased calling in the breeding season represents antagonism between adults and subadult, rather than cohesion between adults (as they redefine their territorial boundaries), because of the way such behaviour can be redirected. Antagonism could be reflected by the greater distance of the subadult from the rest of the group during call bouts, and by the way in; which calls – especially screams – pass to and fro between male and subadult. It has at last been possible to detect the process of group formation in hylobatids, whereby the calls of a 'solitary' male attract a female from her natal group.

The wholly arboreal habits of the siamang give cause for concern about its survival under the present land development policies in Malaya and Sumatra. Although it is effectively adapting to selectively logged forest, this practice is unfortunately infrequent. There must be important consequences to the restriction of gene flow by the division of the siamang's range into countless 'islands' by roads and plantations; these do not provide such a barrier to leaf-monkeys and macaques.

Siamang and the Great Apes

Reference has been made to certain similarities between siamang and the great apes [SCHULTZ, 1933; CHIVERS, 1972a]. The question as to whether the siamang could be more closely related to the great apes than to gibbons had not been fully pursued. GROVES [1972] has suggested that morphologically the siamang represents a conservative stage in hylobatid evolution. This is supported by studies of dentition [FRISCH, 1965], and by the behavioural evidence presented here – less specialised diet, lower level of territorial behaviour, and less intense social behaviour.

Certain morphological features, such as the structure of the thumb, link the siamang closely in phylogeny with the gibbon [GROVES, 1972]. In chromosome number and morphology the siamang resembles the *concolor* more closely than the *lar* gibbon [KLINGER, 1963; BENIRSCHKE, 1969]. Similarities with the great apes must represent the earlier separation from the gibbon stock, and subsequent convergence with the great apes in response to similar environmental pressures. These can mainly be related to the larger body size of the siamang; its movements through trees are very similar to those of chimpanzees and orang-utans.

CROOK [1970] has pointed out the parallel radiations that have occurred among hominoids in Africa and Asia. The chimpanzee and gibbon (and orang-utan) have developed specialised diets which require searching for locally abundant, but irregularly distributed foods. The different social organisation – large loosely organised multi-male groups, compact family groups, and small very loosely organised groups – further reflect the dispersion of food. That is, a specialised diet can involve more widely dispersed food that is either evenly scattered or locally abundant.

In contrast, the siamang and gorilla have become folivore and herbivore, respectively, with food abundant and evenly dispersed. But both are restricted to forests with rich food supplies, although they are not limited by irregularities in the production of more specialised foods. Although the siamang should theoretically be more successful in certain localities, its restricted distribution and small numbers, relative to the *lar* gibbon, reflect its overall limitations.

Hylobatids, however, show greater success in these terms through their adaptations to forest life based on orthogradism and suspensory behaviour, which enables them to exploit the rich terminal branch niche more fully than other primates. The differences between siamang and gibbon are such that until the phyletic distance between gibbon taxa has been accurately measured, it seems advisable to retain the generic distinction originally proposed, while noting the fundamental similarities between them.

Summary

Siamang are restricted to about one fifth of the remaining forested area of West Malaysia, occurring mainly in the western highlands (Chapter 7). There are about eight times more gibbons, which occur throughout the lowland and hill dipterocarp forests of the Peninsula. From observations in three study areas about 100 miles apart – two highland and one lowland – it was confirmed that siamang like gibbons live in small territories in family groups.

Each siamang group called on about 30% of days for about 15 min; 60% of group calls were between 0900 and 1200 h (Chapter 6). Male and female played distinctive parts in the call bout, with the male screaming at the climax of the female rhythmic bark series. The group was alternately restful and mobile during these bouts. Gibbon groups called earlier, more often, and for longer each day.

The behaviour of group members was closely integrated in terms of time and space. All were engaged in the same activity for 75%, and in the same tree for 50%, of the day; individuals were within 10 m of each other on average (Chapter 5). One possible manifestation of this cohesion was the antagonism shown towards other conspecifics. Each group lived in about 23 ha of forest, which was clearly defined topographically by the ranging behaviour of the resident group. This territory was advertised by loud group calls, and defended in occasional border conflicts (Chapter 2).

A siamang family travelled about 700 m daily, increasing to over 1,500 m in the dry season. The pattern of ranging was closely related to the distribution and abundance of food, so that in any 10-day period two thirds of the time was spent in one fifth of the territory. But frequent sorties to the remainder resulted in two thirds of the area consistently being covered in 10 days; these appeared to reflect territorial maintenance as much as food-searching behaviour (Chapter 2). Daily feeding and travel time were increased, and resting time reduced, in warmer, sunnier weather.

The small size of the territory, compared with the gibbon, reflected the less specialised and less frugivorous diet of the siamang. While fruit, especially figs, represented an important part of siamang diet, they mainly ate young leaves throughout the year. During 6 h of feeding each day in about 13 trees, siamang ate young leaves in 67% of trees for 58% of feeding time. Small quantities of flowers and insects (termites and caterpillars) were eaten when seasonally or perennially available (Chapter 3).

Two thirds of feeding time was spent in nearly half the trees in which feeding was recorded, representing about 35 tree species (including 15 species of fig which accounted for 25% of feeding time). Over the year 50% of food trees, and 55% of feeding time occurred in 25% of the territory, although the emphasis changed from one month to the next. Drinking from tree-holes was observed only in the lowlands. The high water content of the diet was supplemented by water obtained from leaves and branches after rain. Suspensory behaviour was as conspicuous during feeding as it was during travel. The subadult and female fed faster and for longer than the male and infant.

Daily activity started at dawn, and the alert period of over 10 h ended 2 h before sunset on average (Chapter 4). The group descended from the emergents to spend most of the day in the main canopy, descending into the under-story for short periods before and after noon. Between 0700 and 1100 h feeding and travel occupied 70 – 85% of each hour, declining to 40% during the afternoon when resting and grooming were more frequent. The secondary grooming peak in mid-morning in the lowlands was the main time of grooming in the highlands. On average resting occupied 27%, feeding 58%, and travel 15% of the alert period, with 30, 44, and 26% of bouts lasting for 66, 72, and 34 min, respectively. There was an average interval of 57 min between feeding bouts, and of more than 100 min between rest and travel bouts.

Two or three days of increased feeding were followed by one or two days of increased travel. There was a marked increase in all activities, except resting – maintenance and social – in the dry season (Chapter 4). Apart from loud concerted group calls, communication within the group was based on subtle gestures in the absence of a varied repertoire of calls, expressions and gestures. Although the male moved first on most mornings, it was the female who led most group progressions followed by the juvenile, then the male carrying the infant, and finally the subadult (Chapter 5).

Although the female initiated many group activities, the adult male maintained a more central position. Young animals followed the female without straying far from the male. The subadult was the focus of frequent, but brief, aggression – mainly from the adults. An important stage in the growth of the infant was the transfer from female to male early in the second year of life. The older infant gained its independence from the adult male, and was thereby more rapidly integrated socially into the group. Play was sedate, and more a solitary than a social activity, characteristic of the infant stage.

The main social interaction was grooming, which occupied 15% of the alert period, in terms of pair association time; bouts lasted for about 13 min, with grooming for 42% PAT (Chapter 5). 58% of bouts and 72% of PAT involved mature animals only (including subadult). Grooming was correlated with increased aggression (and copulation by male and female) between each pairing in the group. Overt sexual behaviour was apparently restricted to several months every three years.

Although this description of siamang behaviour was mainly based on observations at Kuala Lompat, the behaviours in the mountains of Ulu Sempam have special relevance generally – where they are not responses to selective logging. The long-term environmental effects on social organisation are discussed with respect to feeding and breeding behaviour. Discussion also centres on how the siamang is integrated into the tropical rain forest, and on its relationships to other hominoids.

Postscript

Further data were collected on the ecology and behaviour of TS1 at Kuala Lompat during two months in 1971 by Dr. F.P.G. ALDRICH-BLAKE [ALDRICH-BLAKE and CHIVERS, 1973], and during four months in 1972 by the author and several students. It has been mentioned that the TS1 subadult male paired off with a young female (from south of S. Lompat) during 1970, and that an older female joined them for two periods during 1971. Latterly the male was transferring his attention to the old female.

It appears that this pair drifted north during the 1972 dry season, and that a young male crossed north into the area to pair off with the young female (the present TS1A group). The reasons for suggesting this are that the male did not appear to be Sammy, originally of TS1, and that the old female was observed alone once. It may be that Sammy is dead as an adult siamang was found by our local helpers in the middle of TS1's territory in April 1972[1]. But to the north-west of TS1 – in TG1's territory – were two males (one subadult) who were observed twice, and recorded calling together on several occasions.

It seems likely that these two are the oldest off-spring of TS1 – Sammy and Julian – in the 1969–70 study; the appearance of these two males, and the characteristics of their vocalisations support this. TS1 in 1972 contained the same adult male and female, and a juvenile male. The infant born in January 1971 has been lost, but a new infant was born just before Christmas 1972. If Sammy was the first offspring of this pair, they have been together for over ten years, and the one-eyed Murgatroyd must be 20 years old.

The infant was probably mis-sexed in 1970 (so that it is the same individual being observed today as a juvenile male). It was identified as a female as it lacked the preputial tuft present in the RS2 male infant. But GREYSOLYNNE FOX's observations in Milwaukee Zoo [personal commun.] indicate that the tuft may not develop until the third or fourth year of life. The dynamics of

1 The bones were retrieved in 1973 by MACKINNON [personal commun.] and identified as those of *Hylobates lar;* the TS1A male was alone.

these siamang groups will be observed over the next few years, and more fully analysed in future publications.

In a brief visit to Ulu Sempam, it was discovered that the loggers had resumed activity after the end of the 1970 study by penetrating down the RS1/RS2 boundary, and south into RS1's territory. A two-day visit was too short to be certain, but the population may have been seriously disrupted. There was no sight or sound of the adults of either group (not altogether unusual), who might have been ranging quietly in the periphery of their territories. A lone male called near the RS2/RS4 boundary – possibly the RS2 subadult of 1970. A young adult female was seen in the favoured fig, S058, which was fruiting again – possibly the RS1 subadult of 1970.

References

ALDRICH-BLAKE, F. P. G.: Problems of social structure in forest monkeys; in CROOK Social behaviour in birds and mammals, pp. 79–101 (Academic Press, London 1970).

ALDRICH-BLAKE, F. P. G. and CHIVERS, D. J.: On the genesis of a siamang group. Amer. J. phys. Anthrop. *38:* 631–636 (1973).

ALTMANN, S. A. (ed.): Social communication among primates (University of Chicago Press, Chicago 1967).

ALTMANN, S. A. and ALTMANN, J.: Baboon ecology (University of Chicago Press, Chicago 1970).

AMERASINGHE, F. B.; VAN CUYLENBERG, B. W. B., and HLADIK, C. M.: Comparative histology of the alimentary tract of Ceylon primates in correlation with diet. Ceylon J. Sci. *9:* 75–87 (1971).

ANDERSON, J.: Anatomical and zoological researches (Quaritch, London 1878).

BALDWIN, L. and TELEKI, G.: Patterns of gibbon behaviour on Hall's Island, Bermuda. A preliminary ethogram for *Hylobates lar;* in RUMBAUGH The gibbon and the siamang, vol. 4 (Karger, Basel, in press).

BATES, B. C.: Territorial behaviour of primates. A review of recent field studies. Primates *11:* 271–284 (1970).

BENIRSCHKE, K.: Cytogenetic contributions to primatology. Ann. N.Y. Acad. Sci. *162:* 217–224 (1969).

BERNSTEIN, I. S.: Intertaxa interactions in a Malayan primate community. Folia primat. *7:* 198–207 (1967).

BERNSTEIN, I. S.: Activity profiles of primate groups; in SCHRIER and STOLLNITZ Behavior of non-human primates, vol. 3, pp. 69–106 (Academic Press, New York 1971).

BONHOTE, J. L.: Report of the Gunong Tahan Expedition, May–September 1905. Mammals. J. Fed. Malay States Mus. *3:* 1–11 (1908).

BUFFON, G. L., LE CLERC DE: The gibbon (translated by KENDRICK and MURDOCH, 1775). Histoire naturelle, vol. 14, pp. 96–108 (Bell, London 1766).

BURT, W. H.: Territoriality and home range concepts as applied to mammals. J. Mammal. *24:* 346–352 (1943).

CARPENTER, C. R.: A survey of wildlife conditions in Atjeh, North Sumatra. Netherlands Commission of International Nature Protection, vol. 12, pp. 1–33 (1938).

CARPENTER, C. R.: A field study in Siam of the behaviour and social relations of the gibbon, *Hylobates lar.* Comp. Psychol. Monogr. *16:* 1–212 (1940).

CARPENTER, C. R.: The menstrual cycle and body temperature in two gibbons. Anat. Rec. *79:* 294–296 (1941).

CHALMERS, N.: The ethology and social organisation of the black mangabey, *Cercocebus albigena;* unpublished Ph.D. dissertation, Cambridge (1967).

CHANCE, M. R. A.: Social structure of a colony of *Macaca mulatta*. Brit. J. Anim. Behav. *4:* 1–13 (1956).

CHASEN, F. N.: Handlist of birds in Malaysia. Bull. Raffles Mus. *11:* 1–389 (1935).

CHIVERS, D. J.: The monkeys of Barro Colorado Island. Animals *11:* 355–359 (1968).

CHIVERS, D. J.: On the daily behaviour and spacing of howling monkey groups. Folia primat. *10:* 48–102 (1969).

CHIVERS, D. J.: The Malayan siamang. Malay nat. J. *24:* 78–86 (1971 a).

CHIVERS, D. J.: Spatial relations within the siamang family group. Proc. 3rd. Int. Congr. Primat., Zurich, vol. 3, pp. 14–21 (1971 b).

CHIVERS, D. J.: The siamang and the gibbon in the Malay Peninsula; in RUMBAUGH The gibbon and the siamang, vol. 1, pp. 103–135 (Karger, Basel 1972a).

CHIVERS, D. J.: The Siamang in Malaya. Ph. D. dissertation, Cambridge (1972b).

CHIVERS, D. J.: Introduction to the socio-ecology of Malayan forest primates; in MICHAEL and CROOK Comparative ecology and behaviour of primates, pp. 101–146 (Academic Press, London 1973).

CHIVERS, D. J.: RAEMAEKERS, J. J., and ALDRICH-BLAKE, F. P. G.: Long term observations of siamang behavior (in preparation).

CORNER, E. J. H.: Wayside trees of Malaya (Singapore 1952).

CROOK, J. H.: The adaptive significance of avian social organisations. Symp. zool. Soc. Lond. *14:* 181–218 (1965).

CROOK, J. H.: The socio-ecology of primates; in CROOK Social behaviour in birds and mammals, pp. 103–166 (Academic Press, London 1970).

DALE, W. L.: The rainfall of Malaya. I. J. trop. Geogr. *13:* 23–27 (1959).

DALE, W. L.: The rainfall of Malaya. II. J. trop. Geogr. *14:* 11–28 (1960).

DALE, W. L.: Sunshine in Malaya. J. trop. Geogr. *19:* 20–26 (1964).

DEVORE, I.: Primate behavior: field studies of monkeys and apes. (Holt, Rinehart & Winston, New York (1965).

DEVORE, I. and HALL, K. R. L.: Baboon ecology; in DEVORE Primate behavior, pp. 20–52 (Holt, Rinehart & Winston, New York 1965).

ELLEFSON, J. O.: A natural history of gibbons in the Malay Peninsula; unpublished Ph. D. dissertation, Berkeley (1967).

ELLEFSON, J. O.: Territorial behavior in the common white-handed gibbon, *Hylobates lar;* in JAY Primates, pp. 180–199 (Holt, Rinehart & Winston, New York 1968).

ELLIOT, D. G.: A review of the primates, 3 vol. (Amer. Mus. Nat. Hist., New York 1913).

FOODEN, J.: Color-phase in gibbons. Evolution *23:* 627–644 (1969).

FOODEN, J.: Color and sex in gibbons. Field Mus. Nat. Hist. Bull. *42:* 2–7 (1971).

FOX, G.: Behaviour of a siamang family group in Milwaukee Zoo. Folia primat. *18:* 122–139 (1972).

FRISCH, J. E.: Trends in the evolution of the hominoid dentition. Bibliotheca primat., vol. 3, pp. 1–130 (Karger, Basel 1965).

FRISCH, J. E.: Evolution of the siamang in South-East Asia during the Pleistocene. Proc. 3rd int. Congr. Primat., Zurich, vol. 1, pp. 67–73 (Karger, Basel 1971).

GARTLAN, J. S.: Structure and function in primate society. Folia primat. *8:* 89–120 (1968).

GLOGER: Hylobates syndactylus. [RAFFLES, 1821]. Gemeinn. Naturg. *1:* 34 (1841).

GRAND, T. I.: A mechanical interpretation of terminal branch feeding. J. Mammal. *53:* 198–201 (1972).

GROVES, C. P.: Taxonomic and individual variations in gibbons. Symp. zool. Soc. Lond. *26:* 127–134 (1970).

GROVES, C. P.: Systematics and phylogeny of gibbons; in RUMBAUGH The gibbon and the siamang, vol. 1 (Karger, Basel 1972).

GULIK, R. H. VAN: The gibbon in China. An essay in Chinese animal lore (Brill, Leiden 1967).

HALL, K. R. L.: Behavior and ecology of wild patas monkey, *Erythrocebus patas* in Uganda. J. Zool., Lond. *148:* 15–67 (1965).

HAMERTON, J. L.; KLINGER, H. P.; MUTTON, D. E., and LANG, E. M.: The somatic chromosomes of the Hominoidea. Cytogenetics *2:* 240–263 (1963).

HARLAN, R.: Description of a species of orang from Assam (H. Hoolock). Trans. amer. Phil. Soc., n.s. *4:* 52–59 (1834).

HILL, C.: Gestation period for the siamang. Int. Zoo Yb. *7:* 93–94 (1967).

HINDE, R. A.: The biological significance of the territories of birds. Ibis *98:* 340–369 (1956).

HLADIK, C. M.: Surface relative du tractus digestif de quelques primates, morphologie des villosities intestinales et corrélations avec le régime alimentaire. Mammalia *31;* 1: 120–147 (1967).

HLADIK, C. M.: Recherche des caractéristiques histochemiques et cytologiques de la muqueuse intestinale des primates et des corrélations avec le régime alimentaire. Mam. Mus. Hist. nat., Paris, n.s. *A52:* 1–69 (1968).

HOOIJER, D. A.: Quaternary gibbons from the Malay archipelago. Zool. Verh., Leiden *46:* 1–41 (1960).

HOWARD, H. E.: Territory in bird life (London 1920).

JAY, P. C.: The common langur of North India; in DEVORE Primate behavior, pp. 197–249 (Holt, Rinehart & Winston, New York 1965).

JAY, P. C.: Primates: studies in adaptation and variability (Holt, Rinehart & Winston, New York 1968).

JOLLY, A.: Lemur behavior. A Madagascar field study (University of Chicago Press, Chicago 1966).

KAUFMANN, J. H.: Ecology and social behavior of the coati, *Nasua narica*, on Barro Colorado Island, Panama. Univ. Calif. Publ. Zool. *60:* 95–222 (1962).

KAWABE, M.: A preliminary study of the wild siamang gibbon, *Hylobates syndactylus*, at Fraser's Hill, Malaysia. Primates *11:* 285–291 (1970).

KHAN, M.: Distribution and population of siamang and gibbons in the state of Perak. Malay. nat. J. *24:* 3–8 (1970).

KLINGER, H. P.: Somatic chromosomes of some primate species. Cytogenetics *2:* 140–151 (1963).

KLOSS, C. B.: On mammals and birds from the lowlands of Penang. J. Fed. Malay States Mus. *4:* 144–166 (1911).

KLOSS, C. B.: Some remarks on the gibbons, with the description of a new sub-species. Proc. zool. Soc., Lond. *1929:* 113–127.

KOYAMA, N.: Observations on mating behaviour of wild siamang gibbons at Fraser's Hill, Malaysia. Primates *12:* 183–189 (1971).

KUMMER, H.: Social organization of Hamadryas baboons. A field study (Karger, Basel 1968).

KUMMER, H.: Primate societies. Group techniques of ecological adaptation; in GOLDSCHMIDT Studies in cultural ecology (Aldine Publishing, Chicago 1971).

LAWICK-GOODALL, J. VAN: The behaviour of free-living chimpanzees in the Gombe Stream Reserve. Anim. Behav. Monogr. *1:* 161 – 311 (1968).

LIM BOO LIAT: Distribution of the primates of West Malaysia. Proc. 2nd int. Congr. Primat., Atlanta 1969, vol. 2, pp. 121 – 130.

LODGE, G. A. and HARDY, B.: The influence of nutrition during oestrus on ovulation rate in the sow. J. reprod. Fert. *15:* 329 – 332 (1968).

MCCLURE, H. E.: Primates in the Dipterocarp forest of the Gombak valley. Primates *5:* 39 – 58 (1964).

MCCLURE, H. E.: Flowering, fruiting and animals in the canopy of a tropical rain forest. Malayan Forester *29:* 182 – 203 (1966).

MACKINNON, J.: The orangutan in Sabah today. Oryx *11:* 142 – 191 (1971).

MARLER, P.: Aggregation and dispersal: two functions in primate communication; in JAY Primates: studies in adaptation and variability, pp. 420 – 438 (Holt, Rinehart & Winston, New York 1968).

MARSHALL, J.; ROSS, B. A., and CHANTHAROJVUNG, S.: The species of gibbons in Thailand. J. Mammal. *53:* 479 – 486 (1972).

MARTIN, W. C. LINNAEUS: Natural history of man and monkeys. A general introduction to the natural history of mammiferous animals with a particular view of the physical history of man, and the more closely allied genera of the order quadrumana or monkeys, pp. 413 – 446 (Wright, London 1841).

MASON, W. A.: Use of space by *Callicebus* groups; in JAY Primates, pp. 200 – 216 (Holt, Rinehart & Winston, New York 1968).

MATTHEWS, L. H.: Notes on the genital anatomy and physiology of the gibbon *(Hylobates)*. Proc. zool. Soc., Lond. *116:* 339 – 364 (1946).

MEDWAY, Lord: The monkeys of Sundaland; in NAPIER and NAPIER Old World monkeys, pp. 513 – 553 (Academic Press, London 1970).

MEDWAY, Lord: The Gunong Benom expedition, 1967, VI. The distribution and altitudinal zonation of birds and mammals on Gunong Benom. Bull. brit. Mus. nat. Hist. *22:* 103 – 151 (1972a).

MEDWAY, Lord: Phenology of a tropical rain forest in Malaya. Biol. J. Linn. Soc. *4:* 117 – 146 (1972b).

MENZEL, E. W.: Communication about the environment in a group of young chimpanzees. Folia primat. *15:* 220 – 232 (1971).

MILLER, C.: Extract from several letters... giving some account of the interior part of Sumatra. Philos. Trans. *68:* 161 – 179 (1778).

MITCHELL, G. D.: Paternalistic behavior in primates. Psychol. Bull. *71:* 399 – 417 (1969),

MOUSTGAARD, J.: Nutritive influences upon reproduction; in COLE and CUPPS Reproduction in domestic animals, pp. 489 – 516 (Academic Press, London 1969).

NOBLE, G. K.: The role of dominance on the social life of birds. Auk. *56:* 263 – 273 (1939).

OOI JIN-BEE: Land people and economy in Malaya (Longmans, London 1963).

PETTER, J. J.: The lemurs of Madagascar; in DEVORE Primate behavior, pp. 292-319 (Holt, Rinehart & Winston, New York 1965).

POCOCK, R. I.: Observations on a female specimen of the Hainan gibbon now living in the Society's Gardens. Proc. zool. Soc., Lond. *ii:* 169 – 180 (1905).

POCOCK, R. I.: The gibbons of the genus *Hylobates*. Proc. zool. Soc., Lond. *1927:* 719–741.

POORE, M. E. D.: Studies in Malaysian rain forest. I. The forest on Triassic sediments in Jengka Forest Reserve. J. Ecol. *56:* 143–196 (1968).

RAFFLES, T. S.: Descriptive catalogue of a zoological collection made on account of the Honourable East India Company in the Island of Sumatra and its vicinity. Linn. Trans. *13:* 239–274 (1821).

REYNOLDS, V. and REYNOLDS, F.: Chimpanzees of the Budongo forest; in DEVORE Primate behavior, pp. 368–424 (Holt, Rinehart & Winston, New York 1965).

RIPLEY, S.: Leaves and leaf monkeys: the social organisation of foraging in grey langurs. *Presbytis entellus thersites;* in NAPIER and NAPIER Old World monkeys, Evolution, systematics and behaviour, pp. 481–504 (Academic Press, London 1970).

RODMAN, P. S.: Bioenergetics, sexual dimorphism and social groups of primates (in preparation).

SCHALLER, G. B.: The behavior of the mountain gorilla; in DEVORE Primate behaviour, pp. 324–367 (Holt, Rinehart & Winston, New York 1965).

SCHULTZ, A. H.: Observations on the growth, classification and evolutionary specialisations of gibbons and siamangs. Human Biol. *5:* 212–255, 385–428 (1933).

SIEGEL, S.: Nonparametric statistics for the behavioral sciences (McGraw Hill, New York 1956).

SIMONETTA, A.: Catalogo e sinonimia annolata degli ominoidi fossili ed attuali (1758–1955). Atti. Soc. tosc. Sci. nat. Pisa, B *64:* 53–112 (1957).

SOO-HOO, C.: A brief note on the birth of a siamang, *H. syndactylus;* in JARVIS Int. Zoo Yb. *1966:* 147.

STRUHSAKER, T. T.: Ecology of vervet monkeys *(C. aethiops)* in the Masai Amboseli Game Reserve Kenya. Ecology *48:* 891–904 (1967).

STRUHSAKER, T. T.: Correlates of ecology and social organization among African cercopithencines. Folia primat. *11:* 80–118 (1969).

SUGIYAMA, Y.: Social organisation of chimpanzees in the Budongo Forest, Uganda. Primates *9:* 225–258 (1968).

SUGIYAMA, Y.: Social organisation of Hanuman langurs; in ALTMANN Social communication among primates, pp. 221–236 (University of Chicago Press, Chicago 1967).

THOMAS, O.: *Symphalangus syndactylus continentis*. Ann. Mag. nat. Hist. *ii:* 301 (1908).

THOMPSON, N. S.: The motivations underlying the social structure of *Macaca irus*. Anim. Behav. *17:* 459–467 (1969).

WILSON, C.: Spatial factors and the behaviour of non-human primates. Folia primat. *18:* 256–275 (1972).

WYATT-SMITH, J.: Malayan forest types. Malay. nat. J. *7:* 45–55 (1953).

YOSHIBA, K.: Local and intertroop variability in ecology and social behavior of common Indian langurs; in JAY Primates, pp. 217–242 (Holt, Rinehart & Winston, New York 1968).

Appendices

Appendix A: Patterns of Fieldwork and Primate Observation

Monthly details of fieldwork are given below. During seven months in 1968, 166 days were spent in the field – 7 h/day on average. After two months in the forest in Ulu Gombak, most observations were made from 'roads', which include all tracks accessible to a landrover. The mobility afforded by this mode of study was essential for most aspects of the survey. The number of localities visited (Chapter 7), and the mileage covered are indicated. In addition to these measures, contact with each species of primate is given for each month in terms of hours of direct observation (and overall for 1968, as number of sightings). Thus in 1968 contact was maintained with siamang for nearly 1 h/sighting on average, whereas observations of other species could not be, or were not, sustained for long.

Pattern of fieldwork and primate observation in 1968

Month 1968	Number of days	Fieldwork, h forest	road	miles	number of areas	Primate observation, h H. s.	H. l.	P. o.	P. m.	M. f.	M. n.
June	30	93	11	1,190	3	5.3	0.3	0.8	4.0	1.8	0.1
July	25	36	69	1,230	6	3.8	0.5	0.8	2.0	0	0.5
Aug.	24	73	122	1,380	23	3.5	1.0	8.8	0.1	0.1	0.5
Sept.	23	16	176	1,420	22	6.5	0.1	0.2	0	1.3	0
Oct.	24	8	207	1,650	32	6.5	0.3	0.5	0.8	0	0.3
Nov.	18	7	164	1,550	22	6.0	0	0.5	0	5.5	0
Dec.	22	0	190	1,515	26	12.5	1.8	0.5	0.8	0	0
Days	166										
Hours		233 + 939 = 1,172 (+418 driving)									
Hours observation						44.1	4.0	12.1	7.1	8.1	1.4
Number of sightings						45	14	38	27	12	5

$$\frac{\text{Hours of siamang observation}}{\text{Hours fieldwork}} = \frac{77}{1{,}172} = 6.0\%$$

In 1969 and 1970, field work was confined to the three study areas, with few exceptions. Nearly 10 h were spent in the field each day in Ulu Sempam, and over 12 h at Kuala Lompat. Observations were concentrated on siamang so that contacts were sustained for 3.5 h/day, on average, in Ulu Sempam, and 9 h/day in Kuala Lompat, overall. The dusky leaf monkey was ranked second by observation time over the 415 days, but contacts averaged only about 20 min/day.

Pattern of primate observation in study areas

Month		Fieldwork		Observation, h					
		days	hours	H.s.	H.l.	P.o.	P.m.	M.f.	M.n.
1969									
January	R	10	89	0.8	0.3	0.2	0.5	0	0.1
	T	10	77	0	0.2	0.2	0.2	0.1	0
February	R	12	102	0.3	0.3	0.5	0.1	0	0
	T	12	97	13.8	1.8	1.5	0.2	0.2	0
March	R	12	107	0.8	0.5	0.2	0.2	0	0
	T	12	120	25.5	1.0	0.5	0.3	0	0
April	R	12	116	16.3	3.0	5.5	0.3	0	0
	T	12	134	114.0	1.0	3.8	3.3	0	0
May	R	12	103	6.3	2.3	0.8	0.5	0	0
	T	12	146	126.8	3.0	4.0	5.5	0	0
June	R	10	78	2.0	2.5	4.8	0.2	0	0
	T	14	173	156.8	0.3	6.5	4.0	1.5	0.1
July	R	10	78	7.3	2.0	2.5	1.3	0	0.5
	T	14	170	160.0	2.0	0.3	0.3	0.1	0
August	R	12	105	45.0	0	1.5	2.0	0	0.5
	T	12	142	135.5	1.3	0.5	0.5	0.1	0
September	R	10	94	42.5	1.0	1.0	0.1	0	0.3
	T	10	126	116.0	2.5	9.5	6.0	0	0
October	R	10	93	29.3	0.2	1.3	0.1	0	0.3
	T	10	128	118.5	0.3	4.5	3.0	0.1	0
November	R	10	101	48.0	0.5	7.5	3.5	0	1.5
	T	10	123	109.5	5.0	7.0	1.5	0.8	0
December	R	10	96	23.0	0.3	2.3	0.1	0.3	0
	T	10	127	117.0	0.5	10.8	0.1	0.5	0

H.s. = *Hylobates syndactylus*
H.l. = *Hylobates lar*
M.f. = *Macaca fascicularis*
P.o. = *Presbytis obscura*
P.m. = *Presbytis melalophos*
M.n. = *Macaca nemestrina*

R = Ulu Sempam
T = Kuala Lompat
G = Ulu Gombak

Pattern of primate observation in study areas (continuation)

Month		Fieldwork		Observation, h					
		days	hours	H.s.	H.l.	P.o.	P.m.	M.f.	M.n.
1970									
January	R	9	83	44.5	0.2	2.0	0.2	0	0.1
	T	10	143	124.5	2.3	7.0	8.0	0.2	0
February	R	10	114	94.5	0.2	3.5	0.2	0	1.3
	T	10	141	118.5	2.8	18.0	4.0	0.5	0
March	R	12	137	113.5	0.2	7.5	0.2	0	0
	T	10	150	121.0	1.8	1.5	1.5	0	0
April	R	12	155	102.3	0.3	4.0	1.0	0	0.1
	T	10	138	118.3	0.8	3.5	2.5	0	0
May	T	12	183	131.8	2.3	1.8	3.0	1.5	0.2
Total	R	173	1,651	586.4	13.8	45.1	10.5	0.3	4.7
1969+1970	T	190	2,318	1,707.5	28.9	80.9	43.9	5.6	0.3
	G	52	400	33.7	0.3	2.8	2.6	0.2	2.0
		415	4,369	2,327.6	43.0	128.8	57.0	6.1	7.0

Appendix B: Tree Species in Study Areas

Family	Genus or species	Part consumed[1]		
		fruit	leaves[2]	flowers, buds
1. Ulu Sempam, Raub, Pahang				
Anacardiaceae	*Bouea oppositifolia*	S		
	Melanorrhoea			
Annonaceae	*Xylopia*		S	S
Bombacaceae	*Durio*			
Burseraceae	*Canarium*			
	Santiria			
Connaraceae	*Connarus*	S		
Cucurbitaceae	*Macrozanonia macrocarpa*		SD	
Dilleniaceae	*Dillenia reticulata*	DB		DB
Dipterocarpaceae	*Shorea*			
	Shorea curtisii			
Euphorbiaceae	*Baccaurea motleyana*	S		
	Endospermum malaccense			S
Gnetaceae	*Gnetum*	S		
Guttiferae	*Calophyllum*			
	Garcinia	S		
Lauraceae	*Mezzettia leptopoda*			S
Leguminosae	*Cynometra inaequifolia*		SD	
	Dialium platysepalum	SG		
	Saraca thaipingensis			
Melastomaceae	*Memecylon*	S		
Meliaceae	*Aglaia*	S		
Moraceae	*Ficus binnendykii*	S		
	Ficus caulocarpa		SD	
	Ficus consociata	SD		
	Ficus crassiramea	SG		
	Ficus drupacea	SBD		
	Ficus dubia	SGB		
	Ficus heteropleura	S		
	Ficus pellucido-punctata	S		
	Ficus sumatrana	SG		
	Ficus sundaica	S		
	Ficus xylophylla	S		
	Artocarpus	SD		
Myrtaceae	*Eugenia*		S	
Olacaceae	*Ochanostachys amentacea*			
Palmae	*Oncosperma horrida*	P		
	Calamus urens	S		
Rubiaceae	*Wiedelandia ?*	SG		

Tree species in study areas (continuation)

Family	Genus or species	Part consumed[1]		
		fruit	leaves[2]	flowers, buds
Sapotaceae	*Palaquium*		D	
Sterculiaceae	*Scaphium macropodum*	D		
Tiliaceae	*Pentace*			
Vitaceae	*Vitis*	S		
2. Kuala Lompat, Temerloh, Pahang				
Anacardiaceae	*Bouea oppositifolia*	SG		
	Dracontomelum mangifera	S		
	Mangifera indica	SG		
	Pentaspadon velutinum		DB	DB
Annonaceae	*Fissistigma*		S	
	Oxymitra		S	
	Xylopia caudata		S	S
	Xylopia malayana		S	
Apocynaceae	*Alstonia*		S	
	Dyera costulata			
Bombaceae	*Durio iowanus*			
Burseraceae	*Canarium megalanthum*			S
	Santiria			
Celastraceae	*Salacia*			
Combretaceae	*Combretum*		S	
	Terminalia		S	
Connaraceae	*Connarus*	S		
Dichapetalaceae	*Dichapetalum*			
Dilleniaceae	*Dillenium reticulata/ovata*	DB		DB
	Tetracera		S	
Dipterocarpaceae	*Anisoptera laevis*			
	Dipterocarpus grandiflorus			
	Dipterocarpus cornutus			
	Hopea		S	
	Shorea bracteolata			
	Shorea leprosula		S	
	Shorea sumatrana		S	
Ebenaceae	*Dipspyros*		SG	
Euphorbiaceae	*Antidesma coriacea*	SL		
	Cheilosa malayana			
	Endospermum malaccense	S		
	Neoscortechinia kingii		S	
	Sapium baccatum	S		
	Glochidion	S		
Fagaceae	*Lithocarpus lucidus*	S		

Tree species in study areas (continuation)

Family	Genus or species	Part consumed[1]		
		fruit	leaves[2]	flowers, buds
Gnetaceae	*Gnetum*	S		
	Gnetum macrocstachyum	S		
Guttiferae	*Calophyllum*	S	S	
	Garcinia	S		
	Garcinia humbroniana			
Hypericaceae	*Cratoxylon arborescens*			
Leguminosae	*Cassia nodosa*		S	
	Cynometra malaccensis			
	Cynometra inaequifolia			
	Derris		S	
	Dialium platysepalum		S	
	Intsia palembanica		D	
	Koompassia excelsa	SG	SD	
	Koompassia malaccensis			
Leguminosae	*Milletia*			
	Parkia speciosa	DBL		DB
	Saraca thaipingensis			L
	Sindora coriacea			
Linaceae	*Indorouchera*			
Melastomaceae	*Pternandra*			
Meliaceae	*Aglaia*		S	S
	Sandoricum			
Moraceae	*Ficus annulata*	SG		
	Ficus auriantacea	S		
	Ficus benjamina	S		
	Ficus binnendykii	S		
	Ficus bracteata		S	
	Ficus consociata	SGD		
	Ficus heteropleura	S	SL	
	Ficus kerkhovenii		S	
	Ficus pellucido-punctata	S		
	Ficus stupenda	SG		
	Ficus subutata		SD	
	Ficus sumatrana	SGB		
	Ficus virens	SGL		
Myrsinaceae	*Ardisia*	SL		
Myristicaceae	*Horsfieldia*			
	Knema laurina	S		
Myrtaceae	*Eugenia*	S		
	Rhodamnia trinerva			S
Olacaceae	*Ochanostachys amentacea*			

Tree species in study areas (continuation)

Family	Genus or species	Part consumed[1]		
		fruit	leaves[2]	flowers, buds
Oxalidaceae	*Sarcotheca griffithi*	SG		
	Sarcotheca monphylla		S	
Papilionaceae	*Spatholobus*		S	
Polygalaceae	*Xanthophyllum*			
Rhizopheraceae	*Carallia brachiata*	S		S
Rosaceae	*Parinari oblongifolia*		S	
	Maranthes corynbosum	S		
Rubiaceae	*Aegeles*		S	
	Nauclea	S		
	Randia scortechinii	S		
	Uncaria		S	
	Wiedelandia?	S		
Sapotaceae	*Palaquium gutta*			
Ternstroemiaceae	*Adinandra*			S
Tiliaceae	*Grewia blattaefolia*		S	
Ulmaceae	*Gironniera subaqualis*	S		
Urticaceae	*Sloetia elongata*		S	S
Verbenaceae	*Vitex trifoliata*		S	
	Teijsmanniodendron		S	

In the absence of a list of tree species in each area from Dr. SOEPADMO, these lists are based on identifications of siamang food trees (by Dr. SOEPADMO, Dr. WHITMORE, and Prof. CORNER), and of dominant trees species of the forest canopy (more complete for Kuala Lompat).

1 Part consumed by: siamang (S); dusky leaf monkey (D); long-tailed macaque (L); gibbon (G); banded leaf monkey (B); pig-tailed macaque (P).

2 Includes mature and new leaves, stems and shoots.

Appendix C: Pattern of Territory Use by TS1

For each month from April 1969 to May 1970 two measures of daily ranging patterns for TS1 at Kuala Lompat are given for the 10- to 14-day observation periods. On the right, within the dotted line of overall ranging limits, consecutive day ranges are plotted with main food trees (open circles) and sleeping trees (closed circles). On the left is the hectare square grid showing the number of visits to each quadrat, and the proportion of daily activity time spent therein. The five most used squares are outlined in black; the second and third groups of five most used squares are similarly identified with dashed and dotted lines, respectively.

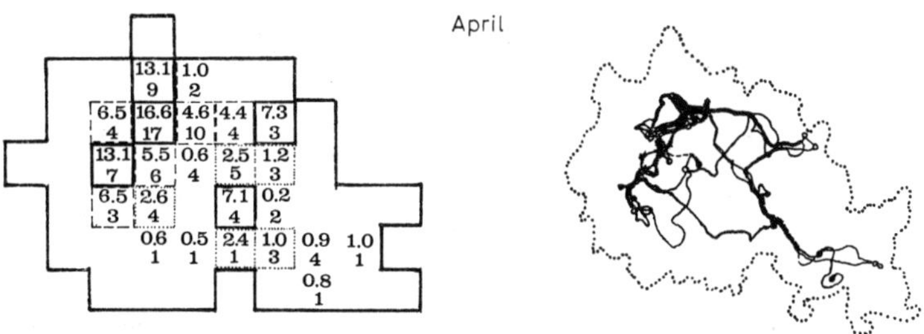

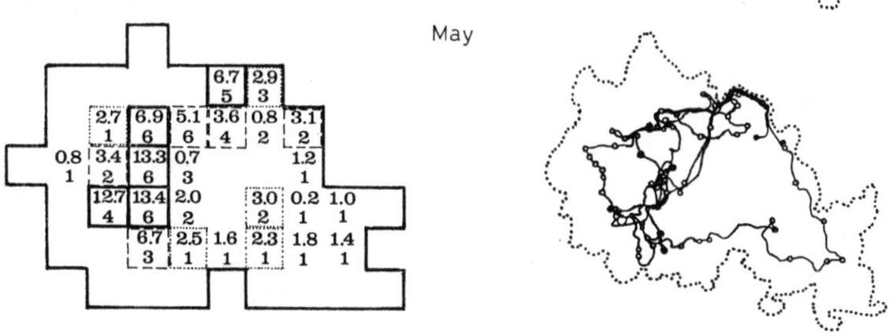

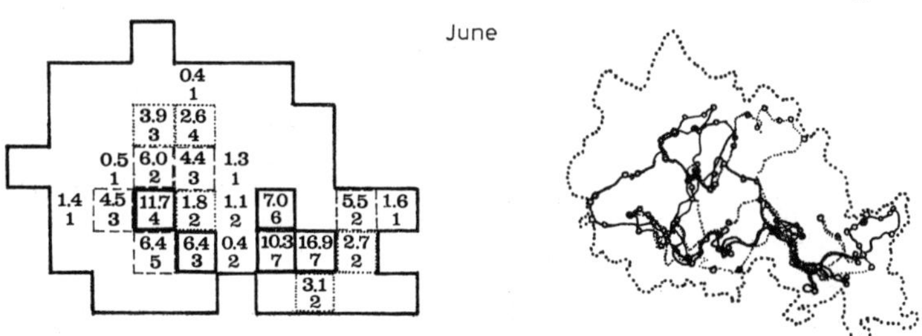

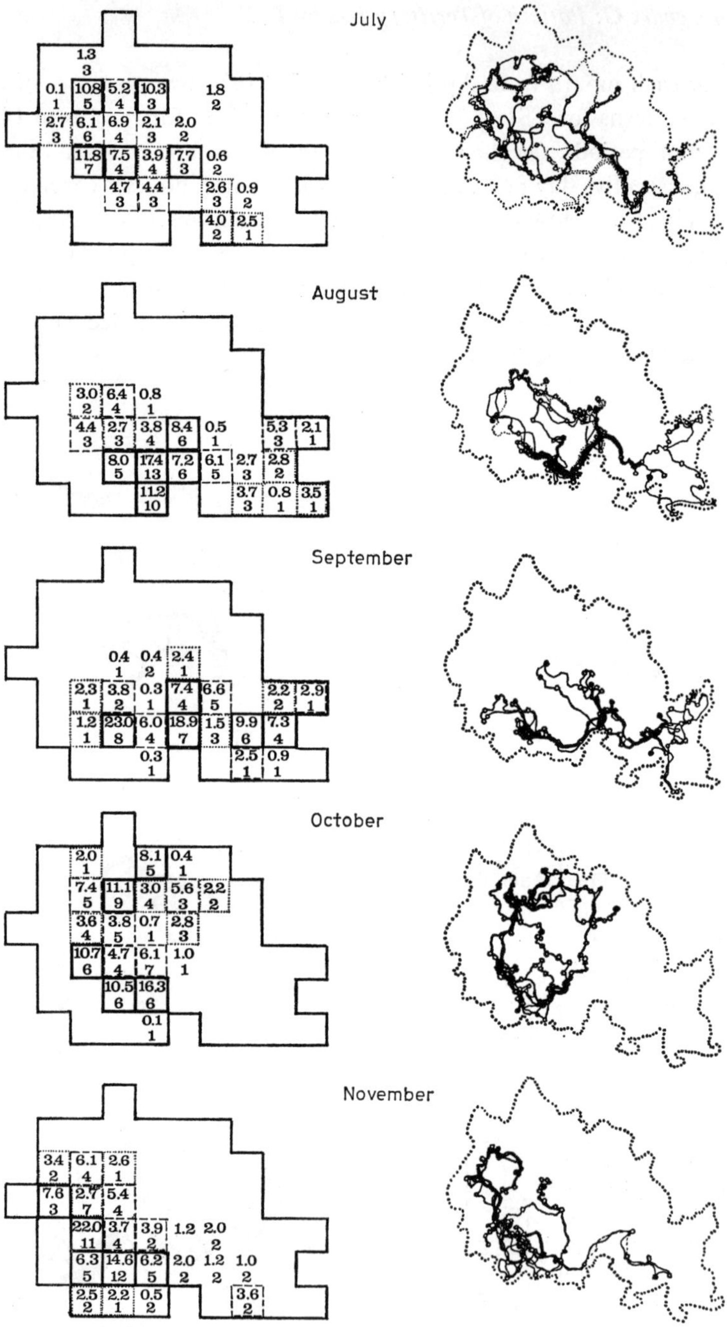
July
August
September
October
November

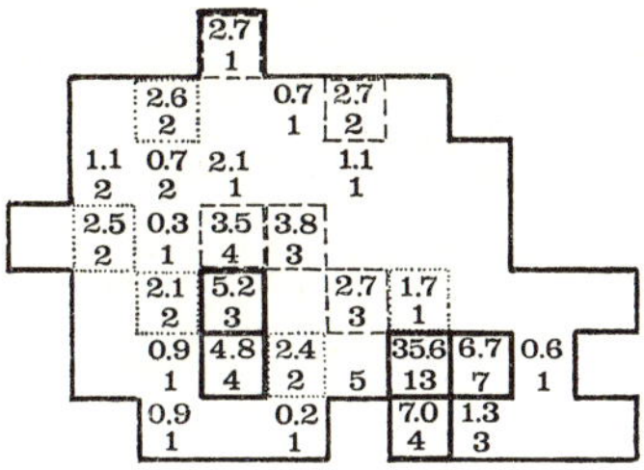

December

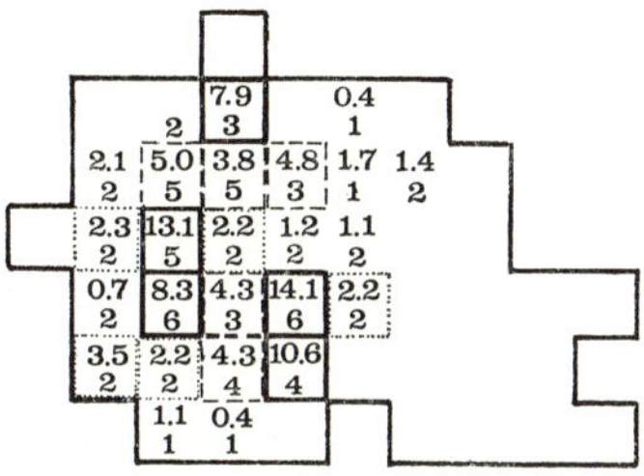

January

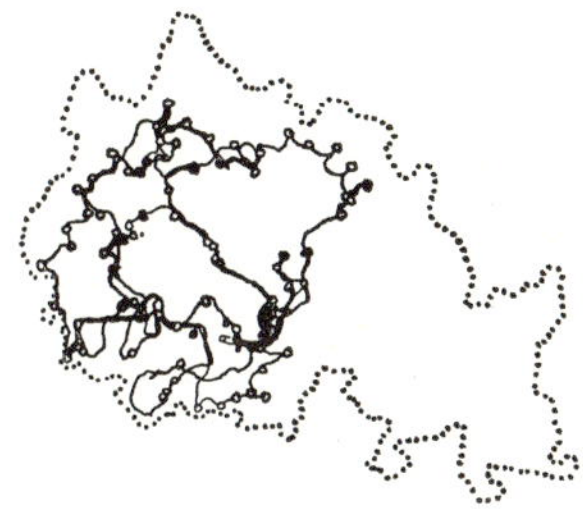

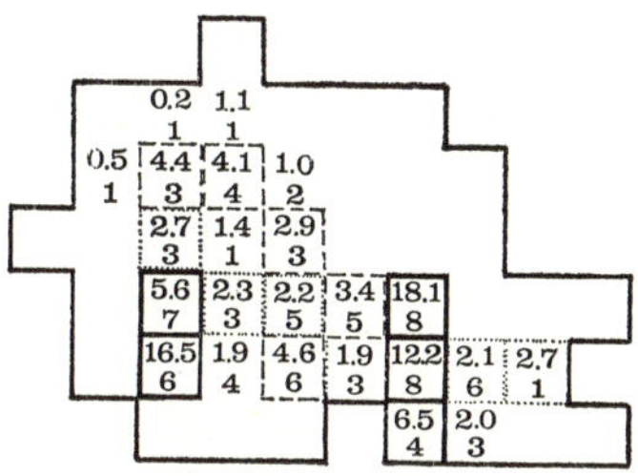

February

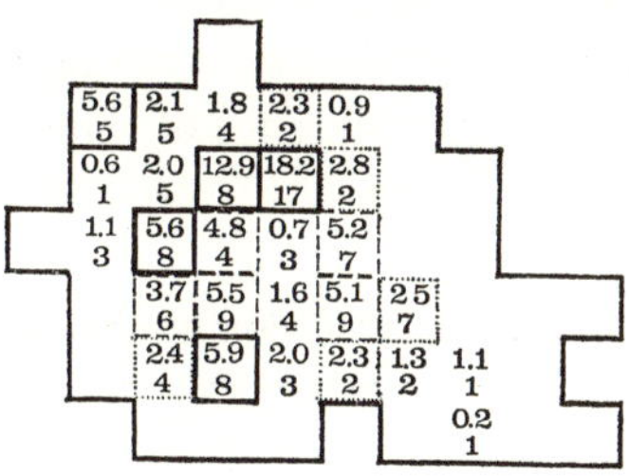

March

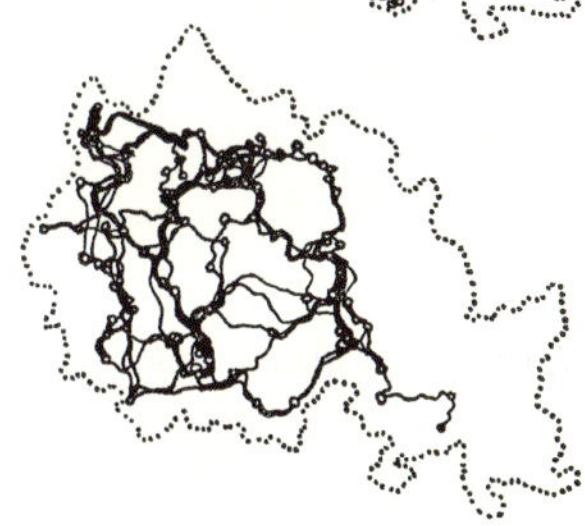

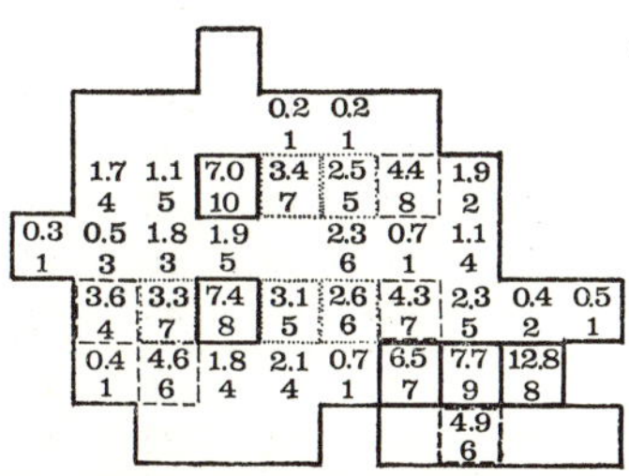

April

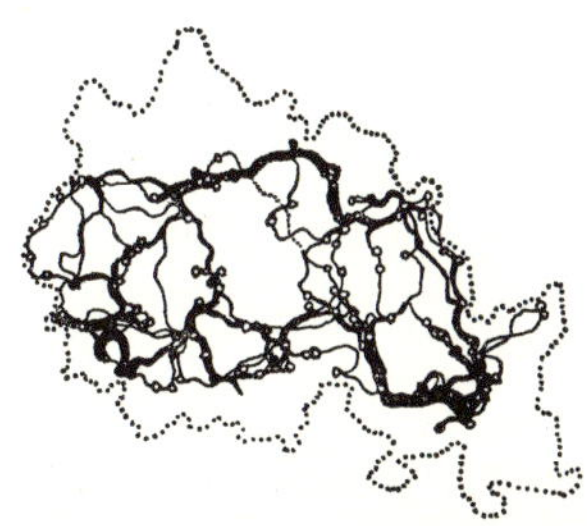

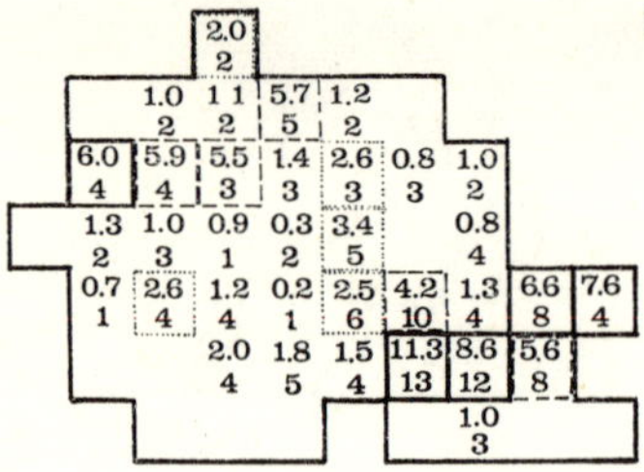

May

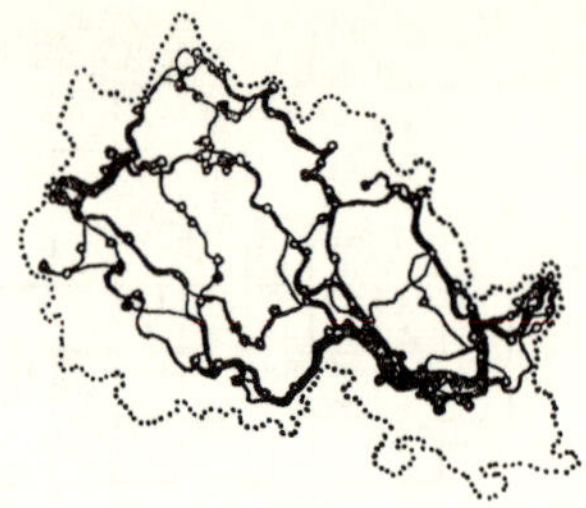

In each month different parts of the territory receive the greatest attention, and there is a concentration of activity about certain food trees, which vary from one month to the next (Chapter 3). In the first few months ranging is concentrated in the west-central area, including the north-west sector, which is then much neglected until the following January (except for October). During the rainy season the group ranges mainly to the south (and west in November). At the end of the dry season (April and May in particular) ranging is most extensive for a 10-day period, almost all parts of the territory being utilised.

The identity of the ten most used hectare quadrats (ranked) in terms of time (monthly and seasonally), and visits (seasonally) for TS1 and RS2 reveals the preferred parts of the territory. The slight difference according to the measure used is also shown. Some hectares might not reach the top ten in any month, and yet still reach the top ten overall. A difference of several rankings for the same square according to time and visits does not represent a very large discrepancy, because of the relatively small range of values.

Territory quadrats ranked for percent of time and number of visits quadrat numbers

	Rank									
Month	1	2	3	4	5	6	7	8	9	10
TS1										
April	16	15	10	33	30	9	11	17	22	28
May	18	17	11	16	27	19	22	28	10	38
June	41	18	36	35	25	19	17	43	11	23
July	11	9	22	30	18	17	10	16	19	25
August	25	26	30	19	31	17	36	43	11	24
Time, %	17	18	11	25	30	16	19	36	41	10
Number of visits	16	17	25	18	36	30	11	22	19	41

Territory quadrats ranked for percent of time and number of visits quadrat numbers (continuation)

	Rank									
Month	1	2	3	4	5	6	7	8	9	10
September	19	31	41	30	44	35	25	18	46	42
October	25	16	11	19	21	9	24	28	18	17
November	11	19	5	12	25	9	17	24	18	42
December	36	37	41	18	19	23	17	27	14	30
January	24	10	25	11	15	9	22	19	18	16
Time, %	19	11	25	36	24	31	18	10	16	9
Number of visits	19	11	25	36	18	10	24	16	9	17
February	35	12	36	37	11	25	9	16	30	23
March	22	16	19	3	10	18	29	30	17	11
April	44	18	41	16	36	42	12	33	35	6
May	36	41	46	43	4	9	21	44	16	35
Time, %	36	35	16	12	22	44	41	18	11	30
Number of visits	35	36	22	41	30	16	18	11	19	29
Overall time, %	11	19	36	25	16	18	17	41	30	10
Number of visits	19	11	36	16	18	25	41	30	22	35
RS2										
February	6	12	17	11	22	16	7	13	18	4
March	6	11	10	15	7	17	22	21	5	3
April	7	17	11	12	21	4	3	22	10	8
Time, %	6	11	12	17	7	22	10	21	15	4
Number of visits	12	11	6	17	7	22	21	15	10	16

Appendix D: Daily Patterns of Weather and Behaviour

Data are presented graphically for each hour of the day for each monthly observation period (ten days) at Kuala Lompat and in Ulu Sempam on:

A. Weather

1. Sun, cloud, rain. The proportion of each weather category in monthly samples for each hour, recorded at 15-min intervals.

2. Ratio of sun : cloud (and sun : cloud + rain). The same data plotted to show the predominance of sun or cloud at each hour of the day.

3. Temperature. A plot of the hourly mean shade temperature from thermohygrograph (Ulu Sempam only, see Chapter 1 for monthly maxima and minima at Kuala Lompat).

At Kuala Lompat the peak of sunshine occurred before noon in July, September, April and May (with a small morning peak in December and January), and after noon in April, May, August, October, February, and March (small in June). November was more cloudy than sunny in all hours except one. Cloudy peaks were distinctive early in the day in April, May, June, August. A little rain sometimes fell at dawn, but most rain fell in the late afternoon and the hours of darkness. (The exclusion of cloud from the rain category from much of this analysis is artificial, but important because of the different effect of each on behaviour.) Daily temperature patterns at Kuala Lompat closely approximate those from Ulu Sempam, but have greater amplitude, especially in the dry season.

B. Behaviour

1. Activity time. From dawn-to-dusk observations (1-min sample) of each individual in the siamang group (excluding the infant) for two consecutive days, it was possible to construct a plot of the proportion of each hour of the day an individual siamang (irrespective of age and sex) spent in each maintenance activity – rest, feed, and travel – in each month of the study. In Ulu Sempam such observations were not attempted, and scores represent group behaviour, in that timings were made from the first group member to change activity; also for April 1969 and May 1970 at Kuala Lompat. The stippled area represents night time resting (sleeping) activity.

2. Activity bouts (Kuala Lompat only). Defining a bout from the time when the first group member changes activity enables a plot to be made of the mean number of bouts of each maintenance activity initiated in each hour of each observation period. (The hatched area represents rest bouts

occurring in or near the night sleeping tree, i.e. after the first movement, but before a delayed first travel or feeding bout).

In most months the relatively small amount of travel was spread throughout the day, although travel bouts were most numerous early or late in the day. Feeding occupied the majority of time in most hours of the day (especially in the morning). Resting increased, in terms of time and bouts, as the day passed. The main exceptions to this pattern were the wet and cloudy months of November and December, when feeding increased as the day passed (see Chapter 4 for similar patterns in Ulu Sempam – sustained to some extent into the dry season).

Month	Peaks					
	early		late		indistinct	
	weather	activity	weather	activity	weather	activity
TS1						
April	c	T	s r	R		F
May	c	F	s	R	r	T
June	c	F	s		r	R T
July	s	T	c		r	R F
August	c	T	s r			R F
September	s	F	c r	R		T
October	c	F	r		s	R T
November	c	R T	s r	F		
December	s		c		r	R F T
January	s		c	R	r	F T
February		F T		R	s c	
March	c	T	s	R	r	F
April		F	r	R	s c	T
May		T		R	s c r	F
RS2						
February		R T		F	s c r	
March	c		s r			R F T
April	s	F T	c r	R		

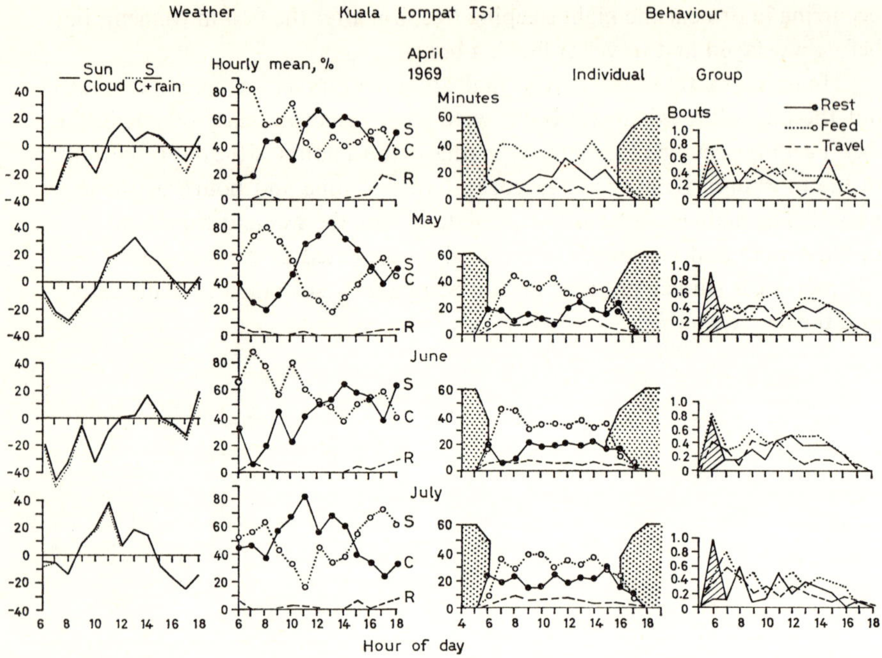
Weather
Kuala Lompat TS1
Behaviour
Sun / Cloud
S / C+rain
Hourly mean, %
April 1969
May
June
July
Individual
Group
Minutes
Bouts
Rest
Feed
Travel
S
C
R
Hour of day

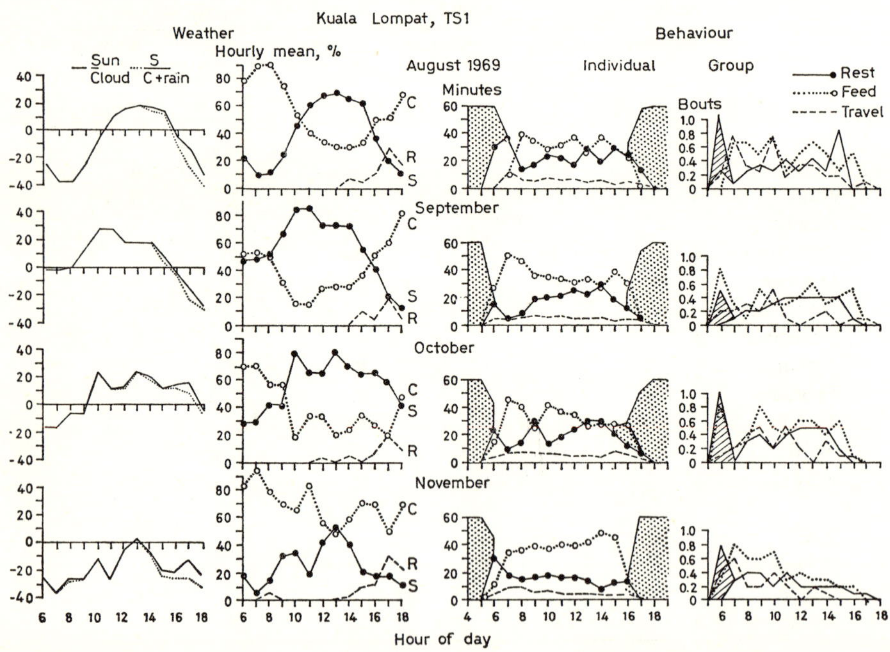
Kuala Lompat, TS1
Weather
Behaviour
Sun / Cloud
S / C+rain
Hourly mean, %
August 1969
September
October
November
Individual
Group
Minutes
Bouts
Rest
Feed
Travel
S
C
R
Hour of day

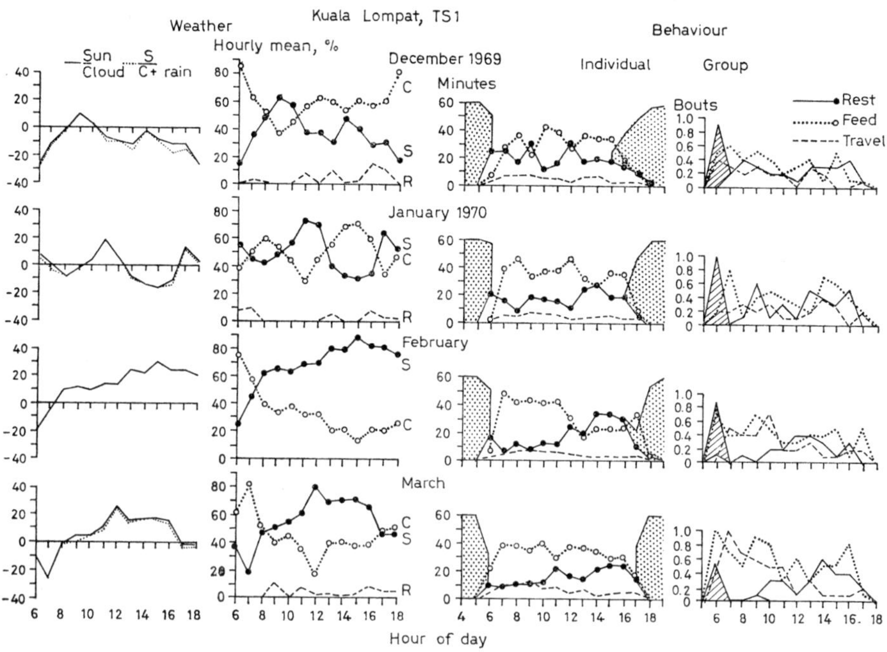
Kuala Lompat, TS1
Weather
Hourly mean, %
Behaviour
Individual
Group
Sun / Cloud
S / C+ rain
December 1969
January 1970
February
March
Minutes
Bouts
Rest
Feed
Travel
Hour of day

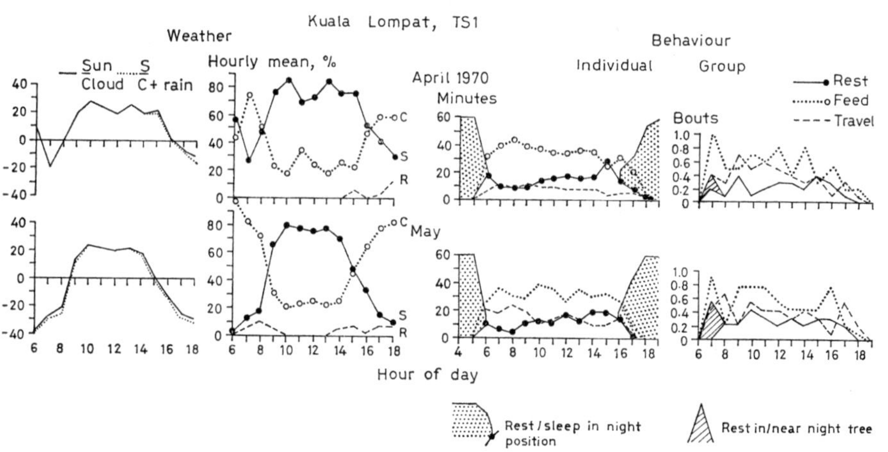
Kuala Lompat, TS1
Weather
Hourly mean, %
Behaviour
Individual
Group
Sun / Cloud
S / C+ rain
April 1970
May
Minutes
Bouts
Rest
Feed
Travel
Hour of day
Rest / sleep in night position
Rest in/near night tree

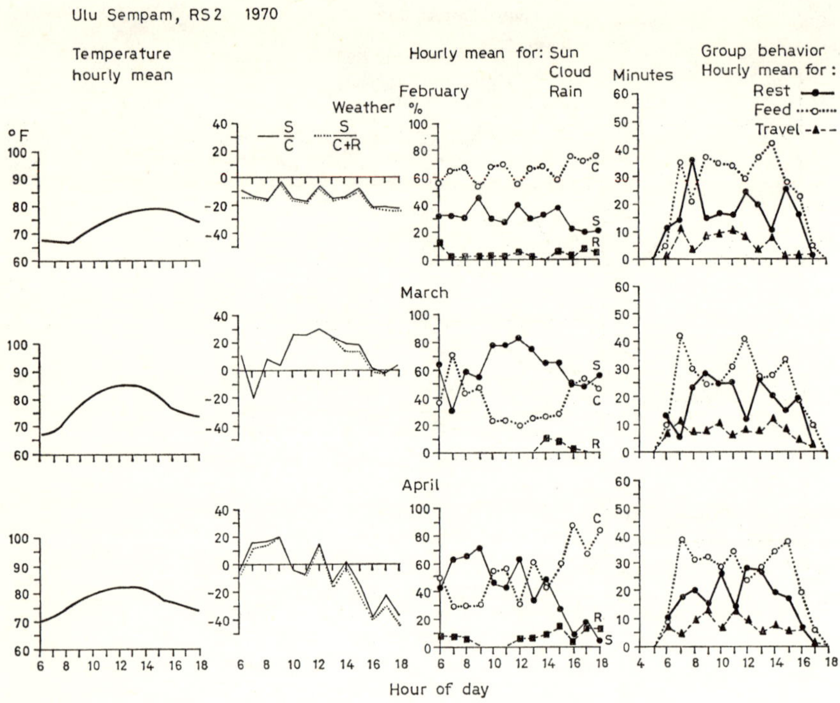

Daily Patterns of Grooming

The daily patterns of activity involving grooming are presented below for each monthly observation period at Kuala Lompat. On the left are the mean number of minutes of group resting, and the cumulative time in pair association, for each hour of the day. On the right are the hourly scores for pair association time as a proportion of resting time to more clearly demonstrate the lack of correlation between the two measures. No scores are given for the last hour or two of the day in most cases, because rest time represents resting during daily activity time only and some grooming occurred in these hours when the siamang were in their night sleeping positions.

Scores exceed 100% when several pairs are grooming during the same hour of resting. In most months there is a clear increase in grooming during the afternoon, when resting was most frequent. There are also indications in all months except June, November, January, March, and April of a marked increase in grooming relative to resting during the morning. In actual time such morning grooming peaks were only apparent in May, July, September, October, and December. Thus grooming was concentrated in the afternoon, although in some months it occurred between feeding bouts in the morning. At other times increased resting was not noticeably associated with grooming.

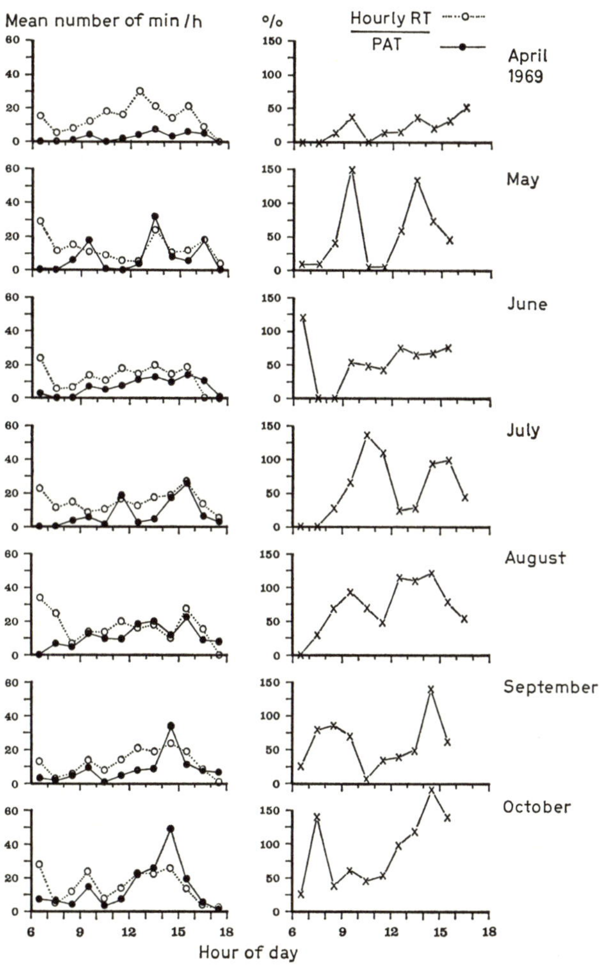

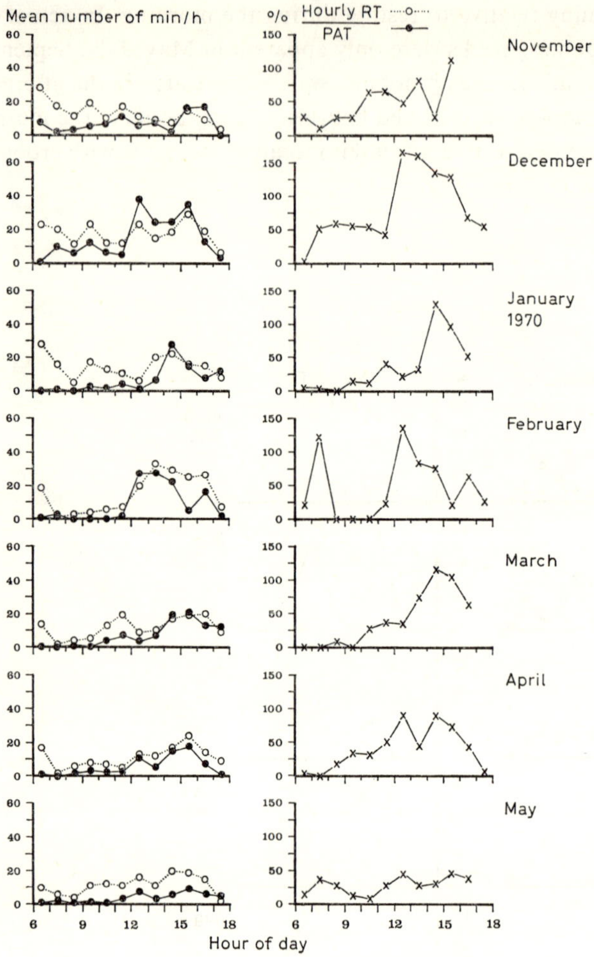
Mean number of min/h
%
Hourly RT
PAT
November
December
January
1970
February
March
April
May
Hour of day

Appendix E: Distribution and Density of Hylobatid Groups in Survey Localities in West Malaysia

State	Locality	Forest/game reserve or district	Siamang			Gibbon			Total hylobatids			
			area[1]	n[2]	lde[3]	area	n	lde	area	n	lde	number of days
Kelantan	Sg. Sok	Sg. Rek FR	–	0	–	34.7	11	11	34.7	11	11	1
	Sg. Pergau	Tanah Merah	17.2	1	1	17.2	2[4]	1	17.2	3	2	2
	Kg. Kalai	Tanah Merah	33.6	3	3	13.8	9[4]	9	33.6	12	12	1
	Bt. Jual	Tanah Merah	13.4	1	1	8.1	4[4]	4	22.2	5	5	1
	Kg. Kuala Tiga	Sokor Taki FR	–	0	–	2.6	8[4]	8	2.6	8	8	1
	Bt. Baka	Ulu Sat FR	–	0	–	33.0	7	7	33.0	7	7	1
	Kg. Tool	Berangat FR	23.5	7	7	2.6	3	3	26.1	10	10	1
	Kg. Amai	Ulu Kelantan	62.6	13	13	–	0	–	62.6	13	13	2
Trengganu	Bukit Bauk	Bt. Bauk FR	–	0	–	9.7	4	4	9.7	4	4	1
	Kg. Keruac	Besut	–	0	–	2.1	2	2	2.1	2	2	1
	Bt. Bintang	G. Tebu FR	–	0	–	7.4	12	12	7.4	12	12	1
	Kg. Langkap	K. Trengganu	–	0	–	9.5	8	8	9.5	8	8	1
	Kg. Sekayu	Ulu Trengganu	–	0	–	42.2	16	12	42.2	16	12	2
	Bt. Kundor	Ulu Chukai FR	–	0	–	6.5	7	7	6.5	7	7	1
	Bt. Berangin	Kemaman	–	0	–	2.7	13	13	2.7	13	13	1
Pahang, east	Bt. Beserah	Beserah FR	–	0	–	6.8	4	4	6.8	4	4	1
	Sg. Lembing	Kuantan	–	0	–	6.1	7	7	6.1	7	7	1
	Kg. Seri Jaya	Kuantan	–	0	–	8.0	25	22	8.0	25	22	1.5
	Bt. Chondong	Lepar FR	–	0	–	7.7	16	16	7.7	16	16	1
	Bt. Beranti	Berkelah FR	–	0	–	5.5	7	7	5.5	7	7	1
	Bt. Galong, S	Berkelah FR	–	0	–	6.8	13	13	6.8	13	13	1
	Bt. Galong, N	Berkelah FR	–	0	–	5.8	11	11	5.8	11	11	1

Distribution and density of hylobatid groups in survey localities in West Malaysia (continuation)

State	Locality	Forest/game reserve or district	Siamang			Gibbon			Total hylobatids			
			area[1]	n[2]	lde[3]	area	n	lde	area	n	lde	number of days
Pahang, central	Gunong Benom	Krau GR	15.1	4	4	4.2	12	8	15.1	16	12	4
	Kuala Lompat	Krau GR	17.6	9	9	4.9	6	6	17.6	15	15	1
	Bukit Besar	Kemasol FR	18.0	3	4	3.8	12	16	18.0	15	20	0.5
	Sg. Muring	Kemasol FR	–	0	–	5.4	8	12	5.4	8	12	0.5
	Bt. Kaling	Lakum FR	16.2	9	10	3.3	8	12	16.2	17	22	0.5
	Bt. Patong	Krau GR	13.5	6	5	2.8	5	5	13.5	10	10	1
	Ulu Dong	Ulu Dong FR	48.4	4	12	19.6	4	8	48.4	8	20	0.5
Pahang, south	Ulu Keratong	Lesong FR	–	0	–	93.1	18	18	93.1	18	18	1
	Ulu Jekatih	Lesong FR	–	0	–	144.9	19	19	144.9	19	19	1
	Ulu Pukin	Lesong / Labis FR	–	0	–	113.8	27	27	113.8	27	27	1
	Ladang Geddes	Palong FR	–	0	–	18.7	6	12	18.7	6	12	0.5
	Tasek Bera	Temerloh	–	0	–	93.1	12	18	93.1	12	18	0.5
	Bt. Bertangga	Temerloh	–	0	–	77.7	10	10	77.7	10	10	1
Johore	G. Besar	Labis FR	–	0	–	5.9	2	6	5.9	2	6	0.5
	G. Bekok	Labis FR	–	0	–	6.3	2	14	6.3	2	4	0.5
	Sungei Selai	Labis FR ext.	–	0	–	72.5	15	15	72.5	15	15	1
	Ulu Lenggor	Lenggor FR	–	0	–	6.3	7	7	6.3	7	7	1
	Bt. Tinggi	Lenggor FR	–	0	–	7.0	5	5	7.0	5	5	1
	G. Sumalayang	Ulu Sedili FR	–	0	–	9.6	1	7	9.6	1	7	0.5
	G. Muntahak	Panti FR	–	0	–	11.4	3	3	11.4	3	3	1
	G. Pulai	Johor Baharu	–	0	–	4.7	6	8	4.7	6	8	0.5

Negri Sembilan	Kg. Beroga	Gapau FR	13.9	1	3	4.3	3	6	13.9	4	9	0.5
	Ulu Bendol	Angsi / Berumbun FR	50.6	5	9	15.5	14	20	50.6	19	29	0.5
	Sri Menanti	Senaling Inas FR	40.6	4	9	9.5	5	20	40.6	9	29	0.5
	Bt. Batu Bras	Pasoh FR	–	0	–	9.4	8	16	9.4	8	16	0.5
	Bt. Batu Elam	Teriang FR ext.	25.4	7	14	7.4	8	16	25.4	15	30	0.5
	Ulu Jempol	Serting FR ext.	22.4	7	14	–	0	–	22.4	7	14	0.5
	Kg. Kemomoi	Jelebu	40.8	1	3	19.4	17	23	40.8	18	26	0.5
	Ulu Kelawang	Lenggeng FR	29.3	1	1	15.3	12	12	29.3	13	13	1
	Sg. Kenaboi	Kenaboi FR	17.6	2	2	17.6	11	11	17.6	13	13	1
	G. Rembau	Johol FR	14.4	5	12	14.4	11	14	14.4	16	26	0.5
	Bt. Besar Jelai	Jelai / Gemas FR	–	0	–	4.4	6	8	4.4	6	8	0.5
	G. Tampin	Tampin FR	–	0	–	–	0	–	–	0	–	0.5
	Ulu Jelai	Johol FR	–	0	–	–	0	–	–	0	–	0.5
Selangor	*Ulu Gombak*	Ulu Gombak FR	15.0	19		4.1	8		15.0			12
	Ulu Rumput	Ulu Gombak FR	4.2	5	3.8	4.2	6	3.5	4.2	11	7.3	12
			29.4	25	9.5	27.1	15	3.0	29.4	40	12.5	12
	Bukit Fraser	Semangkok FR	22.9	11	9	2.3	5	4	22.9	16	13	2
	Ulu Batu	Gombak FR ext.	16.7	16	12	8.7	3	2	16.7	19	13	2
	Ulu Kalong	Batang Kali FR	9.5	6	6	–	0	–	9.5	6	6	1
	Ulu Serendah	Serendah FR	9.0	4	4	6.6	2	2	9.0	6	6	1
	Ulu Gumut	Ulu Selang FR	8.3	5	5	2.5	4	4	8.3	9	9	1
	Ulu Liam	Serendah FR	–	0	–	–	0	–	–	0	–	0.5
	Bt. Lagong	Bt. Lagong FR	–	0	–	–	0	–	–	0	–	0.5
	Ulu Langat	Ulu Langat FR	26.5	5	5	6.8	1	1	26.5	6	6	1
	Sg. Seminyih	Sg. Lalang FR	32.4	1	1	8.1	6	6	32.4	7	7	1
	Bt. Belata	Bt. Belata FR	8.2	1	3	8.2	1	3	8.2	2	6	0.5

Distribution and density of hylobatid groups in survey localities in West Malaysia (continuation)

State	Locality	Forest/game reserve or district	Siamang			Gibbon			Total hylobatids			
			area[1]	n[2]	lde[3]	area	n	lde	area	n	lde	number of days
Pahang, west	Ulu Teranum	Ulu Teranum FR	17.0	2	4	–	0	–	17.0	2	4	0.5
	Sg. Sempam	Raub	43.2	13	16	17.0	6	7	43.2	19	23	0.5
	Sungei Sia	Sg. Sia FR	22.9	4	12	10.9	2	4	22.9	6	16	0.5
	Ulu Tersang	Tersang FR	–	0	–	77.6	3	3	77.6	3	3	1
	Bt. Ulu Telang	Bentong	–	0	–	–	0	–	–	0	–	0.5
	Sg. Perting	Bentong	18.2	1	3	6.1	1	4	18.2	2	7	0.5
	Bt. Tinggi	Bt. Tinggi FR	13.9	10	10	2.7	1	1	13.9	11	11	1
	Bt. Gelong	Bentong	16.1	7	9	3.0	4	12	16.1	11	21	0.5
	Sg. Tanglir	Bt. Tinggi FR	6.0	5	5	3.3	5	5	7.2	10	10	1
	Ulu Kali	Bt. Tinggi FR	12.6	4	3	–	0	–	12.6	4	3	2
	Kg. Janda Baik	Lentang FR	33.4	3	3	3.4	6	6	33.4	9	9	2
	Sg. Dua	Lentang FR	9.5	2	8	3.0	2	6	9.5	4	14	0.5
Perak	Sg. Selim	Selim Hills FR	76.8	6	6	–	0	–	76.8	6	6	1
	Tanah Rata	Tapah Hills FR	101.8	2	2	–	0	–	101.8	2	2	1
	Gunong Besout	G. Besout FR	–	0	–	9.1	2	4	9.1	2	4	0.5
	Sg. Kinta	Kinta Hills FR	–	0	–	4.7	1	3	4.7	1	3	0.5
	Jelapang	Kledang Saiong FR	20.4	6	6	–	0	–	20.4	6	6	1
	Sg. Chior	Chior / Piah FRs	13.6	5	5	–	0	–	13.6	5	5	1
	Kg. Jalong	Korbu FR	18.4	1	3	–	0	–	18.4	1	3	0.5
	Sg. Ibul	Bintang Hijau FR	–	0	–	5.8	6[4]	8	5.8	6	8	0.5
	Kg. Bersia	Grik	9.9	2	4	9.9	6[4]	6	9.9	8	10	1

	Sg. Rui	Lepang Nenering FR	–	0	–	15.4	8[4]	10	15.4	8	10	0.5
	Padang Chong	Padang Chong FR	–	0	–	9.2	6[4]	6	9.2	6	6	1
	Sg. Selama	Bintang Hijau FR	–	0	–	4.0	4[4]	9	4.0	4	9	0.5
	Sg. Ijok	Ijok FR	–	0	–	29.3	5[4]	9	29.3	5	9	0.5
	G. Hijau	Larut Hills FR	–	0	–	10.1	7[4]	10	10.1	7	10	0.5
	Gunong Bubu	Bubu FR	–	0	–	21.1	13[4]	13	21.1	13	13	1
	G.Bujang Melaka	Kampar	–	0	–	–	0	–	–	0	–	0.5
Kedah	Sungei Ketil	Ulu Muda FR	–	0	–	18.4	9[4]	12	18.4	9	12	1
	Sungei Pedu	Padang Terap	–	0	–	4.6	2	2	4.6	2	2	1
	Sg. Padang Sanai	Bt. Perangin FR	–	0	–	29.1	9	9	29.1	9	9	1
	Bt. Kachi	Bt. Perangin FR	–	0	–	13.3	6	9	13.3	6	9	0.5
	Gunong Perak	Bt. Perak FR	–	0	–	8.1	1	4	8.1	1	4	0.5
	G. Jerai	G. Jerai FR	–	0	–	17.3	1	4	17.3	1	4	0.5
	G. Inas	G. Inas FR	–	0	–	6.5	5[4]	8	6.5	5	8	0.5
	G. Bongsu	G. Bongsu FR	–	0	–	3.7	3[4]	6	3.7	3	6	0.5

1 Area of survey in square kilometres, for each locality (by region).
2 n = Number of groups recorded.
3 Ide = Population count or estimate for one day of calling.
4 Black-handed gibbon.
Locality: Sg. = Sungei (river); Bt. = Bukit (hill); G. = Gunong (mountain); K. = Kuala (river mouth); Kg. = Kampong (village).
Situation: Ulu (head of valley around river source); Padang (open space); Ladang (cultivated land). FR = Forest reserve, GR = game reserve (otherwise – district within state).

Appendix F: Notes on Colour Phases and Female 'Great Calls' in Gibbons

Colouration

Although the siamang is monophasic in colouration – all individuals being black – colour in gibbon species varies from sexual dichromatism in Indo-China, through asexual dichromatism in the centre of Sundaland, to monochromatism in the east [FOODEN, 1969; GROVES, 1970]. Thus the white-handed gibbon in Thailand exhibits either a dark-brown/blackish colour or a buff colour. GROVES [1972] asserts that Malayan gibbons show similar extremes of colour, although he notes the relative paleness of the dark phase. But he includes in *Hylobates lar lar* the very pale specimens from southern Thailand, which, in the light of the present study are more likely to be *H.l. entelloides.* The remainder are described as mainly dark phase, with one 'mottled' specimen, and a few pale phase specimens of dubious origin.

The data collected in this study provide additional evidence for the lack of distinct colour phases in the Malayan white-handed gibbon. Of 37 individuals clearly viewed, 18 were classified as dark phase, 8 as pale phase, and 11 (30%) could not be assigned to either. Certainly the colouration was very variable, but the unclassified individuals cannot be explained away by the distorting effects of varied lighting conditions. In addition, individuals of mixed colouration were observed. The TG2 male had golden head, arms, and thorax, and dark brown abdomen and legs; several juveniles in Ulu Sempam were also unevenly coloured with thighs paler than upper parts. Mr. R. TENAZA disagreed with my classification of the TG2 female as a dark phase individual.

It is therefore suggested that not only is the Malayan white-handed gibbon completely separated from its Thai counterpart (Chapter 7), but that it does not exhibit the same distinctive dichromatism. Instead there is evidence to suggest that they are more variable in colour with a continuous gradient from dark-brown (rather than black) to golden (rather than pale buff). No specimens of the rare pale phase in black-handed gibbons were observed.

Great Calls of Female Gibbons

The timing of female great calls in each region was particularly interesting, because of the striking differences in the duration and structure of calls of white- and black-handed gibbons. Although the diphasic 'hoots' of the male black-handed gibbon were qualitatively the most distinctive calls between the two taxa, the female great calls were more highly pitched, rising more rapidly to the climax, and lacking the fluctuations in pitch of the white-handed gibbon call.

Great calls by white-handed gibbons lasted for 19 – 21 sec on average in the west and south, decreasing to 18 in the centre and north-east, and below 16 sec in the north-west (the latter was a very small sample). The sample of 213 calls ranged from 12 to 29 sec – average of 20.0 sec – but the 5% of less than 16 sec appeared to cease before completion. The 49 great calls of black-handed gibbons, however, were consistently shorter, lasting for 12 – 18 sec – 15.2 sec on average (fig. 7.5).

About 50% of black-handed gibbon great calls lasted for 14 or 15 sec, whereas two thirds of white-handed gibbon calls lasted for 18 – 22 sec. These differences are extremely significant (Kolmogorov-Smirnov two-sample test). Although the duration of great calls differed significantly between groups in Ulu Sempam (Chapter 6), there was no significant difference overall between gibbon calls in the two main study areas. The differences between *lar* and *agilis* were significant except in the north-west; the difference therefore between *lar* in the south and north-west was significant.

Thus confirmation is provided that observations in the main study areas are applicable to the country as a whole, except for the north.

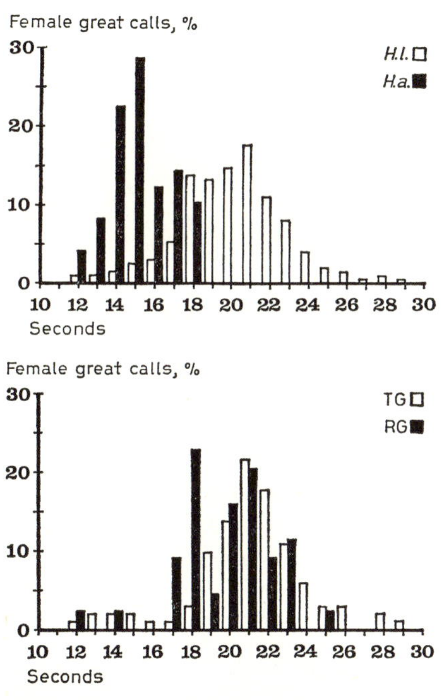

Author Index

Aldrich-Blake, F. P. G. 47, 226, 289, 299
Altmann, S. A. 27, 28, 57, 58, 63
Altmann, J. 27, 28, 58, 63
Amerasinghe, F. B. 63
Anderson, J. 1

Baldwin, L. 129, 223
Bates, B. C. 27
Benirschke, K. 2, 294
Bernstein, I. S. 228, 289, 291, 293
Bonhote, J. L. 273
Buffon, G. L. 1
Burt, W. H. 27

Carpenter, C. R. 3, 4, 120, 172, 219, 220, 228, 237, 274
Chalmers, N. 264
Chantharojvung, S. 5
Chance, MR. A. 228
Chasen, F. N. 5
Chivers, D. J. 3, 24, 27, 29, 47, 59, 60, 125, 143, 148, 163, 173, 177, 185, 226, 254, 264, 278, 279, 286, 291, 292, 294, 299
Corner, E. J. H. 7
Crook, J. H. 289, 290, 291, 295
Cuylenberg, B. W. B. van 63

Dale, W. L. 6, 7, 26
DeVore, I. 27, 57, 63, 163, 264

Ellefson, J. O. 3, 4, 23, 27, 59, 60, 120, 123, 163, 164, 172, 219, 229, 231, 237, 242, 243, 265, 266, 274, 290
Elliot, D. G. 1

Fooden, J. 1, 2, 330
Fox, G. 3
Frisch, J. E. 1, 2, 285, 294

Gartlan, J. S. 228
Gloger, 1
Grand, T. I. 123
Groves, C. P. 1, 2, 285, 294, 330
Gulik, R. H. van 1

Hall, K. R. L. 63, 264
Hamerton, J. L. 2
Hardy, B. 290
Harlan, R. 1
Hill, C. 3
Hinde, R. A. 57
Hladik, C. M. 63, 122
Hooijer, D. A. 2, 285
Howard, H. E. 57

Jay, P. C. 23, 57, 63, 264
Jolly, A. 23

Kaufmann, J. H. 58
Kawabe, M. 3, 27, 172
Khan, M. 3, 285
Klinger, H. P. 2, 294
Kloss, C. B. 1, 273
Koyama, N. 219, 293
Kummer, H. 23, 27, 173

Lang, E. M. 2
Lawick-Goodall, J. van 128
Lim, Boo Liat 242, 273, 285
Lodge, G. A. 290

McClure, H. E. 3, 27, 124, 293
MacKinnon, J. R. 125, 291
Marler, P. 266

Marshall, J. 5
Martin, W. C. L. 1
Mason, W. A. 57, 58, 172, 229
Matthews, L. H. 220
Medway, Lord 5, 7, 25, 125, 273, 291
Menzel, E. W. 227, 229
Miller, C. 1
Mitchell, G. D. 229
Moustgaard, J. 290
Mutton, D. E. 2

Noble, G. K. 57

Ooi, Jin-Bee 5

Petter, J. J. 172
Pocock, R. I. 1, 220
Poore, M. E. D. 86, 125

Raemaekers, J. J. 46
Raffles, T. S. 1
Reynolds, V. & F. 122
Ripley, S. 63, 120, 122, 123
Rodman, P. S. 290
Ross, B. A. 5

Schaller, G. B. 23
Schultz, A. H. 1, 2, 294
Siegel, S. 29
Simonetta, A. 1
Soo-Hoo, C. 3
Struhsaker, T. T. 28, 227, 228
Sugiyama, Y. 63, 122

Teleki, G. 129, 223
Thomas, O. 2
Thompson, N. S. 228

Wilson, C 27
Wyatt-Smith, J. 7, 125

Yoshiba, K. 63, 264

Subject Index

Aggression 222–225
Alert period 129, 188
Altitudinal zonation 272–274

Baboon 63
Botany 14–15
Bouts 137

Callicebus 172, 229
Calling, alarm 242
–, duration 262
–, female 238
–, male 238
Climate 6, 24–26
Conflicts, inter-group 52–53, 59–60, 241–242
–, intra-group 223–225

Daily calls, call hours 258
–, first 256
–, group calls 259
Day length 129, 188
Day range 24, 28–35
Density, siamang 282
–, gibbons 283
Diet 69–86
Distance, inter-individual 181–183
Distribution, siamang 269
–, gibbons 269–272
Drinking 146–147

Facial expression 223
Feeding, habits 114–118
–, movements 118–119
–, rates 116–118
Food categories 69
Food sources 92
Food trees, principal 92–107
–, dispersion 107–113

Geography 6
Geology 5
Gibbon, white-handed 1, 2, 27, 59, 60, 120–121, 163–164, 172, 230, 242–245, 246–249, 250–251, 255, 264–265, 271–284, 292–294, 306, 325–331
–, black-handed 272, 279, 283–286, 328–331
Grooming 197–214
Group cohesion 173–178
Group formation 226
Group size 56
Group spread 178–181

Habituation 20–22
Hylobates 1, 2, 27, 58, 60, 120–121, 146–249, 163–164, 172, 230, 242–245, 250–251, 255, 264–265, 271–284, 292–294, 306, 325–331

Indri 172
Infancy 193–197, 214–219

Langurs 120, 123, 163, 264, 272–273, 292, 306
Leadership 190–192, 229
Leaf-monkeys 120, 123, 163, 264, 272–273, 292, 306
Logging 8–10

Macaca 123, 163, 272–273, 292, 306
Maintenance activities 128, 137–169
Malay Peninsula 5, 7

Maturation 214–219
Maximum radius 28–35
Methods 16–23
Movement, measures of 28–35

Night position shift 28–35

Papio 63
Periodicity of maintenance activities 140–146
Phenology 124
Play 214–219
Population distribution 269–274
–, density 274–284
Presbytis 120, 123, 163, 264, 272–273, 292, 306
Progressions 188–197

Quadrats 28, 41–45

Ranging patterns 35–49

Sampling 128, 164–165
Seasons 24–26, 34, 41
Sexual behaviour 219–222, 231
Siamang 1, 2, 295
Spacing 172–185
Stratification 133–136
Statistics 29
Sumatra 1, 2, 285
Synchrony, activity 137–140, 173–178
Symphalangus 1, 2, 295

Taxonomy 1–2, 295
Territory 52–61
Titi monkey 172, 229
Travel, order 189–192
–, sequence 192–193
Trees 14–15, 49–51, 64–66, 68, 69–73, 86–87, 91–94, 113, 291, 309–312

Vegetation 7, 14–15, 64–69, 309–312
Vocalisations 222, 237–241

Weather 22, 24–26, 34–35, 157–163, 185–186, 263–264